Statistical Analysis of Human Growth and Development

Chapman & Hall/CRC Biostatistics Series

Chapman & Hall/CRC Biostatistics Series

Chapman & Hall/CRC Biostatistics Series

Statistical Analysis of Human Growth and Development

Yin Bun Cheung

Duke-NUS Graduate Medical School

CRC Press
Taylor & Francis Group
Boca Raton   London   New York

CRC Press is an imprint of the
Taylor & Francis Group, an **informa** business

A CHAPMAN & HALL BOOK

CRC Press
Taylor & Francis Group
6000 Broken Sound Parkway NW, Suite 300
Boca Raton, FL 33487-2742

First issued in paperback 2020

© 2014 by Taylor & Francis Group, LLC
CRC Press is an imprint of Taylor & Francis Group, an Informa business

No claim to original U.S. Government works

Version Date: 20130919

ISBN 13: 978-0-367-57627-1 (pbk)
ISBN 13: 978-1-4398-7154-6 (hbk)

Library of Congress Cataloging-in-Publication Data

Cheung, Yin Bun, author.
  Statistical analysis of human growth and development / Yin Bun Cheung.
  p. ; cm. -- (Chapman & Hall/CRC biostatistics series)
  Includes bibliographical references and index.
  ISBN 978-1-4398-7154-6 (hardback : alk. paper)
  I. Title. II. Series: Chapman & Hall/CRC biostatistics series (Unnumbered).
  [DNLM: 1. Growth and Development. 2. Statistics as Topic--methods. WS 103]

  QP84
  612.6--dc23   2013036859

Visit the Taylor & Francis Web site at
http://www.taylorandfrancis.com

and the CRC Press Web site at
http://www.crcpress.com

To Alice

Contents

Preface

For about 15 years I have been involved in the medical and psychological studies of human growth and development in African and Asian countries. I have worked with statisticians and researchers and postgraduate students who were not originally trained in statistics. I realized the need for a textbook on statistical analysis of data from this field for nonstatisticians and also the need for a reference resource for statisticians who are relatively new to this field. This is my attempt to address these needs.

Scope

In this book, *human growth* refers to changes in anthropometric measurements, such as fetal size parameters, height, weight, and body mass index measured by various devices. *Development* refers to changes in cognitive, social-emotional, locomotor, and other abilities, often measured by milestones or multi-item assessment scales. Growth and development are distinct subjects, but they often go hand in hand in research and program evaluation as they may have similar causal factors and they may affect each other. Although the book illustrates with examples from these subjects, the concepts and methods are applicable to studies of many other physical and psychosocial phenomena, such as lung function and depressive symptoms.

The text summarizes existing statistical methods and illustrates how they can be used to solve common research problems. This book is about statistical analysis and not just about statistical methods per se. In addition to describing methods, it deals with how well a method addresses a specific research question and how to interpret and present the analytic results. It covers statistical methods that range from basic to advanced levels. Their inclusion is based on the practical needs for solving common research problems in the field rather than technical considerations. Mathematical details are omitted when possible.

Intended Readers

This book is intended for clinical researchers, epidemiologists, social scientists, statisticians, other researchers, and postgraduate students engaged in the investigation of human growth and development. It is assumed that the readers have attended an introductory course in statistics and know basic statistical methods, such as the t-test and chi-square test.

Organization

Chapters 1 through 3 are background and review materials. Chapter 1 sets the scene by reviewing the issues of human growth and development, and outlining some statistical considerations in relation to the subjects. Chapter 2 reviews the scientific reasoning about causality and study designs. Chapter 3 is a toolbox of statistical concepts and techniques.

Before we can perform the statistical analysis, we need to generate the growth and development variables for the analysis. Chapter 4 discusses the use of existing tools to transform raw data into analyzable variables and back-transform them to raw data. As the development of new tools requires knowledge of regression analysis and model diagnostics, the discussion about new tools is deferred until Chapters 9 through 11. Chapters 5, 6, and 7 cover regression analysis of quantitative, binary, and censored data, respectively. In addition to discussing general statistical concepts and methods, these chapters introduce statistical analyses that are often needed in the studies of growth and development, such as quantile regression and analysis of interval-censored data. Chapter 8 deals with the analysis of repeated measurements and clustered data.

Knowledge of Chapters 5 through 8 is needed to understand Chapters 9 through 11. Chapters 9 and 10 describe the development of new growth references and developmental indices, respectively. Chapter 11 describes the development of new references and the generation of key variables based on longitudinal data. Chapter 12 addresses the processes to verify the validity and reliability of measurement tools.

Finally, Chapters 13, 14, and 15 discuss the bigger and realistic picture of research practice—Chapter 13 is on missing values, Chapter 14 is on multiplicity problems, and Chapter 15 is on what to do when there are many variables available.

Datasets and Software

In addition to using many real examples, this book uses two simulated datasets extensively to illustrate various methods. The description and the Stata® codes that generate these two datasets are included in Appendices A and B.

By using the same seed number that controls the software's (pseudo)random number generator, readers can reproduce exactly the same data and replicate the analysis. These two datasets resemble the longitudinal study of child health and clinical trial of antenatal care discussed in Chapter 2. The advantage of using simulated datasets is that we know the truth. As we set the parameters that control the simulations, we are better positioned to interpret the analytic results and appreciate the properties of the statistical methods.

I have used Stata to implement and illustrate most of the analysis. With the aim of ensuring reproducibility and handling specific issues relevant to the subject matters, I have included some tips on the use of the software in Chapter 3, and also sample codes and outputs in the displays in various chapters. I do not intend to make this book an introductory guide to Stata, and it is not essential for readers to use the same software. A series of good introductory books about the use of several statistical packages are referenced in Chapter 3.

The appendices are available from the Taylor & Francis/CRC Press Web site at: http://www.crcpress.com/product/isbn/9781439871546. These codes and programs are provided to facilitate learning. They have not gone through extensive testing and are provided as is.

Acknowledgments

I thank Tina Ying Xu, Per Ashorn, Lotta Alho, and Mihir Gandhi for their helpful comments on selected chapters of a draft version of this text. I thank Tina Ying Xu and Bill Rising for their comments on the Stata codes. The WMLE program in Appendices D, D1, and D2 are provided by Tina Ying Xu. I also thank Taara Madhavan from the Duke-NUS/SingHealth Academic Medicine Research Institute and William Che for providing editing support for some of the chapters. All remaining errors are mine. This work was supported in part by the Singapore Ministry of Health's National Medical Research Council under its Clinician Scientist Award.

Author

Yin Bun Cheung is a professor in the Centre for Quantitative Medicine, Office of Clinical Sciences at Duke–NUS Graduate Medical School, Singapore, and adjunct professor in the Department of International Health at the University of Tampere, Finland. He has been studying human growth and development in African and Asian countries for about 15 years.

1

Introduction

1.1 Overview

The words growth and development have many meanings. In this book, *human growth* refers to changes in physical size of the body as a person ages. It includes multiple dimensions such as height, weight, and waist-to-hip ratio. *Human development* refers to changes in ability or functional maturation over the life span. It includes the interrelated domains of cognitive, social-emotional, motor, and language abilities.

Many aspects of growth and development are at their highest speed during the first few years of life. Some of them may stop before or at the end of puberty, but some continue throughout the life span. Growth and development are major targets in research and policy concerning children and adults alike. The United Nations' Millennium Development Goals included reduction in the prevalence of underweight children under 5 years of age as a progress indicator (United Nations Development Programme 2003). The estimate that in 2004, more than 200 million children aged under 5 years failed to reach their potential in cognitive ability has stimulated a lot of international policy concerns (Grantham-McGregor et al. 2007). Furthermore, growth and development in early life can have long-term implications that stretch across decades. There has been accumulating evidence that intrauterine growth restriction and accelerated growth in childhood are related to the risk of cardiovascular and metabolic diseases in adulthood. This is known as the *programming hypothesis, fetal origin hypothesis,* or *developmental origin hypothesis* (Barker 1998). All these issues point to the importance of growth and development in early life. However, humans continue to grow—or shrink—physically and mentally during adulthood. Obesity in middle age and cognitive decline in old age are two of the issues important to both individuals and the society. As population aging is occurring in many parts of the world, the next two decades will very likely see intensified research and program activities concerning the improvement and maintenance of cognitive and functional abilities in the elderly.

Human growth and development are distinct fields. This book covers both of them because they are closely related and they often go hand in hand in research and program evaluation. For example, a clinical trial of an infant nutritional supplement may aim to assess its impact on both the growth and development of infants. The research team will need to know the analysis of both types of data. Despite the focus on human growth and development, the statistical considerations and methods are general and can be applied to many other aspects of physical and psychological phenomena, such as lung function and depressive symptoms.

The measurement of growth and development are very different. Growth is objective and usually measured by physical equipment, such as a digital scale for weight, infant length board (infantometer) for supine length, ultra-sonography for fetal size measurements, and skinfold caliper for skinfold thickness. Development is abstract. Different people may have different understanding of what ability means. Multiple indicators are usually needed to measure development so that the measurement covers the breadth of the abstract concept sufficiently, in other words, to obtain sufficient *content validity*. These indicators may appear in a variety of forms, such as:

- Dichotomous items (e.g., whether a child can walk 10 steps without support) on an observation checklist to be administered by a researcher
- Multiple-choice questions in a paper-and-pen assessment test
- Likert-scale items (e.g., 5 points from strongly disagree to strongly agree) on a behavior questionnaire to be filled in by caregivers
- Quantitative measures of performance under purposefully arranged situations (e.g., the number of seconds a person needs to visually search and locate a target among distracters)

Due to the differences in concepts and measurement, there are some differences in the statistical analysis methods.

The main purpose of this book is to explain and illustrate statistical concepts and methods and their applications to research problems commonly encountered in the studies of human growth and development. It focuses on growth and development as responses to exposure—protective or risk factors—or themselves being the exposure that affects aspects of health and well-being. As such, a substantial part of the book is about assessing association and causality. Ultimately, the motivation behind the writing of this book is to facilitate the development of programs and practices that promote human growth and development.

1.2 Human Growth

1.2.1 Fetal Period

The duration from conception to delivery is normally about 38 weeks. The first day of the last menstrual period is on average about 2 weeks before the ovulation. The American Academy of Pediatrics defined *gestational age* (GA) as the time elapsed between the first day of the last menstrual period and the day of delivery (Engle et al. 2004). This is also called gestational duration. This definition is widely used and 40 weeks is considered the average gestational duration of a healthy pregnancy. *Conceptional age* is the time elapsed between conception and delivery, and is about 2 weeks shorter than the gestational age (Engle et al. 2004). In the context of technology-assisted reproduction, precise estimation of conceptional age is possible. Women's self-report of last menstrual period is one way to estimate gestational age. Ultrasonography can be used to estimate or confirm gestational age if it is done early in the pregnancy. This is achieved by comparing the ultrasound estimates of fetal size measurements such as biparietal diameter (width of the skull) and the length of the femur (thighbone) against established fetal growth data (e.g., Hadlock et al. 1982). An estimate of gestational age based on the last menstrual period confirmed by an ultrasound scan is often considered the most reliable estimate. Gestational duration can be partitioned into blocks of 14 weeks starting from the first day of the last menstrual period. They are called the first, second, and third trimester (Rijken et al. 2011).

Repeated ultrasound scans can also be used to measure and monitor fetal growth. Different dimensions of the fetus grow at different rates during different periods. For example, the velocity of growth in body length is higher during the second trimester than the third trimester, but the velocity of growth in weight is higher in the third trimester than the second. Due to such variations, it was suggested that the timing of *intrauterine growth retardation* (IUGR) may determine the type of growth outcome in newborns (Villar and Belizan 1982). If it occurs in the first trimester, a well-proportioned but small baby is expected. If it occurs in the second trimester, the baby is expected to be short. If it occurs in the last month of pregnancy, a thin baby is expected.

1.2.2 At Birth

Size at birth is the combined result of fetal growth and gestational duration. *Birth weight* is most easily measured. *Supine length*, or simply length, is more difficult to measure because babies curl up instead of lying straight for us to measure. Measuring length and other aspects of growth such as head circumferences requires more skills and training, and therefore is less commonly performed. The World Health Organization (WHO) defines *low birth weight* (LBW) as a birth weight less than 2500 grams. Although boys tend to be heavier than girls, the definition of LBW does not take gender into account. At birth, boys and girls on average weigh about 3.35 and 3.23 kg, and measure 49.9 and 49.1 cm in length, respectively (WHO 2006).

When there is no ultrasound data, size at birth measures are often taken as the indicators of fetal growth. In order to separate the effects of gender and gestational duration on size at birth, the measurements are sometimes standardized for gender and gestational duration, that is, growth reference values are developed for each gender and gestational age separately in terms of percentiles or standardized scores, also known as z-scores or standard deviation scores. For example, Niklassson and Albertsson-Wikland (2008) developed the Swedish reference values for weight, length, and head circumference for boys and girls for each gestational age from 24 to 40 weeks (in completed weeks). A negative z-score means that the size at birth measure falls below the median for newborns of the same sex and same gestational age. Short gestational duration and IUGR both can result in a small size at birth. Babies born *small for gestational age* (SGA) is an indication of IUGR. There is no universally accepted definition of SGA. Classifications based on weight/length z-score below −2 or weight/length below the 10th, 5th, or 3rd percentile curves have been used. Since gender is almost always a factor to standardize for, it is common to omit mentioning gender and describe standardized values as weight-for-gestational age values, even though they are actually weight-for-gender-and-gestational age values.

At birth, weight, length, and head circumference tend to be approximately normally distributed given a gestational age (Fok et al. 2003; Wilcox 2001), although some studies have found some skewness in birth weight (Niklasson and Albertsson-Wikland 2008). Figure 1.1 shows the median, 5th, and 95th percentiles of weight-for-gestational age for Swedish girls. There is nonlinearity in the curves. They rise at an accelerating rate in this range of gestational age. The weight-for-gestational age is positively skewed, as reflected by a slightly larger distance between the 95th percentile and the

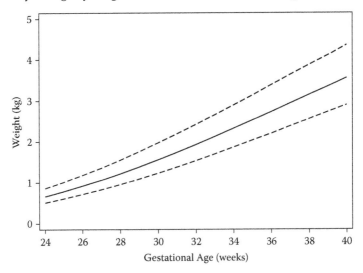

FIGURE 1.1
Weight for the gestational age reference for Swedish girls. Solid line for median and dashed lines for 5th and 95th percentiles. (Data from Niklasson, A., and Albertsson-Wikland, K. 2008, Continuous growth reference from 24th week of gestation to 24 months by gender, *BMC Pediatrics* 8:8, Table 4.)

median than that between the 5th percentile and the median. Furthermore, the variability increases with gestational age, as reflected by the widening distance between the three curves. A newborn at 24 weeks of gestational age falls below the 5th percentile if she is at least 0.15 kg lighter than the median, but a newborn at 40 weeks of gestational age falls below the 5th percentile only if she is at least 0.66 kg lighter than the median. That both the median and variability change in relation to age is a common phenomenon in the studies of growth.

Birth weight is about 5% of adult weight. In contrast, the weight of a newborn's brain is about a quarter of the weight of an adult's (Tanner 1989). The comparison shows the importance of brain development during the fetal period. It has been suggested that fetuses have a brain-sparing strategy, that is, they have a tendency to maintain brain development and head growth even at the expense of slower growth in other parts of the body if the intrauterine environment is not favorable (Barker 1998). Head circumference and ratio of head circumference to body length at birth have been used as indicators of fetal growth.

1.2.3 Infancy to Puberty

In demography and pediatrics, *infancy* usually refers to the first year of life after birth. However, it is common to use this term loosely to refer to the first 2 or 3 years. That is the usage of the word in this book. The terms age, chronological age, and postnatal age refer to the same interval from birth to present. The American Academy of Pediatrics defines *chronological age* as the time elapsed after birth (Engle et al. 2004). In the studies of children up to 2 to 3 years of age who were born preterm, the *corrected age* is sometimes used instead. Corrected age is calculated by subtracting the number of weeks (or days) born before 37 weeks (or 259 days) from the chronological age. Corrected age is used if the researchers aim to study the effect of exposures other than preterm birth and therefore intend to remove the impact of preterm birth from the evaluation. Corrected age and chronological age are synonymous in term infants.

The counting of age in days in daily life and scientific context can differ. In daily life it is common to say that a baby born today is 1 day old and a baby born yesterday is 2 days old today, that is,

$$\text{Age in days} = (\text{Today's date} - \text{Date of birth}) + 1$$

But in the scientific context, as well as some software and formula concerning age standardization, it is usually the time elapsed after birth, that is,

$$\text{Age in days} = \text{Today's date} - \text{Date of birth}$$

This is the way age in days is counted in this book.

From birth to the age of 24 months, the standard anthropometric practice of measuring stature is to measure supine length, meaning the infants are lying on the back. It needs two assessors to hold the infants in a proper posture, with legs extended, in a specially designed measurement board called infantometer. From the age of 24 months onward, it is *standing height* (or simply height) that is measured. As this book will discuss growth both before and after 24 months, it will use the terms length and height interchangeably for convenience.

Weight and height are the two most commonly measured aspects of postnatal growth. Population distribution of postnatal weight tends to be positively skewed whereas that of length tends to be normally distributed. Weight adjusted for height by use of the *body mass index* (BMI) or some other measures is an indicator of short-term growth faltering. A low weight-for-height or BMI is called *wasting*. Length gain is conventionally called *linear growth*. A low height-for-age is an indication of chronic growth faltering and is called *stunting*. Again, gender is usually involved in the standardization, and the term height-for-age usually means height-for-gender-and-age. Epidemiological observations have shown that postnatal wasting and postnatal stunting did not always take place together. The two dimensions of growth faltering appear to represent different biological processes (WHO Working Group 1986).

The *Infancy-Childhood-Puberty* (ICP) model (Karlberg 1987; Karlberg et al. 1994) describes linear growth as three additive and partly overlapping phases (Figure 1.2). According to this model, the *infancy phase* begins before birth and continues at a rapidly decelerating rate after birth. The attained body size is increasing but the growth velocity is decreasing. This phase tails off at about 2 years of age. The infancy phase can be seen as an extension of fetal growth. SGA infants tend to have higher growth rates during this phase. In a study of Swedish and Chinese full-term infants born short for gestational age, 65% to 92% of the infants caught up to normal height range within the first 3 to 6 months (Karlberg et al. 1997). *Catch-up growth* in length could be identified as early as 3 months of age. Thin babies also catch up in weight soon after birth (Ashworth 1998). Such catch-up growth suggests that intrauterine environment instead of inherent health problems in the infants is the main determinant of size at birth. Once released from the adverse intrauterine environment, their "fetal" growth potential starts to pick up.

In the second half of the first postnatal year, the *childhood phase* of the ICP model begins. This is hypothesized to be driven by growth hormone. This phase of growth partly overlaps with the infancy phase. The onset of the childhood growth phase is marked by a sudden increase in growth velocity. An onset after 12 months of age is considered a delayed onset (Liu et al. 2000). The childhood phase of growth is slowly decelerating. Since the partly overlapping infancy phase has a rapidly decelerating growth rate, a delay in the onset of the childhood phase is one of the reasons for growth stunting in young children (Liu et al. 1998).

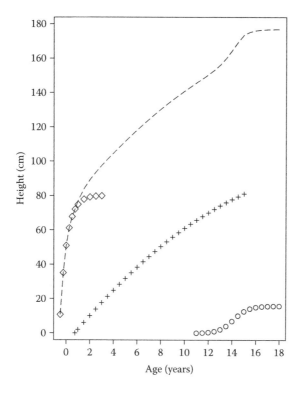

FIGURE 1.2

Illustration of the infancy–childhood–puberty growth model. Height from birth to maturity of a person who enters the childhood phase at age 9 months and reaches peak height velocity at 14 years and 2 months. Attained height (dashed line) and its infancy (◊), childhood (+), and puberty (o) components.

The postnatal pattern of BMI or body fat as percentage of body weight is quite similar across gender and populations (Cole 2004; Fomon and Nelson 2002). In the United Kingdom, median BMI at birth is about 14. It rises sharply and peaks at about 18 at the age of 8 months. Then it drops to about 15 to 16 at about age 6 years. Then BMI rises again (Cole 2004). This second rise in body fat or BMI at about 6 years of age is known as the *adiposity rebound*. It has been an important concern in recent years due to the rising prevalence of obesity and metabolic diseases. There is substantial variation in the age at adiposity rebound. Obese adults tend to have earlier adiposity rebound (Cole 2004; Rolland-Cachera et al. 2006).

There is a wide variation in the onset age of pubertal growth both between the genders and within each gender. It begins at approximately 10 years of age in girls and 12 years in boys. Growth accelerates and reaches peak velocity at about 12 and 14 years of age in girls and boys, respectively (Liu et al. 2000). Further gain in height is limited and uncommon after 16 years in girls and 18 years in boys. It is common to define height measured at age 18 years as *final height*. The final height of girls is on average shorter than that of boys partly because boys continue to grow for a longer period of time during the childhood phase before they enter the puberty phase, which is about 2 years more than that of girls. Similarly, within each gender, those who had a later onset of pubertal growth tend to be taller.

At birth, the length and weight of babies in different countries and socio-economic situations are fairly similar. However, shortly after birth, infants in many developing countries begin to show stunting, wasting, and under-weight. These phenomena continue until about age 24 months and then stabilize (Victora et al. 2010). As such, there have been a lot of international efforts to promote child growth that focused on interventions in the first 1000 days of life, from pregnancy to the child's second birthday (Save the Children 2012). It is believed that a large part of the growth deficits in children in developing countries is the result of a poor nurturing environment rather than ethnic or genetic factors (WHO 2006). Some examples of studies that have provided evidence to support this view include:

- A longitudinal study that showed Pakistani infants on average had poorer postnatal growth than Swedish infants in the first 2 years of life, but Pakistani infants born in upper-middle-class families had growth trajectories similar to the Swedish average (Karlberg et al. 1998).
- A cross-sectional study that showed 12- to 23-month-old Indian children from an affluent area had weight-for-age, length-for-age, and weight-for-length similar to an American growth reference (Bhandari et al. 2002).

Substantial catch-up growth can occur during childhood. A study of Romanian orphans who experienced severe deprivation during institutionalization before adoption by Canadian families at the mean age of 24 months showed considerable gain in height- and weight-for-age z-scores by 10.5 years of age even though there remained a difference from the norm (Le Mare and Audet 2006). It should be noted that catching up in centimeters is more difficult than catching up in z-score. As children get older, the standard deviation in height in centimeters increases due to cumulative exposure to multiple factors, some of which are independent from each other. Therefore, an early deficit in centimeters due to short-term exposure to one risk factor becomes relatively small in z-score at a later time. The impact of an early exposure is relatively speaking diluted by the influences of the subsequent exposures, but its absolute impact can be more persistent.

1.2.4 Adulthood

Although height usually does not change a lot during adulthood, weight and body mass index may change substantially. The WHO defines *obesity* as a

BMI equal to or larger than 30 and *overweight* as a BMI equal to or larger than 25. Table 1.1 shows the BMI cutoff points for various definitions of weight status (WHO Expert Consultation 2004). In some well-off societies, the prevalence of obesity increases sharply from about age 20 to about 50 years and then shows signs of stabilizing (Kwon et al. 2008; Midthjell et al. 1999). There was discussion on whether different definitions are needed for Asian populations, but the WHO definitions have not been modified because there was not sufficient data to indicate a clear BMI cutoff point for either overweight or obesity in Asia (WHO Expert Consultation 2004).

BMI is a crude index of body composition. There are advanced methods to measure body composition, such as magnetic resonance imaging and air displacement *plethysmography*. However, indices based on tape measures are popular because of their affordability and portability. *Central obesity* is also called *abdominal obesity*. A large part of abdominal fat is visceral (intra-abdominal), as opposed to subcutaneous (under the skin). A large waist circumference or a high *waist-to-hip ratio* (WHR) is an indication of central obesity. Women tend to have a lower WHR than men. The range of WHR for healthy Caucasian women and men is about 0.67 to 0.87 and 0.85 to 0.95, respectively (Schernhammer et al. 2007; Singh 2002). Yusuf et al. (2005) defined high WHR as WHR >0.83 for women and >0.90 for men. They found high WHR more strongly predictive of the risk of myocardial infarction than overweight or obesity defined according to BMI. Menopause appears to be associated with gaining fat, especially visceral fat (Toth et al. 2000).

Adult height may decline during late adulthood. Some longitudinal studies estimated that from about 40 years of age height started to decline at an increasing velocity (Leon et al. 1995). From 45 to 69 years, the reduction was

TABLE 1.1

Classification of Adult Underweight, Overweight, and Obesity According to Body Mass Index (kg/m²)

Main Classification	Subclassification	BMI Cutoffs
Underweight		<18.50
	Severe underweight	<16.00
	Moderate underweight	16.00–16.99
	Mild underweight	17.00–18.49
Normal range		18.50–24.99
Overweight		≥25.00
	Pre-obese	25.00–29.99
Obese		≥30.00
	Obese class I	30.00–34.99
	Obese class II	35.00–39.99
	Obese class III	≥40.00

Source: WHO Expert Consultation, 2004, Appropriate body-mass index for Asian populations and its implications for policy and intervention strategies, *Lancet* 363:157–163.

about 1 mm per year. The reduction was over 2 mm per year after the age of 80 years. The causes of the reduction include the loss of vertebral bone mineral density and thinning of intervertebral disks. Furthermore, loss of muscle strength is associated with *kyphosis*, overcurvature of the upper back. Excessive kyphosis is called *hyperkyphosis*. The prevalence in the elderly was estimated at 20% to 40%. It is associated with poor pulmonary and physical function as well as the risk of injuries (Kado et al. 2007).

1.3 Human Development

1.3.1 Fetal Period

Neurons, also known as nerve cells, are the major building blocks of the brain. The generation of neurons, or *neurogenesis*, begins early in the first trimester. By the end of the second trimester, most neurons are already generated and have settled at their final locations to begin the development into a complex structure. The number of neurons hardly increases afterward (Nowakowski 1987). *Synapses*, or connections between nerve cells, begin to form in the second trimester. The formation continues after birth. Around the beginning of the third trimester, the brain and lungs become relatively mature and possibly sufficient for a fetus born at this time to be *viable* (able to survive). Fetus delivered before this point is usually considered *previable*. During the third trimester, the *cerebral cortex*, on which our senses, motor, and cognitive functions depend, develops from a smooth structure to a deep folding structure. Such folding increases surface area and is believed to correlate with later abilities (Dubois et al. 2008).

Although a normal pregnancy is expected to have about 40 weeks in duration, there is substantial variation in gestational duration. A population-based registry showed a standard deviation of 13 days (Bergsjø et al. 1990). A birth before 37 weeks (259 days) of gestation is called *preterm* birth. Preterm birth is a major risk factor of morbidity, disability, and retarded development (Drewett 2007). Birth after 42 weeks (294 days) of gestation is called *postterm*. Postterm birth is also a risk factor of developmental problems. Births between 259 and 294 days are considered normal and called term birth. Gestational duration is negatively skewed. Preterm births are more common than postterm births.

Near the end of the second trimester, the fetus can begin to feel pain and react to sounds; in the third trimester the fetus shows personality and preferences (Berk 2003). Modern imaging techniques have shown that fetuses have facial expressions. Whether that represents emotion is not yet known. There is some evidence that the "programming" of fetus by its environment is dependent on the timing and aspects of development affected. For example, Yehuda et al. (2005) studied cortisol levels in babies born to mothers who developed posttraumatic stress disorder during pregnancy in response to the World Trade Center collapse on September 11, 2001, whereas Davis et al. (2007) studied infant temperament in relation to maternal stress measurements during pregnancy. They found that exposure to maternal stresses during the third trimester had more apparent associations with the offspring's suppression of cortisol levels and temperamental negativity, respectively. In contrast, a study of the Dutch famine during the Second World War showed that maternal malnutrition during (approximately) the first trimester, but not later, contributed to the occurrence of coronary heart disease in the offspring (Roseboom et al. 2000).

1.3.2 Infancy

The brain develops rapidly in the third trimester and in the first few years of life. At birth, a neuron in the cerebral cortex has about 2500 synapses. The number was estimated to increase sixfold by 2 to 3 years of age and then begin to drop (Gopnik et al. 1999). The increase and decrease are referred to as *synaptic blooming* and *pruning*. In the latter process, unused and weak connections are removed but strong connections are maintained. However, different regions have different timing of blooming and pruning. The *prefrontal cortex*, the region that controls higher cognitive functions, shows synaptic blooming at about 3 years of age and the formation of synapses in this region can continue through adolescence (Thompson and Nelson 2001).

Babies are born with not only reflexes but also cognitive abilities. Some reflexes are permanent, such as quickly closing eyelids when exposed to bright light. Some are temporary, such as turning the head toward the source of stimulation when touched near the mouth. These reflexes may be functional to the baby's survival (Berk 2003). *Primary emotions*, or emotions that do not require self-consciousness, like joy and fear also occur early in infancy (Santrock 2008).

Babies demonstrate measurable cognitive abilities soon after birth. For example, newborns can recognize faces and discriminate mothers from unfamiliar females (Pascalis et al. 1995). They can also mimic facial expressions of adults (Gopnik et al. 1999). Measurements based on visual perception are important methods in understanding infant development. The *visual preference* method measures the duration infants look at different stimuli. The pioneering works of Robert Fantz in the 1960s demonstrated that infants as young as only 2 days of age preferred to look longer at visual stimuli that resembled human faces than simple discs (Santrock 2008). *Habituation* refers to a decrease in responsiveness to a repeatedly presented stimulus. Responsiveness is usually measured by the amount of time an infant visually fixates at the stimulus. *Dishabituation* refers to responsiveness to a new stimulus after habituation. These processes involve information processing and memory. Quicker habituation and stronger dishabituation reflect better cognitive abilities. A number of longitudinal studies have assessed the association between infant habituation/dishabituation and later cognitive performance. In an analysis of 38 samples, it was found that habituation/dishabituation had a moderate and stable correlation (average 0.37) with cognitive performance in childhood and adolescence, and the timing of measurement in infancy or in childhood/adolescence did not materially affect the correlation (Kavšek 2004).

Given that dramatic changes in the brain occur during early life, it has long been speculated that there may be a *critical window* for development. One example is that infants under 6 months are able to distinguish native- and foreign-language sounds, but this ability seems to decline sharply between 6 and 12 months unless they are exposed to the foreign sounds (Kuhl et al. 2003). Different parts of the brain have different developmental time courses, and hence there may be no single time period that is critical for all aspects of development.

1.3.3 Childhood to Adolescence

According to the WHO Multicentre Growth Reference Study (MGRS), median head circumference increases from about 34 cm at birth to about 46 cm at 12 months. Then the increase slows hugely to reach about 50 cm at 4 years (WHO 2007). From age 4 to 18 years it merely increases by about 5 cm and is mainly driven by increase in skull thickness, not brain size (Giedd 2004). *Myelination* is the coating of neuronal axons with a layer of fatty cells, which speeds up transmission. *White matter* on magnetic resonance scans indicates bundles of myelinated axons. *Gray matter* contains neuronal cell bodies. Both white and gray matters rapidly increase in volume in the first 3 to 5 years of postnatal life (Tanaka et al. 2012). After that, the amount of white matter continues to increase but that of gray matter declines, similar to synaptic pruning (Giedd 2004; Lebel and Beaulieu 2011). Studies have demonstrated that developmental delays in childhood are associated with reduced levels of white matter maturation (Nagy et al. 2004; Pujol et al. 2004).

By age 6 years, the total size of the brain is approximately 90% that of an adult's (Giedd 2004). Although brain size does not change much during childhood and adolescence, its components do. From about 5 years of age, the amount of white matter increases quite linearly, with a slight degree of slowing down during adolescence. In particular, the *corpus callosum*, which is the white matter structure connecting the similar parts of the left and right hemispheres, grows most linearly from childhood through adolescence into early adulthood (Giedd 2004). It may facilitate coordination of motor activities as well as various aspects of cognitive functions.

In the first few years of life, a test of *intelligence quotient* (IQ) is not feasible. Assessment of development is usually done by observing *developmental milestones*, which are activities that indicate maturity of abilities. Children develop in many aspects, not just intelligence. For example, self-conscious emotions like pride and shame tend to appear during school age. Children also develop cognitive coping strategies to handle stress and suppress negative emotion (Santrock 2008). Developmental assessment usually covers multiple domains, such as gross motor, fine motor, social-emotional, language, and cognitive development. Some milestones may involve multiple abilities and the classification of such a milestone into a single domain involves some degree of arbitrariness. Milestones are usually attained in quite consistent sequence. For example, in the first 2 years of life, infants start to sit without support, stand with assistance, crawl on hands and knees, walk with assistance, stand alone, and walk alone usually in this sequential order. The 10th, 25th, 50th, 75th, and 90th percentiles of the age distributions from the WHO MGRS, a large-scale international study, are shown in Figure 1.3 (WHO MGRS Group 2006b). The performance on developmental milestones may be converted into an overall or domain-specific *developmental quotient* (DQ), or be used to screen for developmental delays or more serious disabilities. There have been questions about whether developmental delay is just a short-term phenomenon that has no lasting consequence, or it will have a long-term impact. Affirmative answers to the former and latter are known as the "lag hypothesis" and the "deficit hypothesis," respectively. Research has suggested that DQ assessed in the first year has weak correlation (<0.3) with IQ at 8 to 18 years of age; but those assessed between 1 to 5 years do have moderate correlation (0.3 to 0.5) with later IQ (Drewett 2007) and with school

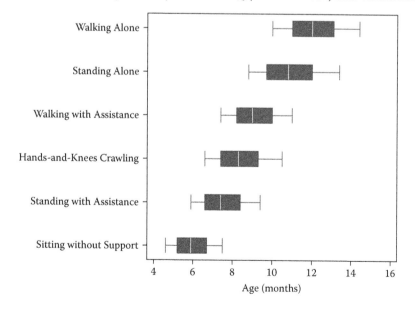

FIGURE 1.3
Distribution (10th, 25th, 50th, 75th, and 90th percentiles) of age at achievement of motor milestones. (Data from WHO Multicentre Growth Reference Study Group, 2006b, WHO Motor Development Study: Windows of achievement for six gross motor development milestones, *Acta Paediatrica* 450:86–95, Table II.)

grade attained by the age of 18 years (Grantham-McGregor et al. 2007). IQ tests can be performed from 4 or 5 years of age. IQ at 8 years of age onward shows stable correlation with adult IQ and predicts educational and occupational success (Drewett 2007; Fergusson et al. 2005).

Adolescence involves intense physical and social changes. On the one hand, adolescents need to continue improving the mastery of abilities they have already acquired. They also need to apply these abilities into practical use in the education or employment arenas. On the other hand, new challenges arise. One needs to adapt to their new gender and social roles. Adolescence comes with a higher mortality and disability rate. This sudden rise in the mortality curve is known as the accident hump and is attributed to risk-taking behaviors. Some research suggested that adolescents showed more extreme emotions. Emotional fluctuations in this period may relate more to social environment than biological changes (Santrock 2008).

Substantial catch-up in cognitive ability during childhood is possible. Studies of Romanian children adopted by British or Canadian families have shown that removal from a depriving environment can lead to significant recovery in cognitive and social functioning, and that the removal was beneficial even if it takes place after the first 2 years of life (Benoit et al. 1996; Rutter 1998; Rutter et al. 2004). They have also shown that major developmental deficits persisted in some of these adoptees, but the developmental outcomes are heterogeneous and therefore suggest a complex pathway instead of simple determinism. A population-based study of Swedish conscripted males showed that being born SGA was associated with deficits in intellectual and psychological performance, but faster growth in height between birth and 18 years cushioned this effect (Lundgren et al. 2001).

Similar to the studies of child growth, studies of development in young children have also demonstrated interesting similarity between populations. For example, in the aforementioned study of Pakistani children up to 24 months, on average the children achieved psychomotor milestones later than European and North American children. But the children with an upper-middle-class family background had development trajectories similar to that of children in Europe and North America (Yaqoob et al. 1993). There are doubts on whether IQ tests developed in Western societies for older children and adults may be culturally biased. There has been concern that some IQ tests may not provide a valid estimate of the intelligence of people from foreign cultures (Shuttleworth-Edwards et al. 2004; Verney et al. 2005). Some countries and studies may specifically develop their own IQ tests, while some tests try to prevent cultural bias by avoiding the use of verbal instructions.

1.3.4 Adulthood

Longitudinal data have shown that the volume of white matter is on the increase, gray matter on the decrease, and total brain size is about stable in people in their early 20s (Giedd 2004; Lebel and Beaulieu 2011). From about 55 to 80 years, total cerebral volume and white matter volume decline by about 0.4% per year, while that of gray matter declines by 0.3% per year (Chee et al. 2009). The decline appears to be quite linear. There are some findings that new neurons can be generated during adulthood, but their functionality is unclear (Santrock 2008). Regardless of changes in size or structure, the aging brain adapts. The brains of younger and older people handle the same tasks in somewhat different ways. For example, young adults mainly show activation of the right prefrontal cortex during memory tasks, but older adults are likely to show activation of both the left and right sides (Madden et al. 1999). This may reflect a compensatory adaptation of the aging brain in handling demanding tasks.

Cognitive function grows in early adulthood and it does not decline until mid-adulthood. The Seattle Longitudinal Study provided excellent data for tracking various aspects of functions across a wide age range (Schaie 2005). Figure 1.4 is based on the mean changes in cognitive function scores over 7 years among people in different age groups at the beginning of the 7-year period. The scores were standardized to have SD 10 before the calculation of mean changes. By piecing the 7-year changes in different age groups together to show cumulative changes, the study demonstrated that most scores increased in early adulthood and peaked at about 53 years of age. Two of the 7 scores increased by about 4 points (0.4 SD) between 25 and 53 years. That was not negligible. Although the scores dropped afterward, they were not lower than their levels at 25 years until the 60s to early 70s. Analysis of changes over 14 years or more among subsamples of participants

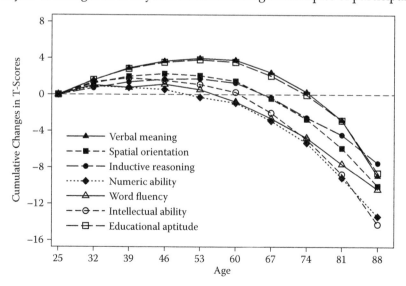

FIGURE 1.4
Mean cumulative changes in cognitive function in adulthood. (Data from Schaie, K. W., 2005, *Developmental Influences on Adult Intelligence: The Seattle Longitudinal Study*, 2nd ed., New York: Oxford University Press, Table 5.1.)

who had been followed for at least 14 years corroborated with the pattern shown in the figure. Analogous to the changes in activation of left and right hemispheres of the brain, problem-solving and decision-making skills and methods also changes as people age. Older adults are more likely to utilize practical experience and they are more effective in solving emotionally laden or interpersonal problems (Santrock 2008). It is important to appreciate that people develop or decline in terms of not only depth but also breadth.

Various aspects of cognitive abilities, processing speed in particular, decline with age from about the mid-50s. Such changes appear to be similar in magnitude across populations (Chee et al. 2009). Deterioration in cognition ranges from *age-related cognitive decline* on the mild side to *dementia* on the severe side. Age-related cognitive decline refers to loss in cognitive abilities that is considered normal for one's age. *Mild cognitive impairment* is a pathological state between normal functioning and dementia (Feng et al. 2009). Dementia is a syndrome with multiple manifestations such as disorientation in time and place, inability to remember relationships, and aggression. *Alzheimer's disease* is the most common type of dementia. The concept of *cognitive reserve* began with the postmortem examination finding that some people with neuropathological features of Alzheimer's disease had good functional and cognitive performance before their death (Katzman et al. 1988). One hypothesis was that these people had larger brain volume and more neurons before the disease started, and such reserve made them more resilient despite neurological damages. Studies have suggested that normal growth in childhood, lifestyle factors such as lifetime bilingualism, and common cognitive activities such as reading may contribute to cognitive reserve and may reduce the decline of cognitive function (Craik et al. 2010; Drewett 2007; Wilson et al. 2002).

Cognitive decline may be slowed or even reversed by interventions. For example, the Seattle Longitudinal Study has demonstrated that cognitive training can improve spatial orientation and reasoning skills (Schaie 2005). About 40% of the receivers of the training restored their performance to a level they had 14 years ago. The training effect was not transient; it lasted for at least 7 years without booster training. The training effect appeared to be specific. Training on spatial orientation did not improve reasoning skill or vice versa. Physical fitness training also benefits the cognitive function of older adults. A review of 18 intervention studies confirmed that aerobic fitness training improves cognitive functioning, mainly in executive-control processes such as planning and switching between tasks (Colcombe and Kramer 2003).

1.4 Statistical Considerations

The characteristics of human growth and development give rise to considerations in statistical analysis, which shaped the content of this book.

1.4.1 Issues Arising from Nature and Nurture

Growth and developmental outcomes naturally change in relation to age, be it gestational, chronological, or corrected age. A person assessed at a younger age is expected to differ from a person assessed at an older age. Not only the level (e.g., mean and median) but also the variability (e.g., standard deviation and range) can change over age. For example, the standard deviation of length is approximately 1.9 cm at birth and 4.7 cm at 5 years (WHO 2006). A difference in 2 cm in length at birth is a big difference, but a difference of 2 cm at 5 years is relatively minor.

Gender is another factor that makes the analysis more complex. It is well known that males and females are different in growth and pubertal development. It is more subtle that some developmental milestones are also dependent on gender or gender-related nurturing. The Denver Developmental Screening Test, Singapore (DDST, SG) is based on the Denver Developmental Screening Test (DDST) of the United States (Frankenburg et al. 1991). One of the test items in the personal–social domain is "comb doll's hair." While 90% of Singaporean girls passed this item at age 22.0 months, it was until 26.9 months that 90% of boys passed this item. If not properly calibrated and interpreted, this item may give rise to an impression that boys are more likely to be delayed in development while in fact it is likely the culture and upbringing that makes the difference. In the development of a screening test for use in Malawi, some initial test items, for example, "wash dishes," were removed from the test because of their gender dependence (Gladstone et al. 2008).

Statistical analysis of data on growth and development usually has to take into account the effects of age and gender. One reason is to make the interpretation of findings easier and independent of age and gender. Another reason is to avoid the results being confounded by age and gender. There are other factors that may affect growth and development, such as parity (number of times a woman has given birth). The first baby a woman gives birth to tends to be lighter than his or her future siblings. There have been calls for taking such additional factors into account. The creation and use of growth reference values (percentiles and z-scores) and the application of multivariable regression analysis play important roles in facilitating interpretation and removing the confounding due to these factors.

1.4.2 Issues Arising from Measurement

Developmental assessment usually requires multiple domains and multiple indicators be measured to sufficiently cover the abstract concepts. Advanced statistical techniques may be needed to combine the data on multiple variables to form summary measures or to perform proper interpretation after multiple analyses.

Since the assessment of some items can involve subjectivity on the part of the test developers as well as the observers, there is a need to establish the validity of the tests and the inter- and intraobserver reliability. Assessment of growth is to some extent easier. However, inaccuracy in measures such as supine length and head circumference is expected. Assessment of reliability is not only part of the data analysis exercise but also part of the observer training.

Measures of growth and development are often done repeatedly, either at regular time intervals or at chosen ages. Age at onset of some events, such as adiposity rebound or cognitive decline, is usually not directly observed. Curve-fitting techniques are needed to find the age and define the trajectory characteristics. In some studies, at the end of the data collection period most participants have experienced the defining events but some still have not. In this case, the age at onset of the event is known to be larger than the age at the last assessment, but its exact timing is not known. This is called *right censoring*. On the other hand, a milestone may be reached between two consecutive assessments. The age at onset is known to be between the ages at the earlier and later assessments, but the exact timing is unknown. This is called *interval censoring*. Statistical methods for analysis of time-to-events, also known as survival analysis, are useful for this kind of data.

1.4.3 Complex Relationship

Growth and development are often measured repeatedly over time. There are complex relationship and interpretation issues in the trend over time and the relationship between variables across time points. For example, a boy may rank number one in his class examination result this school term. In the next school term, at best he remains number one, but chances are he will rank lower, perhaps because of bad luck on his part or good luck on the part of his classmates, or some other reasons. Such changes do not necessarily indicate a change in cognitive ability, it may just be a statistical phenomenon referred to as *regression to the mean*.

Some measures of growth and development are correlated. For example, weight is correlated with height. As such, an apparent relationship (or lack of relationship) between weight and a health outcome may be spurious; it may in fact be attributed to their association with height. Statistical procedures are needed to separate the complex patterns, such as by calculating weight-for-height percentile scores. Some ratio indices have been used to assess body proportionality, such as the waist-to-hip ratio and the ratio of head circumference to supine length. But the use of such ratio indices may produce a spurious correlation as well (Kronmal 1993). We need to know when such measures should be avoided and we need to know alternative analytic methods.

In doing multivariable regression analysis of the impact of an exposure status on an outcome, one should be cautious in including variables that may fall on the causal pathway, or intermediate variable (Victora et al. 1997). The regression coefficient on the exposure variable after adjustment for the intermediate variable does not represent the totality of the impact of the exposure.

There are similar issues in studies of growth and development (Lucas et al. 1999). A later measurement may include an earlier measurement as one of its components. For example, height at age 5 years is the sum of height at 2 years plus height gain from 2 to 5 years. A regression estimate of association between a health outcome and height at 5 years adjusted for height at 2 years requires special care in the interpretation. The use of different types of regression analysis and the interpretation of the findings form a significant part of this book.

1.4.4 Normal versus Abnormal States

In the population, degree of growth and development varies on a broad and continuous spectrum. There may be some subgroups in the population that are distinct from others and their members mostly locate at one extreme end of the spectrum. Furthermore, from a health and social care provider's point of view, it is convenient to have simple decision rules to determine who should be given what type of care. Consequently, there is often a demand to divide the broad spectrum into intervals. The analysis of growth and development involves developing statistical definitions of normal versus restricted intrauterine growth, normal versus impaired cognitive function, and so on. It thus also involves analysis of dichotomous exposure status and outcomes.

Statistically, however, dichotomization of continuous variables creates problems (Royston et al. 2006; Senn and Julious 2009). It is well known that dichotomization leads to a loss of information and statistical power, that is, it is less likely to detect an association even if there truly is an association. Other problems include insufficient adjustment for covariates that have been dichotomized and bias arising from data-driven selection of cutoff points for dichotomization. There is a tension between clinical practice and scientific discovery. The use of dichotomization may sometimes be rightfully justified or dictated by the research context. But we should be mindful of the risk that: "If all you have is a hammer, everything looks like a nail." We will need a range of statistical tools that cover both categorical and continuous variables so that we can view and handle things in flexible ways.

The use of dichotomization also bears the risk of oversimplifying a heterogeneous pattern of outcomes. For instance, while low birth weight is well known to relate to health risk for infants, being heavy may also be a problem. Figure 1.5 shows the chance of survival to discharge without major morbidity in relation to birth weight z-score and gestational age in male infants born under 33 weeks of gestational age in Canada (Shah et al. 2012). Birth weight z-score had a J-shape relationship with the survival outcome given the same gestational age. Had a dichotomy of birth weight z-score been used, the finding might lead the readers to ignore the health risk of being large. A similar J-shape relationship has been reported by other studies of growth and development (e.g., Andersen et al. 2010; Cheung et al. 2001b). Regression analysis methods that allow the exposure variable to be continuous are natural for the exploration of nonlinear

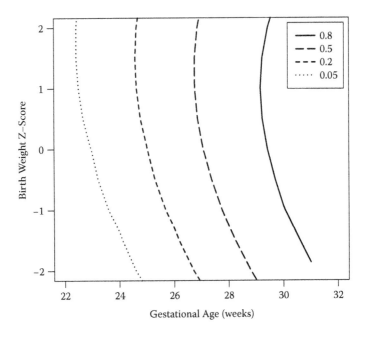

FIGURE 1.5
Contour plot of survival to discharge without major morbidity in relation to gestational age and birth weight z-scores. Contour lines indicate survival probability 0.05, 0.2, 0.5, and 0.8. (Based on Shah, P. S., Ye, X. Y., Synnes, A., et al., 2012, Prediction of survival without morbidity for infants born at under 33 weeks gestational age: A user-friendly graphical tool, *Archives of Disease in Childhood—Fetal and Neonatal Edition* 97:F110–115, Table 4.)

relationships as well as the synergistic/antagonistic (interaction) effect between exposure variables. Indeed, the authors of the Canadian study succinctly estimated and reported the relation between the outcome and the predictors by using quadratic terms in the regression equation

$$\text{log odds of outcomes} = -31.334 + 1.672 \times GA - 0.019 \times GA^2 + 1.107 \times BW$$
$$-0.130 \times BW^2 - 0.029 \times GA \times BW - 0.340 \times \text{Male}$$

where GA and BW denote gestational age in weeks and birth weight-for-gestational age z-scores, respectively, and Male is a binary variable with value 1 for males and 0 for females. Such succinctness and accuracy cannot be achieved by categorizing the continuous variables.

2

Causal Reasoning and Study Designs

The purpose of examining growth and development in relation to exposure status is ultimately to establish cause and consequence. If a causal relation is firmly established, efforts can be made to remove the adversary exposure or introduce the beneficial exposure. However, establishing a causal relationship is not an easy task. A number of conditions need to be considered before making a claim of causality (Babbie 2010; Hennekens and Buring 1987; Hill 1965). This chapter will discuss the key conditions as well as study designs that can facilitate the meeting of the conditions. This book uses the word "subjects" to mean research participants. The word is used in statistics without negative connotations.

2.1 Causality

2.1.1 Association

Association is not causation, but establishing an association between two phenomena is a necessary condition for claiming a causal relationship between them. Therefore, much of empirical research concerns the evaluation of association. To evaluate association, it is essential that the hypothesized cause varies across subjects in the study. For example, we may hypothesize that at least 30 minutes of moderate physical exercise daily prevents cognitive decline in old age. In a study where nobody exercises daily, there is no way to establish the cause and consequence statistically. The cause has to be a *variable*, not a *constant*, in a statistical study of association. Similarly, the consequence also has to be a variable, not a constant.

2.1.2 Temporal Sequence

A cause must occur before its consequence. In order to establish a causal relationship, it is essential to establish the temporal sequence. Some phenomena are short-lived, for example, an episode of an acute disease or a natural disaster. It is easier to examine the temporal sequence involving these phenomena. Some phenomena are slow processes and have long-term presence. If shorter people are more likely to show trait anxiety, could it be that being shorter is the cause of being anxious? The attempt to answer this question will involve the examination of whether being short occurs before having trait anxiety. Since these two are both long-term phenomena, it is not easy to establish the temporal sequence. It requires long-term follow-up data. If the true temporal sequence is the other way around, that is, an anxious child later becomes a short adult, then the hypothesis of shortness causing trait anxiety may commit the fallacy of *reverse causality.*

2.1.3 Nonspurious Association

A claim of causality must not be built on a *spurious association.* The upper panel of Figure 2.1 shows a hypothesized pattern of relationship. Family wealth determines a person's height and degree of happiness. The plus sign and single-headed arrows indicate a positive causal relationship: The wealthier the family is, the taller and happier the person is. There is no causal relationship between height and happiness. However, an association between height and happiness will be observed because they have a common cause, wealth. The plus sign and two-headed arrow indicate a positive association: Taller people are happier, and vice versa. This is an example of a spurious association. This is also called *confounding.* More precisely, this is an example of *positive confounding,* that is, an observed association between two variables due only to their common relationship with a third variable.

We must not forget the possibility of *negative confounding,* that is, the suppression of a true causal relationship from manifesting an association due to a third variable. The lower panel of Figure 2.1 illustrates a hypothesis Stewart et al. (2012) reviewed. Placental polychlorinated biphenyl (PCB) may reduce intelligence quotient (IQ) of the offspring. The minus sign and single-headed arrow indicate this potential negative causal relationship. However,

Positive Confounding

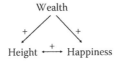

Negative Confounding

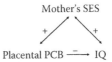

FIGURE 2.1
Illustrations of positive and negative confounding. Single-headed and two-headed arrows indicate causal relationship and noncausal association, respectively. + and – signs indicate positive and negative association, respectively.

mothers with higher socioeconomic status (SES) tend to have more exposure to PCB and their children tend to show higher IQ. Therefore, even if PCB truly reduces IQ, there may be no association observed between them. It is suppressed or reduced by the third variable, mother's SES.

Statistical methods can help to remove confounding and reveal the true relationship or lack of it. However, it requires knowledge on the subject matter to know what is to be included in the statistical analysis.

2.1.4 Observed Association Not Due to Chance

An observed pattern can be the result of chance. It is a well known fact that boys tend to be taller than girls, except around puberty time. According to the World Health Organization (WHO) Multicentre Growth Reference Study (MGRS), the mean (SD) supine length of newborn boys and girls are 49.88 cm (1.89) and 49.15 cm (1.86), respectively, and the distributions are approximately normal (WHO 2006). Despite the truth, if a researcher randomly selects and observes one newborn boy and one newborn girl, it can be shown that there is a 39% chance that the girl is taller than the boy! If the researcher randomly selects and observes two newborn boys and two newborn girls, the probability of finding the mean birth length of girls longer than the mean birth length of boys is 35%. Figure 2.2 shows this probability in relation to sample size. Even if the sample size increases to 20 per gender, there is still an 11% chance of finding girls taller than boys on average, despite the truth.

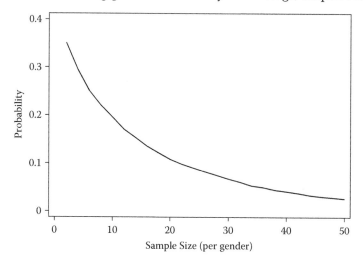

FIGURE 2.2
Probability of finding girls' mean birth length longer than boys in relation to sample size per gender in random samples. (World Health Organization, 2006, *WHO Child Growth Standards. Length/Height-for-Age, Weight-for-Age, Weight-for-Length, Weight-for-Height, and Body Mass Index-for-Age: Methods and Development,* Geneva: WHO Press.)

It is easy to underestimate the impact of chance and overinterpret findings from a sample. Conclusions based on uncritical use of sample observations can be totally wrong. Other factors being the same, the larger the sample size is, the less likely we will observe a pattern that is purely due to chance. The scientific evaluation of a causal relationship requires proof beyond a reasonable doubt that the association is not the result of chance. Statistical inference is responsible for this.

2.2 Study Designs

The focus of this book is on assessing the impact of exposure status on outcomes. Hence, the discussion of study designs is selective. We will discuss two examples in detail: The study of Early Child Health in Lahore, Pakistan, and the Lungwena Antenatal Intervention Study. Appendices A and B are Stata codes that simulate data that resemble the data structure of these two studies. The subsequent chapters at times use the two simulated datasets to illustrate. This chapter does not cover a case-control study, for example. Readers are referred to Hennekens and Buring (1987) and Rothman et al. (2008) for a more general introduction to study designs.

2.2.1 Cross-Sectional versus Longitudinal Studies

From the viewpoint of temporal sequence, there are two major study designs. *Cross-sectional studies* assess participants at a single point in time. In contrast, *longitudinal studies* follow up on participants and collect information from them repeatedly over time. By definition, longitudinal studies collect information on the participants at least at two time points. Longitudinal studies are variably called panel studies and cohort studies in different scientific disciplines.

Both exposure and outcome variables can be simultaneously measured in a cross-sectional study, but there is no way to firmly establish the temporal sequence of the events. For example, a cross-sectional study may show that stunted (exposed) children tend to have poor cognitive ability (outcome). But there is no way to ascertain whether stunting occurred before, alongside, or after the onset of poor cognitive ability. Longitudinal studies are usually needed to establish temporal sequence.

Occasionally, it is possible to make an educated guess about the temporal sequence of events recorded in cross-sectional studies. Suppose a cross-sectional study about the occupational achievement of middle-aged people also measures the height of the subjects. Final height is usually reached by about age 18 and in many countries people begin their occupational journey after this age. Hence in this case, it is reasonable to assume that reaching the present height preceded the present level of occupational achievement. However, if the cross-sectional study also measured IQ, it would be impossible to estimate whether the present IQ of a subject started from early childhood or later. Perhaps variation in IQ leads to variation in occupational achievement, or perhaps the other way around. From the cross-sectional data, there is usually no basis to establish the temporal sequence of this exposure and outcome.

Example 2.1: Early Child Health in Lahore, Pakistan (ECHLP)

Children living in developing countries are exposed to multiple risk factors that can compromise their growth and development. Communities that are undergoing rapid urbanization may face additional problems such as overcrowded housing conditions and lack of hygiene facilities. A community-based longitudinal study of infants born in Lahore was conducted to understand the health situations of infants growing up in this urbanizing area (Karlberg et al. 1993; Yaqoob et al. 1993).

From March to August 1984, an initial house-to-house survey was carried out in three study sites: a typical village about 40 km from central Lahore, Pakistan; a periurban slum; and a typical urban slum. Each study site comprised about 1000 households. All pregnancies were registered during the survey. In addition, a group of upper-middle-class families was recruited through their obstetricians during the second trimester of pregnancy. This group was scattered through the well-developed areas of Lahore. The inclusion of the upper-middle-class families provided sufficient variation in socioeconomic status (SES) for assessment of child health in relation to SES. Without this group, the SES would be too concentrated on the low side and it would be difficult to assess the role of SES in shaping child health. Background information on socioeconomic and health status was collected. Out of 1607 registered pregnancies, 1476 infants born alive were successfully recruited into the longitudinal study. They were assessed at birth or as soon as possible afterward (mostly within 7 days). Then they were followed and assessed for growth and development and other aspects of health at monthly intervals until 24 months of age. The monthly assessment included, among other procedures, anthropometric measures, a set of 10 developmental milestones that were selected from the Denver Developmental Screening Test and the Developmental Screening Inventory, and reports of episodes of diseases.

Among other findings, this study showed that diarrheal incidence in the first 2 years of life had negative associations with weight-for-age and BMI at 24 months, but not height-for-age (Cheung et al. 2001a). The longitudinal design allowed confirmation of the time sequence that diarrheal incidence preceded the growth measures. The study design cannot ensure a nonspurious relationship. The statistical analysis was based on multivariable regression that adjusted for various potential confounders, such as residential district (which reflected not only environmental factors but also socioeconomic status) and mother's education. However, the researchers could only adjust for potential confounders they knew and measured. There was no way to adjust for confounders that the researchers were not aware of or did not measure.

Statistical analysis cannot establish temporal sequence. It is the study design and occasionally additional knowledge that can (e.g., height and occupational achievement in the aforementioned example). Note that it can be tempting to interpret some cross-sectional data as if they are longitudinal data. Figure 1.1 (in Chapter 1) shows the birth weight-for-gestational age of Swedish girls. This is based on cross-sectional data. Every baby was measured for weight only once. One may want to interpret the curves as depicting how the weight of a fetus increases as it ages during the pregnancy. However, it is not the right interpretation. Figure 1.4 (in Chapter 1) shows a partially longitudinal picture. Cognitive functions were measured at the beginning and end of a 7-year interval. The 7-year changes across multiple age groups defined according to age at the beginning of the interval were assembled together to show the trajectories. There is no guarantee that the youngest people would have the same 7-year change that the oldest people in this sample had when they reach the same old age. The interpretation of the figure as long-term trajectories is based on an assumption that the future changes in the younger subjects will be comparable to the changes already observed in the older subjects. When there is further follow-up data available, the researchers can check the validity of the assumption. The reason for making this kind of analysis is that it is difficult or impracticable to follow up on subjects for a very long duration. There is also a timeliness consideration. One may want to provide tentative findings before it is too late.

2.2.2 Randomized Trials versus Observational Studies

Observational studies do not attempt to directly control the exposure status. If the exposure status or putative cause is not variable in the sample, there is no way to formally evaluate association, not to mention causality. A *randomized trial* directly controls the levels of exposure. It randomly allocates subjects to different levels of exposure, usually exposed versus unexposed, and ensures a sufficient number of subjects in each level.

A randomized trial is a special case of longitudinal studies. Usually, the allocated exposure starts at the beginning of the follow-up period and the outcomes are assessed at the end of the study. The temporal sequence of the putative cause and consequence is ensured.

Example 2.2: Lungwena Antenatal Intervention Study (LAIS)

In some sub-Saharan African countries, up to 20% of newborns are delivered preterm. Preterm delivery is a risk factor of low birth weight and growth faltering during infancy, as well as a variety of other developmental and health problems. Maternal malaria and reproductive tract infections are believed to be an important cause of pre-term delivery and low birth weight in sub-Saharan Africa. Some countries offer intermittent preventive intervention for malaria with two doses of sulfadoxine-pyrimethamine (SP) to all pregnant women. However, the efficacy of this infrequent dose schedule has been questioned because the drug does not stay long in the body. Azithromycin is an antibiotic known to have efficacy against multiple common pathogens in the female reproductive tract and it is also expected to have some anti-malaria effect. But the effect of preventive intervention with antibiotics for reproductive tract infections was not well understood.

A randomized trial in Malawi was conducted to examine the potential to prevent preterm deliveries, low birth weight, and postnatal growth faltering through intermittent preventive interventions against malaria and reproductive tract infections (Luntamo et al. 2010, 2013). The eligibility criteria included gestational duration between 14 and 26 weeks at enrollment according to ultrasound assessment. A total of 1320 eligible and consenting pregnant women who received antenatal care between 2003 and 2006 at Lungwena Health Centre, Mangochi District, were randomized to receive two doses of SP (standard care), monthly SP, or monthly SP plus two doses of azithromycin (AZI-SP) during pregnancy. The standard care group and monthly SP group also received two doses of placebo for azithromycin. The primary outcome measures were preterm delivery and frequency of adverse events. The secondary outcome measures include low birth weight and incidence of severe or moderate underweight during infancy. The main study procedures included antenatal care visits and assessments at 4-week intervals; assessment of mothers and infants at delivery (or as soon as possible in the case of home delivery); and postnatal assessment of infants at 4 weeks and 3, 6, 9, and 12 months.

The main results of the study are summarized in Table 2.1. The trend in lower rate of preterm delivery, low birth weight, and stunting (Length-for-age z-score < −2) across the three intervention groups was as hypothesized, but only the differences between the AZI-SP group

TABLE 2.1

Percentage of Preterm Delivery and Low Birth Weight

Outcome	Standard Care	Monthly SP	AZI-SP
Preterm delivery	17.9%	15.4% (P = 0.321)*	11.8% (P = 0.013)
Low birth weight	12.9%	9.1% (P = 0.091)	7.9% (P = 0.021)
Stunting at 4 weeks	24.4%	24.9% (P = 0.932)	14.6% (P = 0.001)

Source: Data from Luntamo, M., Kulmala, T., Mbewe, B., et al., 2010, Effect of repeated treatment of pregnant women with sulfadoxine-pyrimethamine and azithromycin on preterm delivery in Malawi: A randomized controlled trial, *American Journal of Tropical Medicine and Hygiene* 83:1212–1220, Table 2; Luntamo, M., Kulmala, T., Cheung, Y. B., et al., 2013, The effect of antenatal monthly sulfadoxine-pyrimethamine, alone or with azithromycin, on foetal and neonatal growth faltering in Malawi, *Tropical Medicine and International Health*, E-pub ahead of print, Table 4.

* P-values based on Fisher's exact test, using standard care group as reference.

and the standard care group were shown "beyond reasonable doubt" (to be discussed in Chapter 3) that they were not due to chance (P = 0.013, P = 0.021, and P = 0.001, respectively).

Socioeconomic status is associated with utilization of antenatal care and birth outcomes, but it is not causally determined by them. If an observational study reports that the use of an antenatal care package is associated with a smaller chance of preterm birth and low birth weight, it would be difficult to know whether it is the antenatal care package or the socioeconomic status that led to the birth outcomes. A randomized trial can provide strong evidence that the association is not confounded. Regardless of socioeconomic status, every subject has the same chance of being randomly allocated to each intervention group. Therefore, the intervention groups are expected to have no association with socioeconomic status or any other known or unknown confounders. Hence, the association found between randomized exposure status and outcomes is likely to be nonspurious.

A well-designed and implemented randomized trial verifies at least three of the four conditions for claiming a causal relationship described in Section 2.1. If the sample size is large, intuitively we guess that the results are not unduly influenced by the play of chance. Part of the statistical analysis to be discussed in the subsequent chapters is about formal evaluation of the role of chance.

2.2.3 Further Topics in Study Designs

Observational studies often resort to restriction in sample selection, standardization, and multivariable regression analysis to reduce confounding. These methods may also be used in addition to randomization as well. For instance, if sample size is small, randomization may fail to balance the distribution of confounders between intervention groups. In this situation, regression adjustment for the known confounders may be needed even in a randomized trial. Unlike randomization, however, these methods only prevent confounding by known variables and only if the confounders are accurately measured.

2.2.3.1 Restriction

Restriction in sample selection makes the subjects homogeneous and comparable to each other in terms of the confounders. For example, in the examination of birth weight in relation to gestational age, one concern is that those who are born earlier are not comparable to those who are born at an older gestational age in terms of maternal health. Factors such as illness in the mothers may increase the risk of both preterm birth and intrauterine growth restriction. The difference between median birth weight at 35 weeks and 40 weeks may be partly attributable to differences in maternal illness. If the interest is to understand gestational age as an exposure instead of maternal illness, a researcher may opt to restrict the sample by excluding babies whose mothers are known to have significant illnesses. This helps to prevent confounding, possibly at the expense of reduction in *generalizability* of the study findings. Normally, a nonspurious relationship is more important than generalizability as there is no point in generalizing spurious findings.

Note that this approach is to restrict the samples so that they are homogeneous in terms of the potential confounders. No restriction should be made to generate homogeneity in terms of the exposure status. Exposure status has to be variable; otherwise there is no evaluation of association. That is why the LAIS randomized subjects were to receive monthly SP and AZI-SP; otherwise practically nobody in the study area would be receiving these two antenatal care packages. There can be multiple and conflicting purposes in a study. Researchers need to balance the different factors. The ECHLP included a group of upper-middle-class families. That was good for assessing SES as an exposure status and for maintaining generalizability. But it introduces potential confounding by SES when the exposure variables of interest were something else.

2.2.3.2 Standardization

Another approach to deal with confounding is standardization. In fact, we have already seen it. Growth and development data are often standardized for age and gender. There are also growth standards that additionally standardize for parity, ethnicity, or other factors (RCO&G 2002). Suppose a mother's food preference is associated with ethnicity, but it has no impact on fetal growth. But ethnic origins can be associated with birth weight. An apparent relationship between food preference and birth weight may in fact be a result of confounding by ethnicity. By using growth references that are standardized for ethnicity, this confounding can be removed.

2.2.3.3 Multivariable Regression Analysis

Multivariable regression analysis is a powerful way to adjust for confounders. The ECHLP relied on multivariable regression analysis to strengthen its claim of a nonspurious relationship. Chee et al. (2009) observed a negative association between cognitive function and age. Older people, however, had less opportunity to receive education, and education is a predictor of cognitive function. In order to understand whether the observed association was confounded by education, the researchers employed multivariable regression analysis to adjust for education. Regression analysis is an important part of this book and will be discussed in more detail in subsequent chapters.

2.2.3.4 Concluding Remarks

In general, observational studies cannot satisfy the criterion for demonstrating nonspurious association. This is because the possibility of residual confounding due to unknown confounders and inaccuracy in the measurement of confounders always exists to some degree. The aforementioned methods can only reduce confounding and increase the plausibility that the findings are nonspurious. Therefore, well-planned and conducted randomized trials remains the gold standard for establishing a causal relationship. Having said that, we cannot and need not insist on only using randomized trials for the basis of policy and practice. Smith and Pell (2003) put forward an interesting challenge: The benefit of parachute use in the prevention of death and major trauma due to falling from height or jumping from a plane has not been established by randomized trials! Could the death/trauma of a person without a parachute and the unscathed survival of a person with a parachute be confounded by something? If parachute users only jumped from platforms no taller than themselves, the relation between unscathed survival and parachute use could have been confounded. But circumstances (data) allow us to make interpretation informed by subject matter knowledge (common sense in this case). First, we know that jumping from a plane without a parachute would almost certainly mean death. This makes a randomized control group unnecessary because we know what the outcome in the control group would be. Furthermore, suppose our hypothetical data indeed include many parachute users who jump from a trivial height; intuitively we would exercise the restriction method aforementioned to exclude these subjects and only estimate the survival chance of parachute users who jump from lethal heights. This is of course an extreme example. But broadly speaking, although randomized trials provide strong evidence, carefully conducted and analyzed observational studies may provide reasonable evidence.

3

Basic Statistical Concepts and Tools

This chapter begins with reviewing selected basic but important statistical concepts. It also discusses some common misunderstandings about basic statistics. The most exciting part of a soccer game is probably the scoring of goals. However, only a tiny fraction of the time is actually spent on scoring. The players spend most of the 90 minutes trying to organize the attacks. Without such organization, there could be no scoring. Statistical analysis is analogous to scoring goals. Good practice in data organizing and statistical programming is analogous to the organizing of the attacks. The remaining part of the chapter is devoted to these issues.

3.1 Normal Distribution

The *normal distribution* is also known as the *Gaussian distribution,* named after German mathematician Carl Friedrich Gauss (1777–1855). This well-known distribution has two parameters, mean (μ) and variance (σ^2), and is denoted by $N(\mu,\sigma^2)$. Some texts may use standard deviation (σ, or SD) in the notation instead, that is, $N(\mu,\sigma)$. In this book, $N(a,b)$ means a normal distribution with mean a and variance b. $N(0,1)$ is the *standard normal distribution*. A *standard normal deviate*, z, expresses a value as a distance from μ in terms of σ. A normal distribution should have skewness 0 and kurtosis ("pointedness") 3.

Conventionally, statistical tables are used to obtain z values and percentiles. Z values are called *critical values* in the context of statistical inference. Nowadays it is easier to precisely obtain them using statistical packages. Display 3.1 illustrates the Stata codes and outputs for obtaining z values from percentiles and vice versa. Note that Stata's output file adds a dot in front of the commands implemented. In programming or issuing the command, there are no such dots.

```
. ** Obtain critical values for 2.5th and 97.5th percentiles
. display invnorm(.025)
-1.959964

. display invnorm(.975)
1.959964

. ** Obtain percentiles for critical values -1.960 and 1.960
. display normal(-1.960)
.0249979

. display normal(1.960)
.9750021

. * Evaluate the tail of a Chi-square distribution with df = 1
. * Example 3.4
. display chi2tail(1,3.452)
.06317528
```

DISPLAY 3.1
Stata codes and outputs: Examples of obtaining z values (critical values) from percentiles and percentiles from z values based on the standard normal distribution, and obtaining P-values from a chi-square distribution after the likelihood ratio test.

3.2 Statistical Inference and Significance

3.2.1 Population and Sample

Empirical research usually takes a *sample* from a *population* and makes a conclusion about the population. The sample is the means to understand the population. In statistics, a *parameter* is a value that characterizes an aspect of the population, for example, the mean height of the population or the difference in mean height between males and females in the population. The sample provides a window to estimate the parameter. A sample estimate never exactly represents the population parameter. In Chapter 2, Section 2.1.4, we discussed that even if the true difference in mean birth length between boys and girls is 0.73 cm (boys taller than girls), there is an 11% chance that a random sample of 20 boys and 20 girls can give an estimate of the difference that is negative (girls taller than boys). There is always some degree of uncertainty due to sampling or other factors. The uncertainty due to *random sampling* can be quantified by statistical inference.

When there is a need to emphasize the distinction between the sample estimates and the population parameters, different notations are used for them. For example, it is common to use μ and $\bar{x}$ (pronounced x bar) to represent the population and sample mean, π and p to represent the population and sample proportion, and β and b to represent measures of associations in the population and sample, respectively. When there is no need to emphasize the distinction, one may arbitrarily choose one notation to use.

3.2.2 Statistical Inference and Statistical Significance

Statistical inference is the process of using sample data to understand the population. The two major procedures for statistical inference are *hypothesis testing* and *confidence intervals*. This section reviews the concepts, using one sample mean for example.

3.2.2.1 Hypothesis Test

We cannot use a random sample to prove with absolute certainty a statement about the population parameter. One thing we can do is to try to be the devil's advocate, that is, to argue in a way that does not favor the sample estimate. This is making a *null hypothesis*. Then we evaluate how likely it is to obtain the sample estimate or estimates that are even more disagreeable to the null hypothesis purely because of chance if the null hypothesis is true. The degree of how "likely" it is can be quantified as a probability value, denoted as the *P-value*. The smaller the P-value, the less likely the finding was due to chance and the more confident we are in rejecting the null hypothesis and accepting an *alternative hypothesis*. In general, other factors being held constant, the larger the sample size, the more believable the sample estimate. Furthermore, the larger the discrepancy between the null hypothesis and the sample estimate, the less believable the null hypothesis.

> **Example 3.1**
>
> In a study of 1536 Filipino children, the mean weight-for-age z-score (WAZ) at birth was –1.053 and the SD was 0.672 (Cheung and Ashorn 2009), meaning that the Filipino children were about 1 SD lighter than the mean in the reference population. This was a public health concern. But a question was whether the finding was a matter of chance.
>
> As the devil's advocate, we can argue that the population parameter of the mean WAZ of Filipino children was identical to the reference population's, meaning 0 z-score, and the negative sample mean was the result of chance. The null hypothesis is
>
> $$H_0: \mu_{WAZ} = 0$$
>
> The value specified in the null hypothesis is called the null value. This is zero here. The alternative hypothesis is
>
> $$H_a: \mu_{WAZ} \neq 0$$

We asked how likely it was to observe the mean WAZ <−1.053 or >1.053 if the null hypothesis was true. In Stata, it can be answered by plugging the above information to the "immediate" command

"ttesti 1536 −1.053 0.672 0"

The results are shown in Display 3.2.

For the null hypothesis of mean = 0, Stata showed P-value = 0.0000 (middle of the last row), meaning P-value < 0.00005 rounded to 4 decimal places. If we stick to four decimal places, we may report this as P < 0.0001. If the null hypothesis is correct, it would be very unlikely that we observed a sample mean at least as extreme as −1.053; the probability was smaller than 0.0001. So we reject the null hypothesis and accept the alternative hypothesis that the mean WAZ in the Filipino population was not zero.

Stata also shows the P-value for the null hypothesis

$$H_0: \mu_{WAZ} \geq 0$$

and the corresponding alternative hypothesis

$$H_a: \mu_{WAZ} < 0$$

If we only wanted to know whether Filipino children were lighter than the reference population, we should use this one-sided hypothesis. Here, P < 0.0001 again. It is the researcher's responsibility to specify what hypothesis to test. Suppose our concern is whether the mean WAZ is lighter than the conventional cutoff for underweight (WAZ < −2), we may set the null value at −2 and use a one-sided test:

$$H_0: \mu_{WAZ} \geq -2$$

$$H_a: \mu_{WAZ} < -2$$

Statistical packages do not know which hypothesis is relevant. They tend to preset a null value 0. This can be irrelevant to the specific question at hand. Researchers should avoid over-reliance on statistical packages.

Conventionally, P < 0.05 or P < 0.01 lead to the rejection of the null hypothesis and the finding is said to be *statistically significant*. Whether the cutoff point should be 0.05, 0.01, or some other values is a matter of how hard we want to try to be the devil's advocate.

P-value is a function of sample size. In order to ensure that there is evidence on whether a finding is due to chance or not, sample size should be planned. Readers are referred to specialized texts for details on sample size planning, for example, Machin et al. (2009).

```
. ttesti 1536 -1.053 0.672 0

One-sample t test
-----------------------------------------------------------------------
       |    Obs       Mean    Std. Err.   Std. Dev.   [95% Conf. Interval]
-------+---------------------------------------------------------------
    x  |   1536     -1.053    .0171464       .672      -1.086633   -1.019367
-----------------------------------------------------------------------
    mean = mean(x)                                       t =  -61.4122
Ho: mean = 0                                degrees of freedom =     1535

    Ha: mean < 0                Ha: mean != 0                 Ha: mean > 0
 Pr(T < t) = 0.0000     Pr(|T| > |t|) = 0.0000      Pr(T > t) = 1.0000
```

DISPLAY 3.2
Stata codes and outputs: One-sample t-test in Example 3.1.

3.2.2.2 Confidence Intervals

A sample estimate is subject to random sampling (and other) errors. The sample estimate is called a *point estimate*. In Example 3.1, the point estimate of mean WAZ was –1.053. The *confidence interval* (CI) established a range of estimates that is plausible. The 95% CI is the range that will cover the true population parameter 95% of the time if the sampling was repeated many times. Other factors being held constant, the larger the sample size, the smaller the confidence interval. The confidence interval and the hypothesis test are mirror images of each other. A 95% CI that just excludes the null value is equivalent to $P < 0.05$ in the hypothesis test. In Display 3.2, the 95% CI was –1.087 to –1.019, clearly excluding the null value 0.

3.2.3 Clinical Significance and Effect Size

Clinical significance refers to the practical impact on human well-being. A statistically significant relationship may or may not have any clinical significance. Clinical significance is a matter of judgment within a given context. It is important that researchers do not focus exclusively on statistical significance.

Example 3.2

A study of about 430,000 singleton births recorded in the Swedish Medical Birth Registry showed a statistically significant ($P < 0.0001$) difference in the mean gestational duration of boys and girls (Bergsjø et al. 1990). However, the difference was less than one day. Despite the statistical significance, a less than one day difference is probably not considered clinically significant by anyone.

Example 3.3

One of the secondary outcomes in the Lungwena Antenatal Intervention Study (LAIS) introduced in Chapter 2 was mean gestational duration. The AZI-SP group was found to have a mean gestational duration about 3 days (0.4 week) longer than the standard care group (Luntamo et al. 2010). The null hypothesis of no difference in mean gestational duration was rejected (P < 0.01). The 95% CI was 0.1 to 0.7 weeks. The difference was statistically significant. The finding satisfied the conditions of establishing a causal relationship discussed in Chapter 2. Whether this is of clinical significance is a matter of judgment and depends on the context. The important point is that a less than one day difference with extreme statistical significance (P < 0.0001) does not automatically make itself more clinically significant than a three day difference with a lower level of statistical significance (P < 0.01).

There is no universal rule to define the clinical significance of a difference. In the behavioral sciences, it has been common to define an *effect size* as the difference between two groups divided by the SD of one group, assuming equal variance (Cohen 1988). In the case of unequal variance

$$SD_{pooled} = \sqrt{\frac{(n_1 - 1)SD_1^2 + (n_2 - 1)SD_2^2}{n_1 + n_2 - 2}} \tag{3.1}$$

where n_i and SD_i, i = 1 or 2, are the sample size and sample SD for group i.

Example 3.3 (Continued)

In the LAIS, the mean (SD) of gestational duration was 38.4 (2.2) in the standard care group and 38.8 (2.1) in the AZI-SP group. The pooled estimate of SD was

$$SD_{pooled} = \sqrt{\frac{(435 - 1)2.2^2 + (440 - 1)2.1^2}{435 + 440 - 2}} = 2.15$$

and the effect size was

$$(38.8 - 38.4)/2.15 = 0.19$$

An effect size of 0.2 to 0.5, 0.5 to 0.8, and 0.8 or larger are often considered indicative of a "small," "medium," and "large" effect, respectively (Cohen 1988; Machin et al. 2009). However, Cohen (1988) emphasized that there is a risk in using such conventions without consideration of the particular context. He suggested using these definitions only as a last resort.

3.2.4 Parameters

3.2.4.1 Choice of Parameters

Interventions may aim to cause a uniform shift of a distribution, for example from $N(\mu,\sigma^2)$ to $N(\mu+\delta,\sigma^2)$. In this case, the mean is a useful parameter to summarize the intervention effect. However, interventions do not always aim to cause a uniform shift in the distribution. For instance, although preterm birth and underweight are health risks that policy makers want to prevent, postterm birth and overweight are also health risks. A uniform shift in the distribution of gestational duration or weight will prevent health risks for people at one end of the distribution but expose those near the other end to other health risks. Proportion of binary outcomes tends to be a more appropriate parameter in this situation. The primary outcome in the LAIS was the proportion of newborns at gestational age <37 weeks. An alternative is the proportion of term births, that is, the proportion neither preterm nor postterm. The choice of parameters needs to be considered in the context of the subject matter.

3.2.4.2 Parameter Tested by the Mann-Whitney U

There is a common misunderstanding that the Mann-Whitney U test is a test of difference in medians. It actually tests for a null hypothesis

$$H_0: \mathrm{Prob}(A > B) = 0.5$$

where A is an observation randomly sampled from population A, and B from population B (see, e.g., Hand 1994). If the distributions are identical, the probability should be 0.5. Even if the medians are the same across groups, the percentiles above (and/or below) the median can be different between groups, leading to $\mathrm{Prob}(A > B) \neq 0.5$. The Mann-Whitney U test may reject the null hypothesis in this case. Display 3.3 demonstrates a hypothetical situation that the median is 4 in each of the two groups but clearly group 1 tended to have larger values than group 0. The Mann-Whitney U test (implemented as a special case of Kruskal-Wallis test in Stata) rightly rejected the null hypothesis, giving $P = 0.0002$.

3.2.4.3 Clinical "Parameters"

The word *parameter* is sometimes used differently by clinicians and statisticians. For example, obstetricians may call fetal femur length a "fetal size parameter," whereas statisticians call this a "fetal size variable."

```
. tab y group
                 |          group
            y |      0           1 |      Total
    ---------+-------------------+---------
            1 |     10           0 |      10
            2 |     10           0 |      10
            3 |     10           0 |      10
          3.7 |      0          10 |      10
          3.8 |      0          10 |      10
          3.9 |      0          10 |      10
            4 |     10          10 |      20
            5 |     10           0 |      10
            6 |     10           0 |      10
            7 |     10           0 |      10
            8 |      0          10 |      10
            9 |      0          10 |      10
           10 |      0          10 |      10
    ---------+-------------------+---------
       Total |     70          70 |     140

. kwallis y,by(group)

(output omitted)

chi-squared with ties = 14.171 with 1 d.f.
probability = 0.0002
```

DISPLAY 3.3
Illustration of Mann-Whitney U test when both groups have the same median.

3.2.5 Maximum Likelihood

3.2.5.1 Likelihood

The concept of *likelihood* plays a central role in statistics. It is a measure of the support provided by a set of observations for a particular value of the parameter of a probabilistic model (Clayton and Hills 1993). The *maximum likelihood estimation* (MLE) is the estimation of the parameters of interest by finding the value that maximizes the likelihood.

Take a binary outcome variable as an example. A person may succeed ($Y = 1$) or fail ($Y = 0$) in a test. The probability of success is π and the probability of failure is $(1 - \pi)$; π is the unknown parameter of the binary probabilistic model we want to estimate. The relative frequency of observing a particular outcome y in one subject can be represented by

$$\text{Prob}(y) = \pi^y (1 - \pi)^{1-y}, \quad y = 0 \text{ or } y = 1 \tag{3.2}$$

This probability is the likelihood for one observation. The likelihood of a set of observations is the probability of the joint occurrence of the individual observations. Let S and F be the total number of (independent) subjects who succeed and fail, respectively. The sample size is $N = S + F$. The probability of obtaining this set of observations is the product of the likelihoods of the individual observations

$$L = \pi^S (1 - \pi)^{N-S} \tag{3.3}$$

This is analogous to the probability of the joint occurrence of two heads in tossing two fair coins. The probability of getting a head in tossing a coin is 0.5. The likelihood of observing two heads is $0.5 \times 0.5 = 0.25$, or equivalently $0.5^2(1 - 0.5)^{2-2}$.

Example 3.4

Suppose in a random sample of 10 boys we observed 4 successes. Which value is most supported by the data as the parameter value π? Using Equation (3.3), if $\pi = 0.35$, L = 0.0011318; if $\pi = 0.4$, L = 0.0011944; if $\pi = 0.45$, L = 0.0011351, and so on. Figure 3.1 plots the likelihood (multiplied by 1000) in relation to the candidate values for π. It turns out that the likelihood is maximized when π is 0.4. Hence, 0.4 is the most supported value, or the *maximum likelihood estimate*. It is more "likely" than any other candidate value for the parameter of interest.

The MLE may be found by brute force as what we have just done. But there are clever algorithms to find an MLE, although this book will not go into the technical details of maximization algorithms.

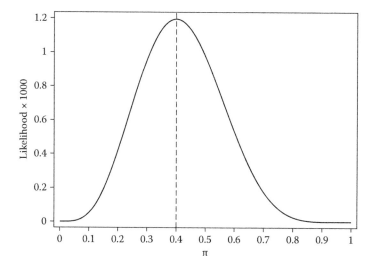

FIGURE 3.1
The likelihood ($\times$1000) for π in the case of S = 4 and N = 10.

Why go to such length to find an estimate of π? Why not just use a sample proportion $p = S/N$ to estimate the probability of success? A reason is that the maximum likelihood estimation method is very general. It can be applied to a simple probabilistic model and it can be applied to much more complicated models. For example, in multivariable regression analysis, several regression coefficients representing the strength of association between the binary dependent variable and independent variables can be estimated simultaneously, adjusting for each other. Maximum likelihood is the major technique to solve such complex problems. An appreciation of the concept of maximum likelihood will help to understand the subsequent chapters.

Likelihood is usually a very small value that is computationally difficult to handle. Since there is a one-to-one correspondence between a value and its log transformation (Section 3.3.1), it is common to work on log-likelihood instead of the likelihood itself. The value that maximizes the log-likelihood is also the value that maximizes the likelihood.

3.2.5.2 Likelihood Ratio Test

The ratio of the log-likelihood values of a model and its nested model can be used to establish the statistical significance of an estimate. A nested model is a special case of a more general, parent model. The nested model usually represents a null hypothesis that there is no association between two variables. That is, the model parameter that represents association is fixed at zero. The *likelihood ratio test* statistics is

$$LR = -2(LL_n - LL_p) \tag{3.4}$$

where LL_n is the log-likelihood of the nest model and LL_p is the log-likelihood of the parent model. The LR follows a chi-square distribution (also called chi-squared distribution) with degrees of freedom (df) equal to the difference in the number of parameters in the two models. The cumulative distribution of a chi-square distribution can be obtained from statistical packages or chi-square distribution tables. The cumulative distribution represents the probability of observing an LR statistic not more extreme than the sample LR statistic if the null hypothesis is true. Therefore, 1 minus the cumulative distribution, or the tail distribution, is the P-value.

Example 3.4 (Continued)

We had observed 4 successes among 10 boys and determined that the MLE for Prob(success) was 0.4. The likelihood was 0.0011944, or equivalently the log-likelihood was −6.730117.

Suppose 10 randomly and independently selected girls were also tested and 8 successes were observed. It can be calculated that the MLE is 0.8 and the log-likelihood is −5.004024 among the girls. The likelihood of the joint occurrence of the two groups (male and female) of observations with parameters described by the regression model

$$\pi = \alpha + \beta \times F$$

where F = 1 for female and 0 otherwise, α is the proportion among boys, and β is the difference in proportion between boys and girls, is the product of the two likelihoods and it is maximized when $\alpha = 0.4$ and $\beta = 0.4$

$$L = \left[0.4^4 (1-0.4)^{10-4} \right] \times \left[0.8^8 (1-0.8)^{10-8} \right] = 8.015 \times 10^{-6}$$

or equivalently the log-likelihood is –11.734141. The shortcut to obtain this log-likelihood value is to use the fact that the log-transformed product of two values equals the sum of the two log-transformed values. Summing the two log-likelihoods –6.730117 and –5.004024 aforementioned gives the same result.

But one might hypothesize that there was actually no gender effect and the difference in the observed proportions was purely due to chance. The null hypothesis is

$$H_0: \beta = 0$$

This means a nested model

$$\pi = \alpha + 0 \times F = \alpha$$

The MLE of α for this model is (4 + 8)/(10 + 10) = 0.6. The log-likelihood is $\ln[0.6^{12}(1-0.6)^8] = -13.460233$.

According to Equation (3.4), the LR test statistic is –2[(–13.460233) – (–11.734141)] = 3.452. The degrees of freedom is 1 because the two models differed by one parameter: the nested model is void of β. Referring to the tail of a chi-square distribution with df = 1 (see Display 3.1 for Stata code), if there truly was no difference between the two groups in the parameter of interest, the probability of seeing a LR test statistic of 3.452 or larger is 0.063, that is, P-value = 0.063. According to the conventional P-value cutoff 0.05, the null hypothesis of no difference could not be rejected. There is not sufficient evidence from the 20 observations to conclude a gender difference.

3.2.6 Derivatives

The *first derivative* (d1) of a function f(x) is the function's rate of change in relation to x. The *second derivative* (d2) is the first derivative of the first derivative, or the rate of change of the rate of change in relation to x. In the MLE context, the x-axis is the candidate values for the parameter to be estimated. The likelihood function is maximized if the d1 is zero (no longer increasing) and the d2 is negative (further change would lead to smaller likelihood value). Statistical packages for MLE have built-in facilities to handle the derivatives.

The derivatives have another usage that concerns us. The same principle for finding the maximum likelihood estimate applies to the search for the peak of a growth or development trajectory, for example, the cognitive function in relation to age in Figure 1.4. Conversely, the trough of a trajectory like an adiposity rebound is found by searching for the point where d1 = 0 and d2 > 0. The analytical approach to find the derivatives can be found in calculus texts, for example, Goldstein et al. (1993). The derivatives can also be numerically approximated, see, for example, Gould (1997). Some of these techniques will be illustrated in Chapter 11.

3.2.7 Normality and Robustness

One common misunderstanding is that statistical inference based on parametric tests is valid only if the distribution of the outcome variable is normally distributed. Some growth data tend to be skewed. Some developmental data are discrete or ordinal (ordered categorical) in measurement level and therefore not normally distributed either. This should not unduly discourage us from using parametric methods. Many parametric methods are to some extent *robust* to violations of model assumptions, meaning the violations of assumption do not seriously affect their performance, especially when sample size is large (e.g., Rice 1995). The degree of robustness of parametric methods and the meanings of "seriously" and "large" are best illustrated by simulation. The robustness of the two-sample t-test is examined here to illustrate.

The histogram in Figure 3.2 describes the body mass index (BMI) distribution of young male adults. This is generated from the parameters reported

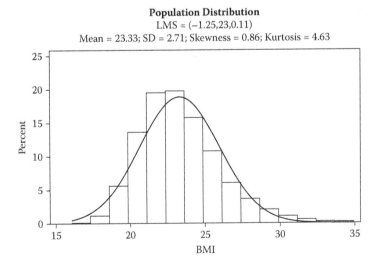

Population Distribution
LMS = (−1.25, 23, 0.11)
Mean = 23.33; SD = 2.71; Skewness = 0.86; Kurtosis = 4.63

FIGURE 3.2
BMI distribution generated according to LMS parameters from a British study.

in a British study (Cole et al. 1995), and has a skewness of 0.86 and kurtosis of 4.63. A normal distribution curve is superimposed on it. It is clear that the distribution is not normal.

The first round of simulation assumes that the two groups concerned have identical BMI distribution as described in Figure 3.2. Figure 3.3a shows the

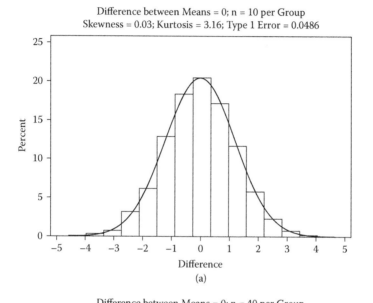

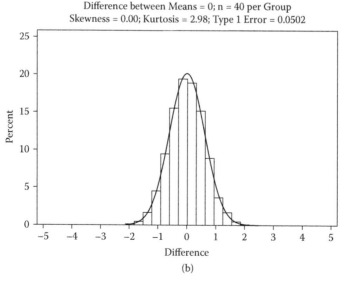

FIGURE 3.3

Sampling distribution of difference in mean BMI between two groups: (a) n = 10 per group; (b) n = 40 per group. True difference is zero; true shape of distribution is nonnormal.

sampling distribution of the 10,000 differences between means estimated from 10,000 simulated samples of 20 persons (10 persons per group). Despite the skewed population distribution, the sampling distribution is similar to a normal distribution. It is the sampling distribution that needs to be normally distributed in the application of the two-sample t-test (e.g., Rice 1995). The skewness is only 0.03 and kurtosis only 3.16. For each simulated sample, a two-sample t-test was applied to test the null hypothesis of the difference between means equaled 0 (which was the truth on which the simulation was based). Out of 10,000 tests, 486 gave $P < 0.05$. So, the type I error rate was 0.0486, slightly smaller than the 0.05 expected.

Figure 3.3b shows the sampling distribution with n = 40 per group. It was almost perfectly normally distributed. The skewness and kurtosis were very close to those of the normal distribution (0 and 3, respectively). Out of 10,000 tests, 502 gave $P < 0.05$. The type I error rate was 0.0502, practically identical to the expected value of 0.05. Note that the sampling distribution in Figure 3.3b is more concentrated around the mean. The larger the sample size, the smaller the *standard error* (SE), which is the standard deviation of the sampling distribution. A smaller SE is an indication of a higher level of precision.

The second round of simulation assumes that the two groups had a difference of one point in mean BMI, everything else being the same. Power calculation (Machin et al. 2009) based on the two-sample t-test suggests that a size of 10 and 40 persons per group has a power of 13.1% and 37.9%, respectively, at 5% two-sided type I error rate. The statistical *power* is the probability of correctly rejecting the null hypothesis. The use of a two-sample t-test in 10,000 simulated samples of 10 and 40 persons per group correctly rejected ($P < 0.05$) the null hypothesis of no difference 13.4% and 37.2% of the times, respectively. Both results are very close to the calculated power despite the nonnormal population distribution.

Readers are also referred to Heeren and D'Agostino (1987) and Sullivan and D'Agostino (2003) for simulation results about ordinal data. An adage in statistics is that "all models are wrong, but some are useful" (Box and Norman 1987). While we should not assume that all parametric methods are robust in all situations, we should also bear in mind that they can be useful at times even if strictly speaking they are "wrong."

3.3 Standardized Scores

3.3.1 Transformation

3.3.1.1 Logarithmic Transformation

It is often possible to transform a skewed distribution toward a normal distribution. The *logarithmic (log) transformation* is a popular choice. A log-transformation with base 10 converts 10 to 1, 10^2 to 2, and so on. A log-transformation with base e (e ~ 2.718) converts e to 1, e^2 to 2, and so on. The use of base e in the transformation is known as the *natural log*. The notations $\ln(y)$ and $\log_e(y)$ both denote the natural log-transformation of y. If $z = \ln(y)$, *exponentiating* z, that is, $e^z = \exp(z) \sim 2.718^z$, back-transforms to the original value y. In the studies of human growth and development, log-transformation usually uses base e. If the $\log_e$ transformed values (z's) follow $N(\mu,\sigma^2)$, the 10th percentile of the population distribution can be obtained by $\exp(\mu - 1.282\sigma)$. Other percentiles can be obtained similarly by replacing the −1.282 with the appropriate critical values in the normal distribution.

3.3.1.2 Power Transformation

The *power transformation*

$$z = y^p \tag{3.5}$$

is another commonly used method to transform nonnormally distributed variables toward normality. The square root transformation is the special case p = 1/2 and is a popular choice for normalization. More generally, the optimal value for p that can make z closest to the normal distribution can be estimated by the maximum likelihood method (Schlesselman 1971).

3.3.2 Calculation and Application

A *standardized score*, or *standard deviation score*, or *z-score* for short, for a value y sampled from a population in which Y follows a normal distribution $N(\mu,\sigma^2)$, is given by

$$\text{z-score} = (y - \mu)/\sigma \tag{3.6}$$

The z-score expresses the sample observation's position in the population in terms of its distance from the mean in standard deviation. Although the population mean (μ) and standard deviation (σ) are usually not exactly known, if the data on which they are based is considered representative enough, the sample mean and SD are treated as if they are the true population parameters. The World Health Organization (WHO) 2006 growth standards were based on a sample of over 8000 children recruited from multiple countries across continents. The sample estimates are generally considered (good approximations of) the true growth parameters of children worldwide. Calculated this way, the z-score refers the observed values in relation to a reference population.

Z-scores can be calculated using the mean ($\bar{y}$) and SD of a particular sample. In this case, the z-scores are only for quantifying the relative positions of the observations within that particular sample. There is no intention to compare with a reference population.

The use of z-scores offers several advantages. First, it makes interpretation easier if Y follows a normal distribution. The z-scores show each observation's percentile distribution, for example, a z-score of 1.96 is higher than about 97.5% of the other observations. If Y is not normally distributed but its log or power transformed variable Z is, Z may be used instead. Even if Y is nonnormal and there is no way to transform it to resemble the normal distribution, z-scores can still be calculated. In that case, they do not have the percentile interpretation. But we may still want to do it because of the other advantages.

Second, the use of z-scores facilitates comparisons of the impact of exposure variables that are on different metrics.

Example 3.5

One may want to study the relationship between systolic blood pressure (SBP), waist-to-hip ratio (WHR), and BMI. A regression coefficient has the meaning of the amount of change in the outcome per one unit change in the exposure (see Chapter 5). Suppose a regression analysis showed that for 1 unit increase in WHR, SBP increased by 50. Also suppose another regression analysis showed that for every 1 unit increase in BMI, SBP increased by 1. At first glance, it appears that WHR is much more influential than BMI, but this interpretation is wrong. It is important to remember that the range of WHR is usually quite small in numbers, for example, from about 0.7 to 1.0. So the regression coefficient 50 represents an effect of an extraordinary difference (1 unit) in WHR. In contrast, a 1 unit difference in BMI is common. The difference in the unit of measurement makes the comparison of the two regression coefficients unfair. One way to make a fairer comparison is to calculate the z-scores for WHR and BMI and use the z-scores for the exposure variables in the regression analysis.

Third, standardization requires centering at the mean. In regression analysis (see Chapter 5), centering a continuous independent variable at the mean has an advantage of making the intercept of a regression model interpretable: the intercept becomes the predicted mean of the dependent variable when the independent variable is "typical" (at the mean). It also reduces the correlation between the estimates of regression parameters and thus makes the findings about nonlinear trends more statistically significant if quadratic or other polynomial terms are used as predictors.

3.4 Statistical Programming

3.4.1 Ensuring Reproducibility

Reproducibility is an essential feature of scientific research. If we cannot reproduce the statistical analysis results we produced 6 months ago, we cannot claim to be doing science. To make statistical analysis reproducible is easier said than done. From time to time, researchers may create some new variables, exclude some observations, or save some new data files, without sufficient documentation. After a while it is difficult to keep track of what has been done. If a few hundred data organizing and statistical procedures were implemented using a mouse to point-and-click a graphical user interface, it is almost impossible to keep track of what has been done and how to reproduce it.

An essential tool in the practice of good statistical analysis is to keep data organizing and statistical analysis commands in *program files*. A program file contains a set of commands to be implemented together and in the order they appear in the file. It is usually in the ASCII or text file format, and is variably called a batch file or script file. In Stata it is called a "do" file. Appendices A and B are examples of Stata do files. Since all procedures are recorded in the do files, we are much less likely to mess up the procedures the next time we want to check or rerun the same procedures. But one also needs to be consistent and disciplined in the creation and naming of computer folders and files. Otherwise one may not be able to identify which program file is the latest version and where it currently is.

Some statistical procedures involve a random process of data generation or a random selection of observations within a sample. The use of bootstrapping techniques in quantile regression is an example (Chapter 5). Statistical packages use *pseudorandom* engines to handle such processes. By specifying a *seed number* for the engines, the results are reproducible.

3.4.2 Date and Age

It is important to include programmable details on the definitions of units of age and time. Some common definitions may include:

- Age in days = Date of assessment – Date of birth
- 1 month = 30.4375 days

Statistical packages can usually store a date as an *elapse date*, meaning the number of days elapsed since an anchor date. Stata uses 1 January 1960 as the anchor date. So, 1 January 1960 is 0; 2 January 1960 is 1, and so on. By saving dates in the elapse date format, calculation about age or duration and sorting by date are easy.

Additional attention is needed for time-to-event analysis (Chapter 7) if the event occurs on the date of birth or date of enrollment to a study. Using the aforementioned definition of age in days, an infant who was born and died on the same day is calculated as having an age at death equal to 0 days. In time-to-event analysis, observations with 0 units of time are excluded from the analysis because that means a person has no exposure time and therefore there is no possibility of the outcome event (Machin et al. 2006). So, either additional precision (e.g., hours) has to be introduced to calculate the age at event, or some assumptions are made to impute the age depending on context. For example, it may be assumed that on average the exposure time begins in the middle of the day and the outcome event occurs at the mid-point between the start of exposure and the end of the day (Machin et al. 2006). Based on this assumption, if a death occurs on the date of birth, the time-to-death is assumed to be 0.25 day.

A study may aim to assess a subject at a specific age. But the assessment may not exactly happen at this age due to participant unavailability or some other reasons. So, the database may need to capture both *target age* and *actual age* at measurement. It is also important to define some age window beyond which the data point is not considered acceptable because it no longer adequately represents the state at the target age. For example, one may define an observation to be missing if the actual measurement date is more than ±7 days different from the target measurement date. The width of the window is to be determined in a way that balances accuracy and practicability.

3.4.3 Long and Wide Formats

Longitudinal studies involve repeated measurements over time. Both longitudinal and cross-sectional studies may involve multiple test items at a single time point. Such data may be arranged either in the *long format* or *wide format*. The long format has one row of data per measure and multiple rows per subject. The wide format has one row of data per subject and multiple columns for the measures. Different statistical approaches may require data in different formats. Therefore, the statistical analysts need to know how to convert data from one format to another.

Example 3.6

Display 3.4 shows the Stata output of a simple example of z-scores at three target ages (28, 56, and 84 days) of two participants in the long format. Both target age (t_age) and actual age (a_age) at measurement were recorded in the data. Note that subject ID 2 did not show up for the second assessment, so the data in the long format consisted of three rows of data for participant ID 1 but only two rows for ID 2. Stata's "reshape" command was used to convert the data to the wide format. The "reshape" is a very important command.

3.4.4 Saved Results

Statistical packages may save estimation results temporarily in the computer memory. In Stata, after general statistics commands and regression analysis

```
. * data in long format
. * t_age is target age in days; a_age is actual age at measurement
. list, noobs clean

id   gender   t_age   a_age   zscore
 1      2       28      30      .03
 1      2       56      60      .05
 1      2       84      83      .06
 2      1       28      27      .13
 2      1       84      88      .04


.
. * reshape to wide format
. reshape wide a_age zscore, i(id) j(t_age)
(note: j = 28 56 84)

(output omitted)

. list, noobs clean
id  a_age28  zscore28  a_age56  zscore56  a_age84  zscore84  gender
 1     30       .03       60       .05       83       .06       2
 2     27       .13        .         .        88       .04       1
```

DISPLAY 3.4
Stata codes and outputs: Conversion between long and wide formats.

commands we can use the "return list" and "ereturn list" commands, respectively, to view what results have been saved in the memory. Using the *saved results* in statistical programming not only improve efficiency but also prevent errors and enhance reproducibility.

Example 3.7

Display 3.5 demonstrates the generation of z-scores using saved results. The data are the Simulated Clinical Trial (SCT) data described in Appendix A. For illustration purposes, a birth weight z-score without adjustment for gestational age or gender was generated. First, the program summarizes the birth weight (bw) variable. Stata recognizes the first three alphabets "sum" as short form of the full command name "summarize." We can, but do not have to, use the "return list" command to review what saved results are available. We see that the mean and SD were part of the saved results. We "generate," or "gen" for short, a z-score variable by subtracting the mean, saved as r(mean), from the data values and then dividing it by the standard deviation, saved as r(sd).

An alternative to do this is to manually enter the mean 2.952817 and SD 0.4360644 into the "gen" command line to generate the z-scores. One advantage of using saved results is that it prevents the risk of typo errors. Another advantage is that even if the data is updated, the do file will still work correctly. Had we manually entered the mean and SD values based on the preupdated data in the do file, we would have errors in the new z-scores.

```
. sum bw
    Variable |       Obs        Mean    Std. Dev.       Min        Max
-------------+--------------------------------------------------------
          bw |      1500    2.952817    .4360644      1.275       4.52
. return list

scalars:
                 r(N)  =  1500
            r(sum_w)  =  1500
             r(mean)  =  2.952816668748856
              r(Var)  =  .1901521590109665
               r(sd)  =  .4360643977796933
              r(min)  =  1.274999976158142
              r(max)  =  4.519999980926514
              r(sum)  =  4429.225003123283

matrices:
                 r(C)  :  5 x 5
. gen bwz = (bw-r(mean))/r(sd)
```

DISPLAY 3.5
Stata codes and outputs: Illustration of generating z-scores using saved results.

In subsequent chapters, more examples of using saved results will be seen.

3.4.5 Computer Software

This text is not intended as a general guide to statistical packages, although it gives some Stata codes for illustration. The codes are included to enhance understanding of the statistical concepts and techniques, and to facilitate the practice of specific methods.

Macros for implementation of the WHO child growth standards in the statistical software packages SPSS, SAS, S-Plus, and Stata are available from the WHO (http://www.who.int/childgrowth; accessed August 17, 2012). The GAMLSS software in R implements a range of statistical models for growth studies (Stasinopoulos and Rigby 2007). Readers are referred to the handbooks written by Everitt and his colleagues on the use of these statistical software packages (Der and Everitt 2009; Everitt 2001; Everitt and Hothorn 2009; Landau and Everitt 2004; Rabe-Hesketh and Everitt 2007). Beddo and Kreuter (2004a, 2004b) reviewed two of these handbooks. The LMSChartMaker software (Pan and Cole 2005) implements the LMS method of Cole and Green (1992). A number of other Stata macros (e.g., for item-response theory) that have been published are referred to in the subsequent chapters when they appear.

4

Quantifying Growth and Development: Use of Existing Tools

Growth and development are changes over time. Anthropometric size and developmental level are states at a point in time. But size and level are the result of cumulative changes. So we use the terms growth and development broadly to include both changes and size and level.

To analyze growth and development as exposure or outcome variables, we must first create the variables that quantify them. The creation of these variables may involve converting raw data to ratio indices, z-scores, percentiles, ability quotients, and so forth. Such a conversion is needed because growth and development usually needs to be interpreted in relation to something. For example, gross motor ability itself does not tell us how well a child is developing. The ability needs to be interpreted in relation to the child's age. The weight of an adult does not easily reveal whether the adult is overweight. The weight needs to be interpreted in relation to the adult's height. This chapter will cover the principles of establishing ratio indices, use of existing growth references (norms), and calculation of change scores. The development of new references and scoring algorithms requires more advanced knowledge about regression analysis and is therefore deferred to later chapters.

4.1 Growth

4.1.1 Ratio Index

4.1.1.1 One Denominator Variable

A *ratio index* is generated by the division of (a function of) one variable by (a function of) another variable. The division serves to standardize the numerator for the denominator. In the studies of growth, ratio indices are usually used to represent body shape or proportionality. Table 4.1 shows some examples of ratio indices. They may be considered *power-type indices*:

$$\text{Ratio index} = \frac{x}{y^p} \tag{4.1}$$

TABLE 4.1

Examples of Power-Type Ratio Indices

Index	Definition	Power Term
Waist-to-Hip Ratio	Waist Circumference/Hip Circumference	1
Body Mass Index	Weight/Height^2	2
Ponderal Index	Weight/Height^3	3
A Body Shape Index*	Waist Circumference/(BMI$^{2/3}$Height$^{1/2}$)	2/3, 1/2

Notes: Length and circumferences in meters; weight in kilograms.
* From Krakauer, N. Y., and Krakauer, J. C., 2012, A new body shape index predicts premature mortality hazards independently of body mass index, *PLOS One* 7:e39504.

where p = 2 for body mass index (BMI) and 1 for waist-to-hip ratio (WHR). The purpose of using these ratio indices is not to study the numerator or the denominator itself, but to study the characteristics they jointly define. A statistical question that need to be asked is: Why is p not the same for all indices?

In 1897, Pearson raised the issue that the use of ratio indices could create a spurious correlation. For over a century, statisticians and psychometricians have reiterated the same concern (Dunlap et al. 1997; Kronmal 1993). It is intuitive and easy to appreciate that even if x, y, and z are independent, randomly generated variables, the ratio variables x/z and y/z are correlated because they share the same denominator. Similarly, if we independently and randomly generate three variables representing weight, height, and head circumference, calculate BMI from the weight and height data, and regress BMI upon head-circumference-to-height ratio, an association will be found even if the three source variables are purely random. The use of an arbitrarily derived ratio index as either an exposure or an outcome variable, or both, may lead to a spurious correlation.

We take BMI for illustration. The power-type ratio index is used to quantify weight relative to height. The index should be independent of height; otherwise any association or lack of association between the index and another variable can be a result of confounding by height. In other words, it is desired that

$$\frac{\text{weight}}{\text{height}^p} = \text{constant} + 0 \times \text{height} \qquad (4.2)$$

The product $0 \times$ height is shown explicitly in Equation (4.2) to emphasize a lack of association between the power-type ratio index and the variable in the denominator. After removing the redundant part about $0 \times$ height and taking log transformation of both sides of Equation (4.2), the equation can be rearranged to

$$\ln(\text{weight}) = \ln(\text{constant}) + p \times \ln(\text{height}) \qquad (4.3)$$

This implies a linear relation of ln(numerator) on ln(denominator) and regression analysis can be used to find the power term p. If p = 1, it implies that the numerator and denominator increase proportionally. If p > 1, it implies that the numerator increases disproportionately more than the denominator does.

Example 4.1

Benn (1971) examined the relationship of ln(weight) and ln(height) in London bus drivers and conductors and found p to be 1.6 and 1.7, respectively. Body weight increased more than proportionately as body height increased. He recommended rounding the power term to the nearest integer for simplicity. So, it turned out to be 2 for BMI. In order to create the optimal ratio index that is not correlated with the denominator, it is best to estimate p from the target population. However, for simplicity and for cross-population comparison, the power term 2 is usually accepted for defining weight relative to height.

Example 4.2

Figure 4.1 shows the waist circumference and hip circumference of 9702 Indonesian adults in the Wave 4 of the Indonesian Family Life Survey (IFLS4) (Strauss et al. 2009) after excluding some outliers. The IFLS4 data is in the public domain (www.rand.org). Figure 4.1a shows the waist circumference plotted against the hip circumference. A 45-degree line was superimposed on it. Waist circumference was smaller than hip circumference, as indicated by many data points falling below the 45-degree line. But waist circumference increased roughly proportionally with hip circumference, approximately parallel to the 45-degree line. Figure 4.1b shows the ln(waist) versus ln(hip) values. Again, the two measurements vary roughly proportionally. The best fitting line in the form of Equation (4.3) fitted using least-squares regression (Chapter 5) was

$$\ln(\text{waist}) = -0.10 + 1.05 \times \ln(\text{hip})$$

The power term p = 1 in the calculation of waist-to-hip ratio is approximately right in generating a waist-to-hip ratio that is uncorrelated to hip circumference.

There are situations where the BMI is too rough for specific purposes. For example, fetal wasting (thinness) and stunting (shortness) are both important concerns of not only perinatal health but also their long-term implications (Barker 1998). The power term usually needs to be larger than 2 in order to make the birth weight index uncorrelated to birth length. The *Ponderal Index*, with p = 3, is often used in studies of size at birth (Beattie and Johnson 1994; Cheung et al. 2002; Cole et al. 1997). To be precise, the power term may be specifically estimated from the target population, as Benn did in 1971.

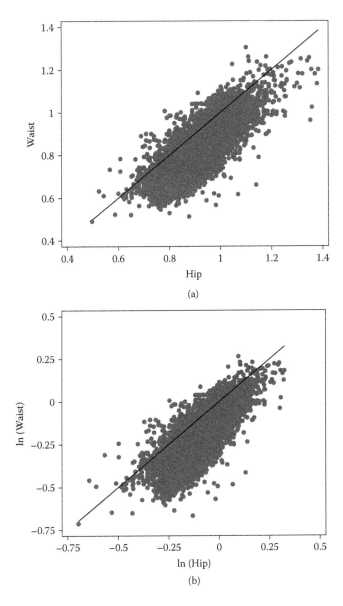

FIGURE 4.1
Scatterplots of waist circumference versus hip circumference in IFLS4 in original metrics (a) and log scale (b). A 45-degree line is superimposed on each panel.

The weight-for-height indices specifically calculated as such are known as the *Benn Index*.

The widespread use of the special cases of BMI in adults and the Ponderal Index in neonates is the result of a balance between the intention to remove the association with height and the intention to maintain simplicity in calculation and facilitate cross-population comparisons. Not all ratio indices have gone through an empirical evaluation. Some are defined arbitrarily and users should beware of the risk of spurious relationships (Kronmal 1993).

4.1.1.2 More Than One Denominator Variables

The same principle can be applied to standardize a numerator variable in relation to more than one denominator variables. This is achieved by extending Equation (4.3) to include more than one log-transformed independent variables.

Example 4.3

Krakauer and Krakauer (2012) developed a new index that characterizes waist circumference in relation to both weight and height. They estimated the relationship:

$$\ln(\text{waist}) = \ln(\text{constant}) + p_1 \times \ln(\text{weight}) + p_2 \times \ln(\text{height})$$

In the American sample they studied $p_1 = 0.681$ and $p_2 = -0.814$. Rounding them to ratios of small integers, $p_1 = 2/3$ and $p_2 = -5/6$. Therefore, they defined *A Body Shape Index* (ABSI) as

$$ABSI = \frac{\text{waist circumference}}{\text{weight}^{2/3}\text{height}^{-5/6}}$$

Since BMI = weight/height^2, or equivalently weight$^{2/3}$ = BMI$^{2/3}$height$^{2\times2/3}$, the ABSI can be expressed as

$$ABSI = \frac{\text{waist circumference}}{\text{BMI}^{2/3}\text{height}^{1/2}}$$

This body shape index is derived in a way such that it is independent of weight and height and BMI. It has been shown to predict premature mortality.

4.1.2 Transformation: Two- and Three-Parameter Models

4.1.2.1 Two-Parameter Models

Chapter 3 has reviewed basic ideas of z-score and transformation. A common practice in growth references is to model the raw data that is normally distributed, or to transform the raw data so that the transformed values are approximately normal. Let T represent the x-axis variable in the standardization. T is age in most growth charts. Assume that data from males and females are to be handled separately. Not performing a transformation can be seen as a special case of power transformation with p = 1. In a model that describes the raw or transformed data as a normal distribution, only two parameters—mean and SD—are needed to fully characterize the distribution for each age and sex combination. Then the following formula is used to convert a raw data value y(t) to a z-score:

$$z(t) = \frac{f\big[y(t)\big] - M(t)}{SD(t)} \qquad (4.4)$$

where f[y(t)] is a transformation of y at time t such that the distribution of y given T = t is normal, M(t) is the mean (and median), and SD(t) is the standard deviation of the transformed values at T = t in the reference population. The z-scores can be converted to a percentile by referring to the cumulative standard normal distribution, that is, $\Phi[z(t)]$. The z-scores can also be converted back to the original metrics, by first calculating the transformed value

$$f[y(t)] = M(t) + z(t)SD(t) \qquad (4.5)$$

and then back-transform f[y(t)] to the raw data.

Example 4.4

According to a Swedish reference for birth-weight-for-gestational age for boys (Niklasson and Albertsson-Wikland 2008), birth weight distribution is approximately normal after log(10) transformation, and the mean and SD at 40 weeks of the transformed values are 0.570 and 0.053, respectively. Using Equation (4.4), for a boy born at 40 weeks with a weight of 4.0 kg, his z-score is

$$z(t) = \big[\log_{10}(4.0) - 0.570\big]/0.053 = (0.602 - 0.570)/0.053 = 0.604$$

The cumulative standard normal distribution $\Phi[0.604]$ can be evaluated using the Stata function "normal(0.604)" (see Display 3.1 for examples). This corresponds to the 72.7th percentile. To back-transform the z-score 0.604 to the raw weight in kg, first use Equation (4.5) to obtain

$$f[y(t)] = 0.570 + 0.604{\times}0.053 = 0.602$$

Then take anti-log(10) to obtain the raw value $10^{0.602} = 4.0$.

4.1.2.2 Three-Parameter Models

Some aspects of anthropometry such as BMI tend to have nonnormal shapes that cannot be easily transformed to normality. The *LMS method* involves three parameters to transform the data toward normality (Cole 1988; Cole et al. 1995; Cole and Green 1992). The acronym stands for the Greek letters lambda, mu, and sigma. It summarizes the distribution of y(t) by a measure of skewness L(t), median M(t), and coefficient of variation S(t). A negative value of L(t) indicates positive skewness. Based on this method

$$z(t) = \frac{\left[y(t)/M(t)\right]^{L(t)} - 1}{L(t)S(t)} \tag{4.6}$$

Given a z-score and the LMS parameters, we can back-transform z(t) to the original data value by

$$y(t) = M(t)\left[z(t)L(t)S(t) + 1\right]^{1/L(t)} \tag{4.7}$$

Quite a few popular growth references use the LMS method and have published the LMS parameters for different ages and genders, including the U.S. Centers for Disease Control and Prevention (CDC) 2000 references and World Health Organization (WHO) 2006 standards.

Example 4.5

According to the WHO 2006 growth standards (WHO 2006), the L, M, and S parameters for the weight of 48-month-old girls are −0.3361, 16.0667, and 0.13884, respectively. Using Equation (4.6), the weight-for-age z-score for a 48-month-old girl who weighs 18.0 kg would be

$$z(t) = \frac{[18.0/16.0667]^{-0.3361} - 1}{-0.3361 \times 0.13884} = 0.803$$

Evaluating $\Phi[0.803]$, this z-score means the girl is at the 78.9th percentile. Conversely, given a z-score 0.803, we can back convert it to raw value in kilograms using Equation (4.7)

$$y(t) = 16.0667 \times \left[0.803 \times (-0.3361) \times 0.13884 + 1\right]^{1/-0.3361} = 18.0$$

4.1.3 Use of Existing References

This section begins with a review of the concepts of references versus standards. Then several popular population-based growth references and standards will be discussed, following a reverse age order, beginning with postnatal growth, size at birth, and fetal growth. The reason for this order of presentation is that there are widely accepted references or standards for postnatal growth, but not prenatal growth. So the discussion is in the order of the likelihood of usage.

4.1.3.1 References and Standards

A growth reference aims to describe the growth pattern in a population. A growth standard aims to prescribe what the normal growth pattern should be. In order to prescribe, it is essential that growth standards are based on data from people without known risk of growth faltering. For example, the WHO 2006 growth standards excluded not only infants with morbidity but also infants whose mothers did not follow recommended breastfeeding and nonsmoking guidelines (WHO 2006).

In practice, most growth references have some criteria to exclude data from people who are at significant risk of growth problems. Furthermore, growth references are often used or misused as growth standards (de Onis and Habicht 1996). The CDC growth charts are developed as references (Kuczmarski et al. 2000). In contrast, the WHO growth charts are purposefully designed to provide growth standards. Both references and standards provide a basis for standardization for age and gender and height (in the case of weight-for-height). They allow for comparing a person's growth status in relation to those of people in the reference population. For the purpose of estimating the prevalence of abnormal growth and determining health care interventions for individuals and communities, the choice between reference and standard has to be made with great care. For the purpose of assessing associations and intervention effects, however, the difference between them is relatively minor. For brevity, this chapter uses the terms *reference* and *standard* interchangeably.

4.1.3.2 Postnatal Growth

The CDC and WHO growth charts are important tools in growth studies. The WHO growth charts have the advantage of being more representative in terms of the data coming from six countries in different continents (Brazil, Ghana, India, Norway, Oman, and the United States). Conceptually, it is more internationally applicable than the CDC growth charts, which are based only on U.S. data. Nevertheless, given the WHO study's finding of "a striking similarity among the six sites" in linear growth (WHO 2006) and other studies that have shown comparable growth patterns across populations given similar socioeconomic status, the advantage of using data from multiple countries need not be overemphasized. The choice between the CDC and WHO growth charts may depend on what parameters and age ranges are being analyzed. Table 4.2 provides a comparison. There are two obvious differences between the two sets of charts. First, the WHO but not CDC growth charts cover arm circumference and skinfold thickness. Second, the CDC weight- and height-for-age charts are up to 20 years, whereas the WHO charts are only up to 60 months. The age range in a study may determine the choice. For instance, in an analysis of height gain from infancy to 18 years,

TABLE 4.2

Comparison of CDC 2000 and WHO 2006 and 2007 Growth Charts

Parameters	CDC 2000	WHO 2006/2007
Length-for-age	Birth to 36 months	Birth to 24 months
Height-for-age	2 to 20 years	24 to 60 months
Weight-for-age	Birth to 36 months	
	2 to 20 years	
	Birth to 20 years*	Birth to 60 months
Weight-for-length	45 to 103 cm	45 to 100 cm
Weight-for-height	77 to 121 cm	65 to 120 cm
BMI-for-age	2 to 20 years	Birth to 60 months†
Head circumference-for-age	Birth to 36 months	Birth to 60 months
Arm circumference-for-age	—	3 to 60 months
Triceps skinfold-for-age	—	3 to 60 months
Subscapular skinfold-for-age	—	3 to 60 months

Sources: Kuczmarski, R. J., Ogden, C. L., Guo, S. S., et al., 2002, 2000 CDC growth charts for the United States: Methods and development, *National Center for Health Statistics, Vital and Health Statistics* 11(246); World Health Organization, 2006, *WHO Child Growth Standards. Length/Height-for-Age, Weight-for-Age, Weight-for-Length, Weight-for-Height, and Body Mass Index-for-Age. Methods and Development*, Geneva: WHO Press; World Health Organization, 2007, *WHO Child Growth Standards. Head Circumference-for-Age, Arm Circumference-for-Age, Triceps Skinfold-for-Age and Subscapular Skinfold-for-Age. Methods and Development*, Geneva: WHO Press.

* Birth to 20 years growth charts are based on weighted analysis of birth to 36 months and 2 to 20 years growth charts.

† BMI from birth to 24 months is length based; BMI from 24 to 60 months is height based.

the CDC growth charts were used so that the z-scores across the whole age range concerned are comparable (Cheung and Ashorn 2009).

Studies that involve children both below and above 24 months of age need to manage the change in length/height measurement method. Measurement protocols may require recumbent length for those below a cutoff age and standing height for others above the cutoff. But in reality, the practice may vary according to field situations, such as inability to verify the exact age of the child on the spot. The measurement method needs to be recorded for the purpose of adjustment from length to height or vice versa. In the WHO Multi-Centre Growth Reference Study (MGRS), some children were measured in both standing and lying positions. The standing height was on average 0.7 cm shorter than the recumbent length (WHO 2006). Analysis of the U.S. National Health and Nutrition Examination Surveys showed an average difference of 0.8 cm (Kuczmarski et al. 2002). The WHO "igrowup" macro (www.who.int/childgrowth/software/en/; accessed January 1, 2011) converts the standing height to recumbent length by adding 0.7 cm for children aged below 24 months (731 days). Similarly, it converts the recumbent length to standing height by subtracting 0.7 cm for children aged above

24 months. Then all the z-scores it produces are length-based for children below 24 months and height-based otherwise.

The Stata macro "zanthro" converts raw data to z-scores using the CDC 2000 growth reference (Vidmar et al. 2004). The Stata macro "igrowup" converts raw data to z-scores using the WHO 2006 growth. Details on how to use the macros can be found from the respective references.

Example 4.6

The CDC 2000 and WHO 2006 growth references were applied to the data on weight at 4 weeks in the Simulated Clinical Trial (SCT) dataset. Note that although the target age for the measure was 4 weeks, the actual age at measurement varied from 16 to 40 days. Comparing the weight in kilograms without any standardization would allow variation in age at measurement to affect the true difference in growth between groups. Using z-scores, that is, standardizing for age and gender, removed this source of variation. In this particular case, the difference was very minor. Note that the prescriptive nature of the WHO versus the descriptive nature of the CDC growth charts resulted in lower mean z-scores and higher prevalence of z-score < –2 according to the WHO standards. Figure 4.2 plots the two sets of measurements against each other. The data points follow closely a curve and the correlation coefficient is high, $r = 0.99$. For studies of association, we can expect similar results no matter which score is used. However, some children (n = 36) who were considered normal in weight (z-score > –2) by the CDC growth charts were considered underweight according to the WHO charts. So, the estimate of the prevalence was more affected by the choice of charts.

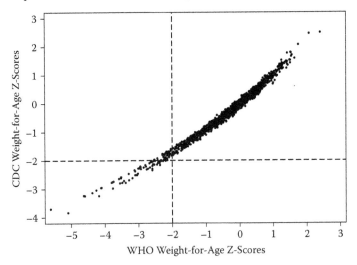

FIGURE 4.2
Comparison of CDC and WHO weight-for-age z-scores as applied to postnatal weight in the SCT dataset.

4.1.3.3 Size at Birth

The CDC and WHO growth charts standardize for chronological age. Perinatal health studies tend to demand standardizing size at birth for gestational age. There is no size at birth-for-gestational age references that are as commonly used as the CDC and WHO references for postnatal growth. Some examples of readily available references include birth weight-for-gestational age reference for Canadian (Kramer et al. 2001) and American babies (Oken et al. 2003) and birth weight-, length- and head-circumference-for-gestational age reference for Swedish babies (Niklasson and Albertsson-Wikland 2008). The Swedish Medical Birth Registry provides a very large sample size for not just birth weight but also length and head circumference as well as estimates of gestational age based on both ultrasound and last menstrual period method. These data characteristics make the Swedish reference very attractive.

To my knowledge, there is no computer software or macro to convert a set of raw data to size at birth-for-gestational age z-score. Most size-at-birth references use two-parameter models and hence the transformation is easy. In preparing your own program to perform the standardization according to existing references, note that the WHO recommendation is to base gestational age on the number of *completed weeks* (i.e., rounding down) instead of rounding to the nearest week (Kramer et al. 2001). So, 279 days should be rounded to 39 completed weeks instead of 40 weeks. Some older references may round to nearest weeks. Users of the references should check the details before application.

In the first few days of life, neonates become lighter than they were at birth because of a change in hydration status. The loss in weight can be up to about 300 grams (Greenwood et al. 1992; Michaelsen et al. 1991; Rijken et al. 2011). By the time they are about 1 week old, their weights return or exceed their birth weights. Although this phenomenon is commonly known to obstetricians and pediatricians, the exact degree and time course of weight loss in the first week of life has not been well documented. Postnatal growth charts ignore this and smooth this out. This has an implication to the studies of birth weight where home birth is common. In such a setting, the first anthropometric measurement tends to not happen on the day of birth and hence birth weight and prevalence of low birth weight could be wrongly estimated. Clinical trials should aim to measure birth weight within 24 hours of birth. The timing of measurement also needs to be comparable between intervention groups.

Table 4.3 shows the mean weight as percent birth weight in a community-based study of 902 Gambian infants, which were used to adjust the estimate of birth weight (Greenwood et al. 1992). For example, the estimate of birth weight for a child who was first weighed when 3-days-old equaled the weight on that day multiplied by 100/96. The accuracy of such a blanket adjustment is questionable, but not doing any adjustment is known to cause a bias. In the study of associations, the researchers may include day of measurement as a covariate in multivariable regression analysis.

TABLE 4.3

Mean Weight as a Percentage of Birth Weight in the First 6 Days of Life in Gambia

Age in Days	0	1	2	3	4	5
% Birth Weight	100	98	96	96	98	99

Source: Greenwood, A. M., Armstrong, J. R., Byass, P., et al., 1992, Malaria chemoprophylaxis, birth weight and child survival, *Transactions of the Royal Society of Tropical Medicine and Hygiene* 86:483–485 (supplementary figure).

4.1.3.4 Fetal Growth

The body size of fetuses cannot be directly observed. The estimation of fetal growth is usually based on ultrasonography. Not surprisingly, there are not many fetal growth references that are estimated from a large sample size. Furthermore, some studies pooled both sexes in the same set of references. Based on a relatively large sample size (634 pregnancies, 1799 ultrasound assessments), Johnsen et al. (2006) provided gender-specific percentiles for estimated fetal weight from 20 to 42 weeks of gestation.

Although there is a fair amount of evidence to argue that the intrinsic postnatal growth pattern is universal and differences between populations are largely due to extrinsic factors, there is not sufficient data to verify this point in fetal growth and size at birth.

4.2 Development

4.2.1 Standard Use of Multi-Item Inventories

Human development is a multidomain concept. There is not a universally agreed definition of what it is and what it should comprise. Although there is no standard measure of human development, there are some widely used measures. In the assessment of infants and young children, it is common to use inventories of *developmental milestones* that cover domains such as gross motor, fine motor, personal, social, and language development. Some milestones can be observed without the infants' conscious cooperation with the assessors, for example, whether a child can walk without support. In the assessment of older children, adolescents, and adults, measurements that require the participants to follow instructions and perform tasks may be employed.

It is common to use a multitude of test items to assess different aspects of development. The resultant large amount of data is then summarized to represent a smaller number of developmental domains or the overall level of development. The summary scores are then used to calculate a developmental quotient or intelligence quotient in relation to the norms previously constructed for a reference population. Making reference to a previously examined population is comparable to the age-standardization of growth variables described in the previous section.

Some examples of measures that are based on multiple test items are shown in Table 4.4. Some assessments concern developmental milestones, that is, tasks that normal children are expected to encounter and able to manage eventually, such as walking without support and taking off shoes and socks. For instance, the Griffiths Mental Development Scale–Revised: Birth to 2 Years (GMDS 0-2) includes 276 milestones, covering five domains of development (Griffiths and Huntley 1996). Some assessments comprise test items purposefully developed to assess specific aspects of development. For example, the Raven's Standard Progressive Matrices (SPM) is a test of reasoning ability. It consists of 60 multiple choice items (Raven, Raven, and Court 2003). Each test item involves showing the respondent a graphical pattern that has a missing part. The respondent is required to identify one piece of graphics, among the given choices, to complete the pattern.

TABLE 4.4

Examples of Development and Intelligence Assessment Scales

Scales	Age Range	Domains
Bayley Scales of Infant Development III (BSID III)	1–42 months	Cognition, language, motor, social-emotional, general adaptive
Denver Developmental Screening Test–Revised (DDST-R)	Birth–6 years	Gross motor, language, fine motor-adaptive, personal-social, behavior
Griffiths Mental Development Scales–Revised: Birth to 2 years (GMDS 0-2)	Birth–2 years	Locomotor, personal-social, hearing and language, eye and hand coordination, performance
Griffiths Mental Development Scales–Extended Revised: 2 to 8 years (GMDS-ER 2-8)	2–8 years	Locomotor, personal-social, language, performance, practical reasoning
Raven's Standard Progressive Matrices (SPM)	6–16 years	Nonverbal reasoning
Wechsler Intelligence Scale for Children, Fourth Edition (WSIC-IV)	6–16 years	Verbal comprehension, perceptual reasoning, processing speed, working memory, intelligence quotient*
Wechsler Adult Intelligence Scale, Fourth Edition (WAIS-IV)	16–90 years	Verbal comprehension, perceptual reasoning, processing speed, working memory, intelligence quotient,* general ability*
Repeatable Battery for the Assessment of Neuropsychological Status (RBANS)	Adults[†]	Immediate memory, delayed memory, visuospatial/constructional, language, attention

* Composite scores summarizing results of multiple domains.
† Initially developed for older adults. From Randolph, C., Tierney, M. C., Mohr, E., et al., 1998, The Repeatable Battery for the Assessment of Neuropsychological Status (RBANS): Preliminary clinical validity, *Journal of Clinical and Experimental Neuropsychology* 20:310–319.

It is important that users of the existing assessment scales study and follow the respective manuals, as different assessments may have their own ways of administration, scoring, and interpretation. For example, some instruments require administering all items to each assessee. But some others may order the test items according to their difficulty level, and then apply standardized rules to skip items that are too easy for a child at a particular age and also skip the more difficult items after a child has failed a number of consecutive items. So, what appears in a raw data file as missing values or "not done" may have special meaning that must be interpreted according to the user manual.

It is common that a raw score is calculated by summing the number of items the assessee has passed. Many well-established assessment scales provide age-specific norms so that a raw score can be compared with the reference population. This is similar to the use of z-scores and percentiles in growth reference. However, oftentimes there are two differences. First, growth references usually allow the use of exact age, or age in decimal places. But norms for development are usually prepared for discrete age intervals. Second, while growth references often convert raw values to z-scores that center at zero, developmental references tend to convert raw scores to quotients that center at 100. Assessees with a *developmental quotient* (DQ) or *intelligence quotient* (IQ) above 100 are considered above average, and vice versa. The concept of DQ is similar to IQ, but has broader contents covering locomotor ability for example. Recall that in Example 3.1 we tested for a null hypothesis of mean weight-for-age being 0 z-score in order to assess whether Filipino children were underweight as compared to the reference population. In the studies of DQ and IQ, a null value of 0 is usually meaningless. The null value more likely should be 100.

The differences between growth and development references are to some extent a matter of difference between conventions in scientific fields. The former usually more involves statisticians and the latter more involves psychologists/psychometricians. However, there is also some substantive difference in terms of measurement. Growth is measured objectively and the distribution may be nonnormal. A lot of statistical efforts are made to transform the data into normal distribution, including in the smoothing of the transformation parameters across ages. In contrast, the distribution of scores on a developmental assessment can to some extent be manipulated by including, excluding, or modifying test items. For instance, if the distribution of the raw score is bimodal, it may suggest two groups of items that are centering at different developmental levels. There is an insufficient number of items that target the intermediate levels. By modifying the test, the test developers can shape the distribution of the raw scores. As such, the focus in the test development is not the same as that in the measurement of growth. However, note that not all well-established developmental assessment scales show a normal distribution. For example, despite the Raven's SPM score sometimes presented as having mean 100 and SD 15, the distribution can be bimodal and skewed (Raven 2000). In such cases, it is not valid to estimate percentiles based on z-scores. Furthermore, developmental assessments may use different sets of items for subjects at different ages. Therefore, it is sometimes natural to develop references in discrete age intervals.

Example 4.7

The Bayley Scales of Infant Development II (BSID-II) manual provides norms for 38 age intervals (Black and Matula 1999). The norms were based on a stratified sample of about 1700 healthy American infants/children. The intervals include 1-month intervals for the first 36 months of age, and 3-month intervals for the age range from 37 to 42 months. The 1-month interval for age T centers at T and covers age (T–1) months and 16 days to T months and 15 days, inclusive. Similarly, the 37- to 39-month interval covers 36 months and 16 days to 39 months and 15 days, and so on. Norms are provided for the mental scale raw score and for the motor scale raw score separately, resulting in a Mental Development Index (MDI) and a Psychomotor Development Index (PDI). The indices have mean 100 and SD 15. So, a child with an MDI value of 115 is one SD above average, or according to the normal distribution, at the 84th percentile, as compared to children in the same age interval.

A score can be converted to a *developmental age*. If the mean raw score is Y for children at age T in the reference sample, a child who has raw score Y is said to have developmental age T.

In developmental screening practice, it is common to use test items or scores to classify individuals into categories, such as normal or delayed in development. However, in research it is usually not advisable to collapse quantitative scores into categories because that results in a loss of information and statistical power.

Example 4.7 (Continued)

The BSID-II manual provides a table to convert mental scale and motor scale scores into developmental ages. For example, a mental scale raw score of 140 is mapped to a developmental age of 26 months. The MDI and PDI can also be used to classify individuals to categories. Infants who score 115 or above (≥1 SD), 85 to 114, (–1 to 1 SD), 70 to 84 (–2 to –1 SD), and below 70 (<–2 SD) are in the Accelerated Performance, Normal, Mildly Delayed Performance, and Significantly Delayed Performance classification, respectively (Black and Matula 1999).

The use of developmental age has the advantage of relative ease in appreciating the analysis results. However, developmental age, DQ with mean 100, or percentiles are basically different ways of presenting the same information. In terms of comparison between exposure groups and in terms of estimation of associations, the different representations should give practically the same result.

4.2.2 Use of Selected Items

There are problems when an existing scale is not used according to its manual. One of the common reasons why a scale is not used according to its manual is cultural inappropriateness of some items. Many developmental assessment scales were developed and validated in Western societies. Not all items are valid for all societies. Some invalid items may have to be excluded from the administration. Then there is a problem in using the scoring systems and reference norms in the user manuals as the present data is not compatible with the original designs.

> **Example 4.8**
>
> In a clinical trial of nutritional supplements in Malawi, southern Africa, the Griffiths scales of development were used (Phuka et al. 2012). However, 9 of the 276 items were excluded from the administration to adapt to the local circumstances. For example, "tries to turn doorknob or handle" and "uses spoon and fork together without help" were excluded from the 58-item Personal-Social scale due to cultural inappropriateness.

In such circumstances, the conversion to z-scores or DQ according to a user manual is problematic. A simple approach to handle this problem is to use the raw score as the dependent variable and include gender and age at assessment as the independent variables in multivariable regression analysis in addition to the main exposure variable(s) of interest. Recall that a main purpose of standardization for age and gender is to avoid confounding. The use of regression adjustment for covariates serves the same purpose. A recent investigation of different scoring methods showed that the use of standardized scores and raw scores with adjustment for age and gender as covariates gave practically equivalent results in terms of effect size per unit increase in the exposure status (Cheung et al. 2008). However, there is no way to make a valid comparison of developmental level between the present sample and the reference population if the original set of items is not used in their totality. It may be tempting to adjust the raw scores obtained from the subset of items by computing

$$\text{adjusted raw score} = \text{raw score} \times \frac{\text{total number of items in test}}{\text{number of items administered}} \quad (4.8)$$

But this adjustment is not generally valid because it assumes that the difficulty levels of the items excluded are comparable to that of the items administered. As will be discussed in Chapter 10, it is possible to solve the problem of comparing individuals assessed on different items if the items are well calibrated.

4.3 Change Scores

Given measures at two static time points, or *point measures*, the degree of change between the two time points can be calculated. It is common to use "gain" (or change) over a follow-up period, such as weight gain (in kilograms) or height gain (in centimeters), as an exposure or outcome variable. The gain in Y may be denoted as

$$\Delta Y = Y_t - Y_0 \tag{4.9}$$

where the subscript 0 denotes study baseline. The gain divided by the duration of follow-up gives an estimate of the growth rate. The results in terms of gain and growth rate are easy to interpret if all participants are at the same age when the follow-up starts and ends. When duration of follow-up varies, the gain values for a study participant may be adjusted by

$$\Delta Y_{adjusted} = \Delta Y \times (T_{target} / t_{actual}) \tag{4.10}$$

where ΔY is as in Equation (4.9), T_{target} is the duration of follow-up the study targets, and t_{actual} is the actual duration of follow-up of a participant. However, if the follow-up is at a period that growth is nonlinear in relation to age, variation in age at starting or ending of follow-up time can introduce a bias. The change in age standardized z-scores

$$\Delta z = z_t - z_0 \tag{4.11}$$

or a similar change in DQ or IQ may be considered instead. They facilitate comparisons as properly developed z-scores and developmental quotients are expected to have no relationship, linear or nonlinear, with age.

5

Regression Analysis
of Quantitative Outcomes

We have seen regression equations in daily life. For instance, the maximum attainable heart rate (MAHR) of a person at a certain age can be estimated by

$$MAHR = 220 - 1 \times Age$$

The "$1 \times Age$" is usually written simply as "Age." You may have seen this equation on treadmills in gymnasiums. This is a simple form of linear regression equation. A person who is 40 years old is accordingly estimated to have MAHR = 220 − 40 = 180. The left-hand side of the equation is the *dependent variable* (MAHR). The right-hand side includes an *intercept* (220) and the product of a *regression coefficient* (1) and the *independent variable*. In some scientific disciplines, it is common to use the phrases *outcome variable* and *exposure variable* instead of dependent and independent variables, respectively. Furthermore, in the statistics literature, independent variables are sometimes referred to as *covariates*. We use these phrases interchangeably.

The study of relationship may involve only one independent variable or multiple independent variables. It is straightforward to extend the right-hand side to include multiple variables. We will first review the analysis of a single independent variable. We will later discuss multivariable analysis.

5.1 Least-Squares Regression

5.1.1 Model, Estimation, and Inference

Let y_i and x_i be the observations on the dependent variable (outcome) and independent variable (exposure) of the *i*th subject, i = 1, 2, ..., n. A simple linear relationship between them is obtained by finding a regression line in the form of

$$\hat{y}_i = b_0 + b_1 x_i \tag{5.1}$$

where $\hat{y}_i$ is the predicted outcome value for the ith subject. The equation can be equivalently expressed as

$$y_i = b_0 + b_1 x_i + e_i$$

where y_i is the observed outcome value of the ith subject and e_i is the prediction error, or residual. The coefficients on the right-hand side of Equation (5.1) are estimated in a way that minimizes the *residual sum of squares*

$$SS_{Residual} = \sum_{i=1}^{n} e_i^2 = \sum_{i=1}^{n}(y_i - \hat{y}_i)^2 \qquad (5.2)$$

The sum of squares

$$\sum_{i=1}^{n}(y_i - v)^2$$

is minimized if v is a value that equals the arithmetic mean of y. Instead of the usual way to calculate a sample mean, one can obtain the mean by finding a value that minimizes the expression. The least-squares regression finds the regression coefficient that minimizes the residual sum of squares. Therefore, it estimates the conditional mean given the covariate value. The predicated value on the left-hand side of Equation (5.1) is the conditional mean of y given the value in x. As will be seen in Section 5.2, not all regression methods relate the conditional mean to the independent variables.

The regression coefficient b_1 represents the amount of difference in the predicted outcome per one unit increase in the exposure variable. The intercept b_0 represents the predicted outcome when x = 0. They can be calculated by

$$b_1 = \frac{\sum(x_i - \bar{x})(y_i - \bar{y})}{\sum(x_i - \bar{x})^2} \qquad (5.3)$$

and

$$b_0 = \bar{y} - b_1 \bar{x} \qquad (5.4)$$

where $\bar{y}$ and $\bar{x}$ are the sample means of the dependent and independent variables, respectively (Montgomery et al. 2001).

The model assumes that the (linear) relationship specified is correct. The estimation of the regression coefficients does not require assumptions about the distribution of Y. It is the hypothesis testing and the construction of confidence intervals that do make distributional assumptions, including

- For each level of X, Y is normally distributed.
- The variance of Y is constant across levels of X.

The standard error (SE) of the regression coefficient b_1 is

$$SE(b_1) = \sqrt{\frac{\sum \frac{(y_i - \hat{y}_i)^2}{(n-2)}}{\sum (x_i - x)^2}} \qquad (5.5)$$

The numerator involves the residual sum of squares, $SS_{Residuals}$, in Equation (5.2). The smaller the residual sum of squares, the smaller the SE. The strength of the relationship can be tested by comparing the following test statistics to the critical values in a t-distribution with (n – 2) degrees of freedom

$$t = \frac{b_1 - \beta_1}{SE(b_1)} \qquad (5.6)$$

where β_1 is the null value specified according to the null hypothesis of interest. When $\beta_1 = 0$, Equation (5.6) tests for the presence of a relationship. While computer packages usually by default test the null hypothesis of $\beta_1 = 0$, researchers may set β_1 to other values according to the specific scientific questions.

The $100(1 - \alpha)\%$ confidence interval (CI) can be calculated by

$$b_1 - t_{df,1-\alpha/2} \times SE(b_1) \text{ to } b_1 + t_{df,1-\alpha/2} \times SE(b_1) \qquad (5.7)$$

Example 5.1

Display 5.1 shows the results of least-squares regression analysis of weight at 4 weeks in relation to birth weight in the Simulated Clinical Trial (SCT) dataset. In this analysis, $b_1 = 0.981$, meaning weight at 4 weeks was estimated to be 0.981 kg higher for every 1 kg heavier at birth. The SE was 0.020; $t = 0.981/0.020 = 48.97$. The null hypothesis of $\beta_1 = 0$ was clearly rejected (two-sided $P < 0.0001$).

The null hypothesis H_0: $\beta_1 = 0$ concerned whether there was a presence or absence of a relationship. This is what researchers often but not always want to know. For illustration purposes, let's consider the null hypothesis H_0: $\beta_1 \geq 1$. Babies who are smaller at birth tend to catch up. So, we may expect that the difference in weight would be somewhat smaller at 4 weeks than at birth. If that is true, b_1 should be smaller than 1. So, the aforementioned null hypothesis can be seen as a test of whether there is catch-up growth or not. After running a regression analysis, Stata temporarily keeps the regression coefficient of *varname* and its standard error as _b[*varname*] and _se[*varname*] and the sample size as e(N) in the memory (see "ereturn" in the help menu). The "manual" calculation in Display 5.1 used Equation (5.6) to calculate the t-value −0.970. Stata recognizes "disp" as the short form of the "display" command. Comparing this with a t-distribution with df = n − 2 gave a one-sided P-value of 0.166. This did not reject the null hypothesis $\beta_1 \geq 1$. The infants were maintaining their difference in weight from birth to 4 weeks of age. In this population, babies light at birth did not appear to be catching up.

Stata's postestimation command "test" makes testing against null values other than the default value of 0 easy. The "test _b[bw] = 1" tested the two-sided null hypothesis $H_0 : \beta_1 = 0$, using the saved results from the regression. Dividing the resultant P-value by 2 gave the same one-sided P-value that the manual operation did.

```
. * Simple regression
. regress weight4wk bw

    Source |       SS       df       MS              Number of obs =    1500
-----------+------------------------------           F( 1,  1498) = 2398.40
     Model | 274.071823      1   274.071823          Prob > F      =  0.0000
  Residual | 171.18063     1498   .114272784         R-squared     =  0.6155
-----------+------------------------------           Adj R-squared =  0.6153
     Total | 445.252453    1499    .29703299         Root MSE      =  .33804

------------------------------------------------------------------------------
  weight4wk |      Coef.   Std. Err.      t    P>|t|     [95% Conf. Interval]
-----------+------------------------------------------------------------------
        bw |   .9805748   .0200226    48.97   0.000     .9412996    1.01985
     _cons |    1.11706   .0597638    18.69   0.000     .9998302   1.234289
------------------------------------------------------------------------------

. * manually test b1 = 1
. disp (_b[bw]-1)/_se[bw]
-.9701627

. disp ttail(e(N)-2,abs((_b[bw]-1)/_se[bw]))
.16606101

. test _b[bw] = 1

 (1)  bw = 1

    F( 1,  1498) =   0.94
         Prob > F =  0.3321
```

DISPLAY 5.1
Stata codes and outputs: Least-squares regression of weight at 4 weeks in relation to birth weight in the SCT dataset.

If there is only one binary independent variable, the least-squares regression, the two-sample t-test, and the analysis of variance (ANOVA) are equivalent and they give identical P-values.

Example 5.1 (Continued)

In Display 5.2, the "xi:" in front of the regression command and the "i." prefix for the group variable jointly generated indicator variables to represent levels of the categorical variable. Since the group variable has three levels, two indicator variables were created: _Igroup_1 = 1 if group = 1, _Igroup_1 = 0 otherwise; and _Igroup_2 = 1 if group = 2, _Igroup_2 = 0 otherwise. The indicator variables contrasted the group concerned against the control group, which was the group with the smallest value (group = 0 in this case). Nevertheless, the "if" option only included two groups for the present illustration of the equivalence between least-squares regression, two-sample t-test, and ANOVA.

The mean weight at 4 weeks was 0.06 kg higher in the intervention group than the control group. All three methods gave P = 0.0808. The main body of the regression output rounded the P-value to 0.081. If necessary, users can "manually" plug the saved results into the t-distribution function to show the precise P-value, which was 0.08077691. The three pieces of information needed—degrees of freedom, regression coefficient, and its SE—were stored after the regression command as e(df_r), _b[*varname*], and _se[*varname*].

The three procedures are the same in a two-group comparison. The least-squares regression is more general in that it can handle quantitative and multiple exposure variables and is easier for robust inference for repeated measurement and cluster data (Section 8.2).

Version 12 Stata allows using the "i." prefix without the "xi:" in front of the command. Users may want to explore this new syntax. The analysis is the same using the new or old syntax.

5.1.2 Model Diagnostics

5.1.2.1 Linear Relationship

The estimation of the regression equation assumes a linear relationship. When there is only one right-hand side variable in the equation, this assumption can be graphically checked by plotting either the y-axis variable or the regression residuals against the x-axis variable. In order to highlight the signal and not be inundated by the noise in the data points, we can superimpose a *locally weighted regression smooth*, or *lowess* (Cleveland 1979), on the graph. The procedure provides the smoothed mean values of y in relation to x, by using the following basic steps:

1. For a data point (y_i, x_i), perform a weighted linear regression of the y-axis variable upon the x-axis variable. Data points closer to i,

```
. xi: regress weight4wk i.group if group==0 | group==1
i.group              _Igroup_0-2      (naturally coded; _Igroup_0 omitted)
note: _Igroup_2 omitted because of collinearity

      Source |       SS       df       MS              Number of obs =    1000
-------------+------------------------------           F( 1,  998)   =    3.06
       Model | .908619566        1  .908619566         Prob > F      =  0.0808
    Residual | 296.787217      998  .297381981         R-squared     =  0.0031
-------------+------------------------------           Adj R-squared =  0.0021
       Total | 297.695836      999   .29799383         Root MSE      =  .54533

------------------------------------------------------------------------------
    weight4wk |      Coef.   Std. Err.      t    P>|t|     [95% Conf. Interval]
-------------+----------------------------------------------------------------
    _Igroup_1 |   .0602866   .0344895     1.75   0.081    -.0073937    .127967
    _Igroup_2 |          0  (omitted)
        _cons |   3.981311   .0243878   163.25   0.000     3.933454   4.029169
------------------------------------------------------------------------------
. disp 2*ttail(e(df_r),_b[_Igroup_1]/_se[_Igroup_1])
.08077691

. ttest weight4wk,by(group), if group==0 | group==1

Two-sample t test with equal variances
------------------------------------------------------------------------------
    Group |     Obs       Mean   Std. Err.  Std. Dev.  [95% Conf. Interval]
---------+--------------------------------------------------------------------
C (Contr |     500   3.981311  .0255026    .570255    3.931206   4.031417
       A |     500   4.041598  .0232195   .5192043    3.995978   4.087218
---------+--------------------------------------------------------------------
combined |    1000   4.011455  .0172625   .5458881     3.97758    4.04533
---------+--------------------------------------------------------------------
    diff |          -.0602866 .0344895   -.127967   .0073937
------------------------------------------------------------------------------
    diff = mean(C (Contr) - mean(A)                        t =  -1.7480
Ho: diff = 0                              degrees of freedom =      998

  Ha: diff < 0                 Ha: diff ! = 0                 Ha: diff > 0
Pr(T < t) = 0.0404       Pr(|T| > |t|) = 0.0808         Pr(T > t) = 0.9596

. oneway weight4wk group if group==0 | group==1

   Analysis of Variance
    Source              SS          df       MS            F     Prob > F
------------------------------------------------------------------------------
Between groups      .908619566       1   .908619566      3.06     0.0808
 Within groups      296.787217     998   .297381981
------------------------------------------------------------------------------
     Total          297.695836     999    .29799383

Bartlett's test for equal variances: chi2(1) = 4.3783 Prob>chi2 = 0.036
```

DISPLAY 5.2
Stata codes and outputs: Equivalence of least-squares regression, two-sample t-test, and ANOVA for a binary exposure variable.

according to the distance $|x_j - x_i|$, j ≠ i, are given heavier weight and data points farther from i are given lighter weight.

2. Use the resultant linear regression equation to obtain a predicted value for this data point i only, $\hat{y}_i$.

3. Repeat the procedure for every other data point and graphically connect all the predicted values.

The *weight function* and the *bandwidth* control the degree of smoothing. The tricube weighting function is a popular choice (Cleveland 1979). The bandwidth specifies the proportion of data points to be included in the prediction of each data point. For example, bandwidth = 0.8 means that the 20% data points most far away from *i*th data point are not included in the prediction of $\hat{y}_i$ (i.e., the weights are zero). The smaller the bandwidth, the less smoothed the line is.

Example 5.1 (Continued)

Figure 5.1a,b shows the scatterplots and lowess functions of the weight at 4 weeks and the regression residuals versus birth weight. The linear relationship appeared to be reasonable but not strictly correct. There was some curvature in the smoothed values. The residual plot showed this curvature more clearly than the plot of the original variable, because the range of residual values was smaller than the range of the original values and therefore the y-axis scale could be more precise. Indeed, fitting a second-order natural polynomial to the data would give a statistically significant coefficient for the polynomial term. However, researchers should appreciate that no assumption is strictly correct. There is always some degree of departure from model assumptions. The question is whether the departure is clinically significant enough to justify a more complicated analysis model. This is a judgment call and depends on the specific research purpose.

5.1.2.2 Normal Distribution

To assess the normality assumption, the *quantile-normal plot* (Q-normal plot, a form of the Q-Q plot) can be applied to the regression residuals. A *quantile* is a data value that divides the cumulative distribution into subsets. A *percentile* is a special case of the quantile that divides the cumulative distribution into two subsets with a specific percentage of the probability below and the remaining probability above the value. The quantile-normal plot plots the quantiles of the observed distribution against the corresponding quantile of a theoretical distribution, the standard normal distribution in the present context. The plotting begins with sorting the residuals in ascending order, such that $e_{(1)} < e_{(2)} < \ldots < e_{(n)}$, where n is the sample size and $e_{(n)}$ is the largest residual value. Then $e_{(i)}$ is taken as the quantile below which $100[i/(n+1)]\%$ of

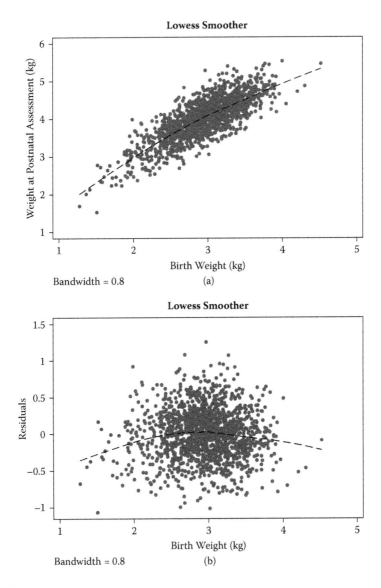

FIGURE 5.1
Scatterplots and locally weighted regression smooth of weight at 4 weeks (a) and regression residuals (b) versus weight at birth.

the residuals fall. The corresponding quantile, $q_{(i)}$, in the theoretical (normal) distribution is calculated by using the observed mean, μ, and observed SD, σ, of the residuals, such that

$$\Phi\big[(q_{(i)} - \mu)/\sigma\big] = i/(n+1) \tag{5.8}$$

where $\Phi(\bullet)$ is the cumulative standard normal distribution. If the normality assumption holds, the pairs of $e_{(i)}$ and $q_{(i)}$ should appear as a straight line and show a slope of one, overlapping with a 45-degree line. A histogram of the residuals may serve a similar function with an intuitive interpretation but it tends to draw attention to the center of the distribution, while the Q-normal plot draws more attention to the tails.

Example 5.1 (Continued)

Figure 5.2a shows the Q-normal plot of the 1500 residuals, whose mean was 0 and SD was 0.3379. Apart from the largest residuals, $e_{(1498)} < e_{(1499)} < e_{(1500)}$, all the data points fell almost exactly on the 45-degree reference line. The proportion of residuals smaller than, for example, $e_{(1498)} = 1.0716$, is calculated as $1498/(1500 + 1) = 0.9980$. In a standard normal distribution, the z value 2.8782 is the quantile below which 99.8% of the data points fall, or $\Phi(2.8782) = 0.998$. Therefore, from Equation (5.8), $(q_{(i)} - 0)/0.3379 = 2.8782$, or $q_{(i)} = (2.8782 \times 0.3379 + 0) = 0.9725$. As such, this data point had (x,y) coordinate $(0.9725, 1.0716)$ in the plot. The coordinates for the other pairs were calculated similarly (by a statistical package). Figure 5.2b shows a histogram of the residuals. The symmetry and bell-shape again suggest an approximately normal distribution.

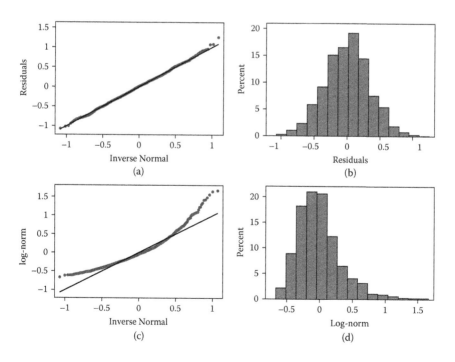

FIGURE 5.2
Quantile-normal plot (a) and histogram (b) for the regression residuals and quantile-normal plot (c) and histogram (d) for a shifted log-normal variable.

For illustration purposes Figure 5.2c,d show the Q-normal plot and a histogram of a (shifted) log-normal distribution. The Q-normal plot clearly deviated from a linear pattern and from the 45-degree line. The histogram clearly showed a positively skewed distribution.

If the residuals are always skewed toward one direction, a single Q-normal plot of the residuals would show this nonnormality. If, however, for some x values the residuals are positively skewed but for some others the residuals are negatively skewed, a single plot may fail to detect the nonnormality. When there are such concerns, the residuals may be grouped, for example, as tertiles according to their corresponding x values, and one plot is examined for each group. A more elaborate diagnostic plot extended from the Q-normal plot will be discussed in Chapter 9.

5.1.2.3 Equal Variance

Similarly, to assess the assumption of equal variance of y for each x, the residuals can be grouped according to their corresponding x values, and the variance calculated for each group and then compared. Another way of examining the unequal variance assumption is to calculate the *scaled absolute residual* (SAR) and check whether the SAR is related to the x-axis variable. The SAR has an expected value equal to the SD (Altman 1993)

$$\text{SAR}_i = \sqrt{\pi/2}\,|e_i| \qquad (5.9)$$

where $|\bullet|$ means absolute value. Then, perform a regression analysis of the SAR in relation to the independent variable. The presence of an association would suggest unequal variances. The SAR is also useful in the development of age-related references (Chapters 9 and 10).

Example 5.1 (Continued)

The SDs of the residuals in the lowest, middle, and upper tertiles of birth weight values were 0.121, 0.108, and 0.112, respectively, demonstrating approximately equal variances. A regression analysis of the SAR versus birth weight (kg) gave a regression coefficient of −0.030 and P = 0.053, again showing no evidence of unequal variance.

5.1.2.4 Residuals Explained

The *coefficient of determination*, commonly denoted by R^2, or R-squared, is calculated by

$$R^2 = 1 - \frac{\text{SS}_{\text{Residual}}}{\text{SS}_{\text{Total}}} \qquad (5.10)$$

where

$$SS_{Total} = \sum (y_i - \bar{y})^2$$

is the *total sum of squares*. The R-squared is also called the proportion of variance explained by the regression model. A smaller R-squared is usually preferred.

The coefficient of determination is sometimes considered a tool for model diagnostics. The validity of this depends on the specific analysis aim and application. For instance, in the development of new growth reference (Chapters 9 and 10), the focus is on finding regression equations that fit the data well. Usually it requires estimation of a nonlinear relationship. A nonlinear regression equation with a larger R-squared value captures the shape of the growth pattern better than a linear regression equation with a smaller R-squared. The R-squared is a measure of fit and therefore is important in this context. However, in the context of assessing relationship, whether a regression equation fit the data well and therefore has a large R-squared value has no direct implication to the specific research aim. Suppose the research purpose is to assess whether IQ is related to head circumference, and the regression analysis gives a statistically and practically nonsignificant regression coefficient, a confidence interval that is precise, and an R-squared that is close to zero. The research question is answered: No, IQ is not related to head circumference. That the R-squared value is very small has no direct impact on the validity of the answer to the specific research question. That is the reason why there is a saying that "the most important thing about R-squared is that it is not important" (as quoted in Gujarati 1995, p. 211).

The indirect implication of R-squared will be discussed later.

5.1.3 Robustness

The least-squares method is to some extent robust to the violation of the distributional assumptions, especially when sample size is large. Display 5.3 shows the Stata codes and results of simulating a relationship between X, which follows a flat (uniform) distribution and Y, which follows a Poisson distribution with mean $= \mu = 1 + 0.5x$. The Poisson distribution is nonnormal, with (positive) skewness $= \mu^{-1/2}$, violating the assumption of normality for each X. Furthermore, its variance is equal to its mean, violating the assumption of equal variances for each X (see, e.g., Rice 1995).

As shown in Display 5.3, for a small sample size $n = 5$, the mean of regression coefficient b_1 in the 1000 simulation replicates was 0.444, lower than the true value 0.5 in the simulation parameter. Furthermore, in 936 of 1000 (93.6%) simulation replicates the lower and upper boundary of the 95% CI contained the true value 0.5. This is slightly lower than the expected 95% coverage of the 95% CI. However, as sample size increased to 50, the mean of the 1000 estimates of b_1 was 0.497, very close to the true value 0.5. The 95% CI

```
. *** Robustness of ordinary least squares
.
. * define similation program with true relation y=1+0.5x
. capture program drop sim_ols

. program define sim_ols, rclass
  1.              version 12.1
  2.              syntax [, obs(integer 1)]
  3.              drop _all
  4.              set obs `obs'
  5.              tempvar x y
  6.              gen `x'=2+uniform()*3
  7.              gen `y'=rpoisson(1+0.5*`x')
  8.              regress `y' `x'
  9.              return scalar b1 =_b[`x']
 10.              return scalar low=_b[`x']-invttail(e(N)-2,0.025)*_se[`x']
 11.              return scalar up =_b[`x']+invttail(e(N)-2,0.025)*_se[`x']
 12.      end

.
. * small sample size n=5
.
. simulate b1=r(b1) low=r(low) up=r(up), reps(1000) nodots nolegend ///
> seed(123): sim_ols, obs(5)

. sum b1

    Variable |      Obs        Mean     Std. Dev.         Min        Max
-------------+-----------------------------------------------------------
          b1 |     1000     .4443102     1.197788    -6.244461   6.567042

. count if low<0.5 & up>0.5
936


.
. * moderate sample size n=50
.
. simulate b1=r(b1) low=r(low) up=r(up), reps(1000) nodots nolegend ///
> seed(123): sim_ols, obs(50)

. sum b1

    Variable |      Obs        Mean     Std. Dev.         Min        Max
-------------+-----------------------------------------------------------
          b1 |     1000     .4968459     .2698566    -.2778136    1.60049

. count if low<0.5 & up>0.5
943
```

DISPLAY 5.3
Stata codes and outputs: Simulation of the robustness of ordinary least-squares regression.

covered the true value in 943 out of 1000 (94.3%) replicates. This shows that the ordinary least squares (OLS) method is robust to violation of distributional assumption. The commonly asked question is how large a sample size is large enough for it to be robust to violation of distribution assumptions. There is no fixed answer; it depends on how serious the violation is.

5.1.4 Multivariable Least-Squares Regression

5.1.4.1 Estimation and Testing

From a practical point of view, it is straightforward to extend the aforementioned single variable least-squares regression to a multivariable regression. The right-hand side of the regression equation extends from having one variable as in Equation (5.1) to having multiple independent variables

$$\hat{y}_i = b_0 + b_1 x_{1i} + b_2 x_{2i} + \ldots + b_p x_{\text{p}i} \tag{5.11}$$

where $p < n - 1$ is the number of independent variables. The regression coefficients b_j, $j = 1, 2, \ldots, p$, represent the expected change in the outcome variable y per unit change in the covariate x_j when all of the other $(p - 1)$ covariates in the equation are held constant. For this reason, the regression coefficients are often said to be "adjusted" for the other covariates. If the covariate concerned is a binary exposure variable, the coefficient is often called the "adjusted difference" between the two groups. The number of covariates p has to be smaller than $(n - 1)$. Otherwise there is not enough data for the estimation.

It can be shown that the variance (squared SE) of a regression coefficient is (Montgomery et al. 2001)

$$SE(b_j)^2 = \frac{\sigma^2}{(n-1)\text{Var}(x_j)} \times \frac{1}{(1 - R_j^2)} \tag{5.12}$$

where R_j^2 is the proportion of variance of x_j that can be explained by the other covariates x_k ($k \neq j$) and σ^2 is the variance of the residuals. R_j^2 is an application of the coefficient of determination in Equation (5.10). It is the proportion of variance in covariate x_j explained by the other covariates x_k ($k \neq j$) in a least-squares regression model that does not involve the original outcome variable y

$$\hat{x}_j = b_0 + b_1 x_1 + b_2 x_2 + \ldots + b_{j-1} x_{j-1} + b_{j+1} x_{j+1} + \ldots + b_p x_p$$

Having obtained the SE, we can test the statistical significance or construct the confidence interval for a regression coefficient b_j as in Section 5.1.1, using $df = [n - (p + 1)]$. A smaller SE means a better level of precision. Equation (5.12) is important in understanding regression analysis. It separates the variance of a least-squares regression coefficient into four components: $(n-1)$, $\text{Var}(x_j)$, σ^2, and $1/(1-R_j^2)$. First, it is clear that the larger the same size, the smaller the SE. Second, the larger the variance of the exposure variable, the smaller the SE. Hence, in the design of a study, it is desirable to ensure heterogeneity in the key exposure variables. For instance, if a researcher aims to study the impact of socioeconomic status on IQ, it is desirable to recruit a heterogeneous sample of participants instead of sampling from only one residential community where the people are homogeneous in terms of socioeconomic status. The inclusion of an upper-middle-class sample in the Early Child Health in Lahore, Pakistan (ECHLP) study discussed in Chapter 2 is a case in point.

5.1.4.2 Linear Combinations and Multicollinearity

The multivariable regression equation in (5.11) involves (p + 1) parameters to be estimated. The number of parameters must be smaller than the total sample size, otherwise there is not enough data to estimate or test the parameters. Furthermore, the covariates cannot be linear combinations of each other. For example, if $x_3 = x_1 + x_2$, at most two of these three variables can be included as covariates in the regression model. It is not uncommon to encounter this kind of combinations in the studies of growth and development. If x_1 is birth weight, x_2 is gain in weight from birth to 28 days of age, and x_3 is weight at 28 days, by knowing two of the three variables the third can be fixed. Although there are three variables, there are only two pieces of information. The regression model cannot simultaneously estimate the parameters for all these three linearly dependent covariates.

If some of the covariates are highly correlated but not exactly linearly dependent, they may be referred to as showing near-linear dependencies or multicollinearity. Although it is possible to estimate the regression coefficients for these covariates, multicollinearity leads to imprecise estimation of individual regression coefficients. The extent of correlation between the covariates is indicated by R_j^2 in Equation (5.12). The stronger the correlation between x_j and the other covariates is, the larger the R_j^2 is. The quantity $1/(1-R_j^2)$ in Equation (5.12) is called the (jth) *variance inflation factor* or VIF (Marquardt 1970). A strong correlation between the covariates results in a large VIF, which in turn results in a large and imprecise SE for the estimate of the individual parameters. A rule of thumb is that VIF >10 indicates an unacceptable level of multicollinearity, which corresponds to $R_j^2 = 0.9$ (see, for a debate, O'Brien 2007). Multicollinearity also tends to result in estimates that are erratic. An addition or deletion of a small number of observations may make big changes to the regression coefficients. If there is a strong multicollinearity in the data, a large sample size will be needed.

Centering the variable Z means generating a new variable Z-centered = (Z – Mean of Z). The original variable Z tends to be strongly correlated with the square of Z. The centered variable tends to be less correlated to its square. In the estimation of a nonlinear relationship one may want to include a variable and its square in the regression equation. The correlation between the two terms makes the estimation of the nonlinear relation less statistically significant. That is one of the reasons why the use of z-scores, which is centered, can be desirable even if the independent variable is not normally distributed.

5.1.4.3 Explained and Unexplained Variation

The precision of the estimation is linked to the variance of the residual, σ^2, as shown in Equation (5.12). The larger the proportion of the variation in the outcome explained, the smaller the variance of the residuals and therefore the more precise the estimates are. A larger R-squared in the least-squares regression model (5.1) or (5.11) leads to a smaller σ^2. From the viewpoint of precision of a regression coefficient, a larger R-squared is good. Therefore, sometimes researchers may want to include covariates in the regression model even if they do not confound the relationship between the outcome and the key exposure variable. If the covariates are predictive of the outcome, they can improve the R-squared, reduce the σ^2, and make the regression coefficient on the key exposure variable more precise.

5.1.4.4 Multivariable Least-Squares Regression in Matrix Notation

Although from a practical point of view it is not essential to know the model in matrix notation, such knowledge will help understand more advanced methods. In matrix notation, Equation (5.11) can be written as

$$\mathbf{Y} = \mathbf{X}\mathbf{B} + \mathbf{e} \tag{5.13}$$

where

$$\mathbf{Y} = \begin{bmatrix} y_1 \\ y_2 \\ \vdots \\ y_n \end{bmatrix} \quad \mathbf{X} = \begin{bmatrix} 1 & x_{11} & x_{21} & \cdots & x_{p1} \\ 1 & x_{12} & x_{22} & \cdots & x_{p2} \\ \vdots & \vdots & \vdots & & \vdots \\ 1 & x_{1n} & x_{2n} & \cdots & x_{pn} \end{bmatrix}$$

$$\boldsymbol{\beta} = \begin{bmatrix} b_0 \\ b_1 \\ \vdots \\ b_p \end{bmatrix} \quad \mathbf{e} = \begin{bmatrix} e_1 \\ e_2 \\ \vdots \\ e_n \end{bmatrix}$$

and n is the number of observations. $\mathbf{X}$ is called a design matrix and the column of 1's in $\mathbf{X}$ is for the intercept b_0. It can be shown that the estimates that minimizes the sum of squares are

$$\hat{\beta} = (\mathbf{X}'\mathbf{X})^{-1}\mathbf{X}'\mathbf{Y} \tag{5.14}$$

When the design matrix only has two columns (one for the intercept and one for the single explanatory variable), Equation (5.14) boils down to Equations (5.3) and (5.4). The variance–covariance matrix of the regression coefficients are

$$\text{Var}(\hat{\beta}) = \text{Var}\left((\mathbf{X}'\mathbf{X})^{-1}\mathbf{X}'\mathbf{Y}\right)$$

$$= (\mathbf{X}'\mathbf{X})^{-1}\mathbf{X}'\text{Var}(\mathbf{Y})\mathbf{X}(\mathbf{X}'\mathbf{X})^{-1} \tag{5.15}$$

$$= \sigma^2(\mathbf{X}'\mathbf{X})^{-1}$$

where σ^2 is the variance of $\mathbf{Y}$ given $\mathbf{X}$ and is estimated by the residual sum of squares divided by the residual degrees of freedom. The second line of Equation (5.15) is simplified to the third line because some of the elements involving $\mathbf{X}$ divide themselves out. Similar to Equation (5.5), the residual sum of squares

$$\sum (y_i - \hat{y}_i)^2$$

divided by the degrees of freedom is an estimate of the variance of the residuals σ^2. The df is $[n - (p + 1)]$.

5.2 Quantile Regression

5.2.1 Single Quantile

5.2.1.1 Model and Estimation

As mentioned earlier, the sum of squares

$$\sum (y_i - v)^2$$

is minimized if v is a value that equals the arithmetic mean of the y's. The ordinary least-squares regression finds the regression coefficients that minimize the residual sum of squares and therefore it estimates the conditional mean given a set of covariate values. In contrast, the sum of absolute deviation

$$\sum |y_i - v|$$

is minimized if v equals the median. The *median regression* finds the regression coefficients that minimizes the sum of absolute values of the residuals

$$\sum |e_i| \tag{5.16}$$

The median regression is a special case of *quantile regression*, with $q = 0.5$. In the median regression, it is the sum of the unweighted absolute values of the residuals that are minimized. By more heavily weighing the absolute values of the residuals above the center, that is, $e_i > 0$, the regression line would shift upward to fit the higher quantile ($q > 0.5$) rather than the median, or vice versa. Define the weight w_i

$$w_i = \begin{cases} 2q & \text{if } e_i > 0 \\ 2(1 - q) & \text{otherwise} \end{cases} \tag{5.17}$$

The quantile regression minimizes the sum of the weighted absolute residual values

$$\sum |e_i| w_i \tag{5.18}$$

For $q = 0.5$, the weight is constant and it reduces to the median regression.

Quantile regression is a nonparametric procedure to estimate a specific quantile in relation to covariates. Under quantile regression, the predicted values on the left-hand side of regression Equations (5.1) and (5.11) are the conditional quantiles instead of conditional means. The method does not assume normality in the data.

5.2.1.2 Statistical Inference

Following Koenker and Bassett (1982), Rogers (1994) proposed a method for estimating the SE of the regression coefficients in a quantile regression equation. However, this method tends to underestimate the true SE if there are unequal variances. Hence, Gould (1993, 1998) proposed the use of the *bootstrap method* for estimating the SE in quantile regression.

Eforn and Tibshirani (1993) offer a comprehensive explanation of the bootstrap method. The essence of the method is described here. The standard error refers to the standard deviation of the sampling distribution. Knowing the true population distribution, in Chapter 3, Section 3.2.7, we obtained the sampling distribution by simulation and calculated the SE from the simulated data. In real life, we do not know the true population distribution. So the SE is usually calculated according to statistical theory. The bootstrap method is a resampling method that resembles the simulation. Assuming that the observed distribution defined by a sample of size n is a good approximation of the true population distribution, the bootstrap procedure randomly selects with replacement n observations from the sample. If the sample size is not large, one would not like to consider the bootstrap method because the assumption of the observed distribution being a good approximation of the true population distribution is difficult to substantiate. To select with replacement means that a randomly selected observation is put back into the pool and is available for the next selection again. Hence, a bootstrap sample may include multiple copies of the same observation. This bootstrap sample is used to calculate the statistics concerned, such as the quantile regression coefficient. This process is repeated k times, each time providing a statistics $\hat{\theta}_i$, i = 1, 2, ... k, calculated using the *i*th bootstrap sample. The bootstrap standard error is

$$SE_{bs}(\hat{\theta}) = \left\{ \frac{1}{k-1} \sum_{i=1}^{k} (\hat{\theta}_i - \bar{\theta})^2 \right\}^{1/2} \tag{5.19}$$

where $\hat{\theta}$ is the sample statistics and

$$\bar{\theta} = \frac{1}{k} \sum_{i=1}^{k} \hat{\theta}_i$$

Let $\hat{\theta}$ be the quantile regression coefficient estimated from the true sample, the 100(1 − α)% CI can be calculated as

$$\hat{\theta} - z_{1-\alpha/2} SE_{bs}(\hat{\theta}) \text{ to } \hat{\theta} + z_{1-\alpha/2} SE_{bs}(\hat{\theta}) \tag{5.20}$$

Similarly, one may test the null hypothesis that the true regression coefficient is β_1 by calculating

$$z = \frac{\hat{\theta} - \beta_1}{SE_{bs}(\hat{\theta})} \tag{5.21}$$

and comparing this to the critical value in the standard normal distribution.

The larger the number of replicates, k, the more precise the SE is. But it will take more computer time and the incremental improvement in precision will eventually level off. Since the bootstrap method involves random sampling, it is important to set the seed number so as to ensure reproducibility next time the same analysis is run.

In Chapter 3 we discussed that, despite popular misunderstanding, the Mann-Whitney U test does not test the null hypothesis of equal median. The quantile regression offers a means to assess relationships between not only the median but also other quantiles of a dependent variable in relation to an independent variable. Use the quantile regression to analyze the median in the data shown in Display 3.3, that is, to fit the regression equation

$$\text{median of } y = b_0 + b_1 \text{ group}$$

The coefficient b_0 is found to be 4, showing the median in the reference group. The coefficient b_1 is 0, rightly showing no difference in the median between the two groups. Table 5.1 shows the standard errors of b_0 and b_1 using the method of Rogers (1994) and bootstrap methods with 100, 1000, and 2000 replicates. The SE for the coefficient b_1 is clearly very different between the analytic method of Rogers and the bootstrap method (1.45, 1.50, and 1.42), suggesting a significant degree of underestimation by Rogers. There is no major difference between the bootstrap estimates for the SE of b_1 based on 100, 1000, and 2000 replicates.

Example 5.2

In a cohort study of women, Terry et al. (2007) used the quantile regression to assess whether body mass index (BMI) at age 20 years and 40 years were associated with perinatal factors. They examined the 10th, 50th, and 90th percentiles of BMI at age 20 in relation to maternal weight gain during pregnancy (pound), adjusting for maternal prepregnancy BMI. The regression coefficients were −0.01 ($P > 0.05$), 0.03 ($P > 0.05$), and 0.22 ($P < 0.01$), respectively. For every one pound more in maternal weight gained during pregnancy, the 90th percentile of the offspring's BMI at age 20 years was 0.22 points higher. The effect of maternal weight

TABLE 5.1

Median Regression Using Four Estimates of Standard Errors: Data as in Display 3.3

Method	SE(b_0)	SE(b_1)
Rogers (1993)	0.39	0.57
Bootstrap 100 replicates	0.34	1.45
Bootstrap 1000 replicates	0.47	1.50
Bootstrap 2000 replicates	0.47	1.42

gain appeared to affect only the upper end of the BMI distribution. The study did not report how they calculated the standard errors.

5.2.2 Simultaneous Quantiles

Sometimes there are specific hypothesis about an exposure status that affects only the upper (or lower) percentiles of the distribution of the outcomes. For example, in the Lungwena Antenatal Intervention Study (LAIS), the rationale of enhancing preventive treatment for infectious diseases during pregnancy was that they may prevent preterm birth without increasing the risk of post-term birth. In other words, the interventions were expected to raise the lower percentiles of gestational duration without affecting the upper percentiles. Terry et al. (2007) found some signs of differential impact of perinatal factors on some but not all BMI percentiles in adulthood. However, they did not formally test the hypothesis of the differential impact. Let b_1 and b_2 be the regression coefficients of two quantile regression models that regress two different percentiles upon an exposure variable x. Say b_1 and b_2 are the regression coefficients of the analysis of the 10th and 90th percentile of the outcome variable, the testing of the null hypothesis $b_1 = b_2$ can be of interest. In the LAIS context, it would be hoped that $b_1 > 0$ and $b_2 = 0$ and therefore $b_1 > b_2$.

In order to test a hypothesis that compares the quantile regression coefficients for two different percentiles, the covariance between the two coefficients is needed. The bootstrap method can be used to obtain the estimate of covariance. In each of the k bootstrap samples, both regression models are estimated, giving the pair of statistics $(\hat{b}_1, \hat{b}_2)$. Similar to Equation (5.19), which calculates the square root of the variance of one regression coefficient, the covariance between two regression coefficients is estimated as

$$\text{COV}_{bs}(b_1, b_2) = \frac{1}{k-1} \sum_{i=1}^{k} (\hat{b}_{1i} - \bar{b}_1)(\hat{b}_{2i} - \bar{b}_2) \tag{5.22}$$

where $\bar{b}_1$ and $\bar{b}_2$ are the mean of the k bootstrap estimates of the two regression coefficients. The bootstrap standard error of the difference $(b_1 - b_2)$ involves the covariance of the two coefficients

$$\text{SE}_{bs}(b_1 - b_2) = \left\{ \left[\text{SE}_{bs}(\hat{b}_1) \right]^2 + \left[\text{SE}_{bs}(\hat{b}_2) \right]^2 - 2 \times \text{COV}_{bs}(b_1, b_2) \right\}^{1/2} \tag{5.23}$$

The confidence interval for this difference can be estimated and the hypothesis with specific null value β can then be tested by plugging the estimated difference and bootstrap SE into Equations (5.20) and (5.21).

Example 5.3

Display 5.4 shows a simultaneous quantile regression of the SCT dataset, which resembles the LAIS trial. The effects of the two interventions (group 1 and group 2) versus the control (group 0) on the 10th and 90th percentile of the gestational duration were estimated. The bootstrap method with 1000 replicates was used to estimate the variance and covariance of the regression coefficients. The 10th percentile in group 2 was 0.77 weeks higher than group 0 (P = 0.015), whereas the 90th percentile was only 0.018 weeks higher (P = 0.873). The null hypothesis of no difference in the regression coefficients was tested by using Stata's

```
. *** quantile regression
. set seed 123

. xi: sqreg gestweek i.group,q(.1 .9) reps(1000) nodots
i.group           _Igroup_0-2        (naturally coded; _Igroup_0 omitted)

Simultaneous quantile regression                  Number of obs =    1500
  bootstrap(1000) SEs                              .10 Pseudo R2 =  0.0098
                                                   .90 Pseudo R2 =  0.0000
```

gestweek	Coef.	Bootstrap Std. Err.	t	P>|t|	[95% Conf. Interval]	
q10						
_Igroup_1	.6103096	.3123521	1.95	0.051	-.0023847	1.223004
_Igroup_2	.7699432	.3173645	2.43	0.015	.147417	1.392469
_cons	35.45163	.2377517	149.11	0.000	34.98527	35.91799
q90						
_Igroup_1	.0122185	.1082905	0.11	0.910	-.2001987	.2246356
_Igroup_2	.0181351	.1132303	0.16	0.873	-.2039718	.2402419
_cons	41.08407	.0841543	488.20	0.000	40.919	41.24914

```
. test [q10]_Igroup_2 = [q90]_Igroup_2

 (1)  [q10]_Igroup_2 - [q90]_Igroup_2 = 0

       F( 1, 1497) =      5.31
            Prob > F =    0.0213

. lincom [q10]_Igroup_2 - [q90]_Igroup_2

 (1)  [q10]_Igroup_2 - [q90]_Igroup_2 = 0
```

gestweek	Coef.	Std. Err.	t	P>|t|	[95% Conf. Interval]	
(1)	.7518082	.3261933	2.30	0.021	.1119637	1.391653

DISPLAY 5.4
Stata codes and outputs: Simultaneous quantile regression of the SCT dataset.

postestimation command "test." The null hypothesis of equal coefficients was rejected (P = 0.021). The postestimation command "lincom"—linear combinations of estimators—was used to estimate the confidence interval of the difference between the coefficients on the 10th and 90th percentiles. The 95% CI was 0.112 to 1.392. It excluded the null value 0. There was statistically significant evidence to demonstrate different intervention effects on different percentiles. Relative to group 0, group 2 had a higher 10th percentile but not 90th percentile.

5.2.3 Multivariable Quantile Regression

Similar to the extension of single variable to multivariable least-squares regression, from a practical point of view it is straightforward to add covariates to the right-hand side of a quantile regression equation. Technically, there are a lot of differences between quantile and least-squares regression, such as the estimation of quantile regression equation is by the simplex method for linear programming (Armstrong et al. 1979) and estimation of the quantile regression standard errors is (better) by bootstrapping (Gould 1993). However, the aforementioned general concepts about the least-squares equation are equally applicable to multivariable quantile regression: The regression coefficients represent the unit change in the outcome variable per unit increase in the exposure variable having controlled for the covariates; the number of covariates must be smaller than $(n - 1)$. The inclusion of highly correlated covariates affects model performance. More comprehensive discussion of quantile regression can be found in Hao and Naiman (2007).

5.3 Covariate-Adjusted Variables

The result of a regression analysis of Y upon X can be used to produce a new variable "adjusted-Y." First, estimate the regression equation

$$Y = b_0 + b_1 X$$

The adjusted-Y can be calculated as

$$\text{adjusted} - Y = Y - b_1(X - C) \tag{5.24}$$

where C is a constant to be specified by the researchers such that the adjusted Y represents the distribution of Y at $X = C$. Usually, C is the mean of X. If X and Y are positively correlated ($b_1 > 0$), Equation (5.24) adjusts down those that have a high value on the X variable, and vice versa. This removes the influence of X on Y and makes the adjusted-Y values comparable across observations that have different X values. This operation is based on the assumption that the regression model fits the data well, for example, a valid assumption of linear relationship. Model diagnostics should be applied before calculating the adjusted-Y.

Example 5.4

We use the SCT dataset to illustrate the calculation of a variable that represents birth weight adjusted for gestational age, or adjusted birth weight in short. Least-squares analysis of the data gave the following result:

$$\text{Birth weight} = -2.024 + 0.128 \times \text{Gestational age}$$

The intercept –2.024 represents the predicted birth weight if gestational age (GA) was zero. The intercept was meaningless here for two reasons. First, it is meaningless to talk about birth weight for a baby who had not been conceived. Second, the range of GA in the data does not cover zero. To try to interpret the intercept here is to extrapolate too much out of range.

Since 40 weeks is the expected duration of a healthy gestation, we set C = 40 and calculate for each observation

$$\text{Adjusted birth weight} = \text{Birth weight} - 0.128 \times (\text{Gestational age} - 40)$$

The adjusted birth weight represented the birth weight if every subject was born at 40 weeks of gestation. The correlation between gestational age and birth weight was 0.602. The correlation with adjusted birth weight must be 0. Further analysis of adjusted birth weight in relation to other variables would be free of the influence of gestational age.

The new variable adjusted-Y can be used in further analysis as an exposure variable in relation to another outcome variable Z. The Pearson's correlation coefficient between Z and adjusted-Y is the *partial correlation*, in the sense that the effect of X has been removed.

Example 5.4 (Continued)

The SCT dataset includes a 10-item developmental test at age 36 months. (Ignore items 11 and 12 for the time being.) A score that ranges from 0 to 10 can be created by summing the 10 item scores. The correlation between the score and birth weight was 0.133. The partial correlation between the score and adjusted birth weight was 0.102. Part of the correlation between developmental test score and birth weight was due to the effect of gestational age on the scores. The partial correlation coefficient 0.102 was free of that effect.

A reason for deriving and using a covariate adjusted variable is to remove the association between X and Y so that X cannot be a confounder in the studies of the association between Y and Z. For this purpose alone, there is no difference from performing a regression analysis of Z on Y with adjustment for X

$$Z = b_0 + b_1 X + b_2 Y$$

The use of external references to generate z-scores that represent Y given X also serves the same purpose of preventing confounding, such as using birth-weight-for-gestational age references to generate birth weight z-scores. The use of covariate-adjusted variables is often for the purpose of ease of presentation and interpretation. For instance, for most people it is easier to appreciate an association by seeing a scatterplot instead of reading a multi-variable regression equation. It is also easier to appreciate (adjusted) birth weight in grams or kilograms than birth weight z-scores. The use of adjusted variables facilitates such presentations.

6

Regression Analysis of Binary Outcomes

This chapter will begin with a review of the measures of association based on risk and odds. It will also discuss the concepts of sensitivity and specificity and their relevancy in studying binary outcomes. In Chapter 5 we saw that the left-hand side of a regression equation can be the (conditional) mean, median, or some other statistics. In the analysis of binary outcomes, the most well-known regression method is probably the logistic regression, which has the log odds for the left-hand side of the regression equation. We will discuss a class of regression models, called generalized linear models, that includes the logistic regression model. In this chapter, the terms *risk, proportion,* and *probability* are synonymous.

6.1 Basic Concepts

6.1.1 Risk and Odds

We use p to represent the sample estimate of the true risk in the population, π; p = k/n, where k is the number of occurrences of the outcome and n is the sample size. *Odds* is denoted by Ω and the sample estimate is denoted by ω. It can be calculated using the sample estimate of the risk

$$\omega = \frac{p}{1-p} \qquad (6.1)$$

Alternatively,

$$\omega = \frac{k}{n-k} \qquad (6.2)$$

Equation (6.2) is preferred because it avoids a rounding error in p. Conversely,

$$p = \frac{\omega}{1+\omega} \qquad (6.3)$$

If the outcome concerned is rare, say, p = 0.001, (1 − p) is very close to 1. Then the odds and the risk are approximately equal. In epidemiology, it is

TABLE 6.1

Notation for Entries in a 2-by-2 Table

Group	Outcome		Risk of Failure	Sample Size
	Success (y = 0)	Failure (y = 1)		
Unexposed	a	b	$p_0 = b/(a+b)$	$n_0 = a + b$
Exposed	c	d	$p_1 = d/(c+d)$	$n_1 = c + d$
	s = a + c	f = b + d		$n = n_0 + n_1$

said that the odds and risk parameters are approximately equal under the *rare disease assumption* (Clayton and Hills 1993).

Risk ratio (RR), *risk difference* (RD), and *odds ratio* (OR) are complementary measures of the difference in proportions between groups. Table 6.1 shows the notations for entries in a 2-by-2 table. The rows denote two groups defined by different exposure status. The two columns denote the binary outcome. The cell frequencies are represented by a, b, c, and d. The sample estimates of the risk of failure in the two groups are p_0 and p_1. The risk ratio

$$RR = \frac{p_1}{p_0} = \frac{d/(c+d)}{b/(a+b)} \tag{6.4}$$

and risk difference

$$RD = p_1 - p_0 = \frac{d}{(c+d)} - \frac{b}{(a+b)} \tag{6.5}$$

indicate the relative impact and absolute impact of the exposure, respectively. In regression analysis, it is common to estimate the logarithmic transformation of RR. The estimated ln(RR) is then back-transformed to RR if needed. The RR is also variably called *relative risk* or *relative risk reduction*. The RD is sometimes called *absolute risk reduction*. The *odds ratio* is estimated by

$$OR = \frac{p_1/(1-p_1)}{p_0/(1-p_0)} = \frac{d/c}{b/a} = \frac{ad}{bc} \tag{6.6}$$

The last decade has seen a lot of research and concerns about which of the three measures of associations is easier to interpret by lay people. It appears that in the communication of treatment benefits or harms to patients the use of risk difference is more appropriate (see, for example, Akl et al. 2011; Stovring et al. 2008). The odds ratio tends to be larger in numerical value than the risk ratio. So, it may give a false sense of strong association. Nevertheless, when the risk is small, the odds ratio and risk ratio are approximately the same. Figure 6.1 illustrates the odds ratio value if the risk ratio is 2.0, at various levels of risk in the

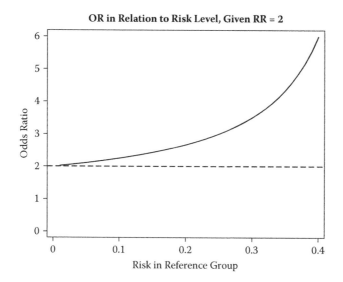

OR in Relation to Risk Level, Given RR = 2

Risk in Reference Group

FIGURE 6.1
Odds ratio (solid line) in relation to the risk in the reference group (p_0), given that the risk ratio is 2.0 (dashed line).

reference group. When the risk level is close to zero, the OR is approximately equal to the RR. As the risk level increases, the OR increases in numeric value in an accelerating manner even though the true risk ratio is still 2.0. The use of the odds ratio has been popular yet controversial. Many previous researchers have criticized the uncritical use of the OR. For instance, Sinclair and Bracken (1994) maintained that "because the control group's risk affects the numerical values of the odds ratio, the odds ratio cannot substitute for the risk ratio in conveying clinically important information to physicians" (p. 881).

On the other hand, in the context of equivalence studies, Cheung (2007) pointed out that the risk ratio has a property that makes it problematic. That is, the RR for $Y = 1$ does not equal the inverse of the RR for $Y = 0$. In equivalence studies, it is common to define the equivalence zone as 0.8 and $1/0.8 = 1.25$. If the confidence interval (CI) of the comparative parameter totally falls within this zone, equivalence is concluded. If the CI totally falls outside this zone, nonequivalence is concluded. If the CI overlaps with the zone, there is no clear conclusion. The use of RR in an equivalence study can give different and counterintuitive conclusions according to whether the risk ratio for $Y = 1$ or $Y = 0$ is the parameter. In equivalence studies, the use of risk difference was recommended (Cheung 2007).

Example 6.1

Consider the hypothetical example described in Table 6.2. The RR for $Y = 1$ (diseased) is $0.40/0.25 = 1.60$. It can be shown, using generalized

TABLE 6.2

Demonstration of Risk Ratio for Y = 0 Does Not Equal the Inverse of the Risk Ratio for Y = 1

	Y = 0 (Nondiseased)	Y = 1 (Diseased)	Prob(Y = 1)	Prob(Y = 0)
Unexposed	750	250	0.25	0.75
Exposed	600	400	0.40	0.60
RR			0.40/0.25 = 1.60	0.60/0.75 = 0.80

linear models described in the next section, that the 95% CI is 1.40 to 1.82. This totally falls above the equivalence zone of 0.8 to 1.25. Therefore, it can be concluded that the two groups are not equivalent as far as the risk of disease is concerned. However, the RR for Y = 0 (nondiseased) is 0.6/0.75 = 0.80, which does not equal the inverse of the RR for Y = 1, that is, 1/1.60 = 0.625, not 0.80. The 95% CI is 0.75 to 0.85, which overlap with the equivalence zone. So there is no conclusion about whether the groups are equivalent as far as the risk ratio of Y = 0 is concerned.

6.1.2 Sensitivity and Specificity

Sensitivity refers to the probability that a diseased subject be correctly diagnosed as diseased. *Specificity* refers to the probability of a nondiseased subject being correctly diagnosed as nondiseased (Hennekens and Buring 1987). Table 6.3 illustrates the concept of sensitivity and specificity. The following example illustrates the relative importance of sensitivity and specificity in the estimation of association. The planning of studies usually gives priority to using highly specific measures of clinical outcomes (Nauta 2010; Orenstein et al. 1988).

Example 6.2

Table 6.4 demonstrates the impact of insufficient sensitivity and insufficient specificity in outcome assessment. The true diseased risk was 20/100 = 0.2 and 40/100 = 0.4 in the unexposed and exposed group, respectively. The true RR of disease is 2.0 and true RD is 0.2. If the diagnostic test of the disease has 100% sensitivity but only 50% specificity, a number of nondiseased people would be misdiagnosed as diseased: 0.5 × 80 = 40 in the unexposed group and 0.5 × 60 = 30 in the exposed

TABLE 6.3

Notation for Sensitivity and Specificity for Diagnosing a Disease

	Diagnosis			
Truth	Nondiseased	Diseased	Sensitivity	Specificity
Nondiseased	a	b		a/(a + b)
Diseased	c	d	d/(c + d)	

TABLE 6.4

Illustration of the Relative Impact of Imperfect Sensitivity or Specificity

Group	True Nondiseased	True Diseased	Observed Diseased (Sens = 100%; Spec = 50%)	Observed Diseased (Sens = 50%; Spec = 100%)
Unexposed	80	20	$20 + 0.5 \times 80 = 60$	$0.5 \times 20 + 0 = 10$
Exposed	60	40	$40 + 0.5 \times 60 = 70$	$0.5 \times 40 + 0 = 20$
RR		$0.4/0.2 = 2.0$	$0.7/0.6 = 1.17$	$0.2/0.1 = 2.0$
RD		$0.4 - 0.2 = 0.2$	$0.7 - 0.6 = 0.1$	$0.2 - 0.1 = 0.1$

Notes: Sens and Spec stand for sensitivity and specificity, respectively.

group would be misdiagnosed as diseased. More important, the degree of misclassification was not the same across the two groups. Both the RR and RD would be biased toward the null values of RR = 1 or RD = 0. In this example the RR and RD estimates are 1.17 and 0.1, respectively.

In contrast, if the diagnostic test has 100% specificity but 50% sensitivity, the estimate of RR is not affected. Nevertheless, the estimate of RD is still biased toward the null value.

6.2 Introduction to Generalized Linear Models

The *logistic regression* model is a member of a bigger family called *generalized linear models*, or GLMs. Note that some authors may use the acronym GLMs to stand for *general linear models*, which is a different thing. One reason why logistic regression is important in the learning of statistics is that it is a gateway to the learning of the GLMs. The seminal paper of Nelder and Wedderburn (1972) first demonstrated an underlying unity to a group of regression models. Among others, Hardin and Hilbe (2001) gave a comprehensive review of GLMs.

A GLM consists of three components:

- A random component that specifies the probability distribution of the outcome variable Y;
- A systematic component that specifies the exposure variables in a linear predictor equation; and
- A *link function*, g[E(Y)], that specifies how the expected value of the outcome variable on the left-hand side of the regression equation relates to the systematic component on the right-hand side.

The logistic regression model has the Bernoulli distribution, which specifies $P(Y = 1) = \pi$ and $P(Y = 0) = (1-\pi)$, for the random component. It uses the *logit link*, that is, the log odds, for the link function

$$g\left[E(Y)\right] = \ln\left[\frac{E(Y)}{1 - E(Y)}\right] \tag{6.7}$$

The systematic component is up to the analyst's ideas about what variables are related to the log odds and in what manner.

Table 6.5 shows three popular link functions for modeling binary outcomes. The following sections will discuss the three models. Computationally, the logistic regression model is easier to estimate. Historically some statistics software implemented only the logistic regression. That is one of the reasons why the logistic regression has been popular. Some general issues about the GLMs are discussed in Section 6.3. So the subsequent sections will be briefer.

6.3 Logistic Regression

6.3.1 Model and Estimation

The logistic regression uses the logit link function, meaning the left-hand side of the regression equation is the log odds of the outcome. It may be more intuitive to call this the log odds regression. The coefficients, β's, represent the amount of difference in the log odds per one unit difference in the exposure. As such, exp(β) represents the odds ratio in the population (Table 6.5). The regression equation to be estimated from the sample is

$$\ln\left(\frac{p}{1 - p}\right) = a + b_1 x_1 + \ldots + b_p x_p \tag{6.8}$$

For brevity, we use matrix notation $\mathbf{Xb}$, where $X = (1 \; x_1 \; \cdots \; x_p)$ and $b = (a \; b_1 \; \cdots \; b_p)'$, to represent the systematic component $a + b_1 x_1 + \ldots + b_p x_p$. The model gives the predicted probability of a positive outcome ($Y = 1$)

TABLE 6.5

Generalized Linear Models for Binary Outcomes Using One of Three Popular Link Functions

Model	Link Function	LHS	RHS	Interpretation of β's
Logistic	Logit	$\ln[\pi/(1 - \pi)]$	$\alpha + \beta_1 x_1 + \ldots + \beta_p x_p$	log OR; exp(β) is OR
Log-binomial	Log	$\ln(\pi)$	$\alpha + \beta_1 x_1 + \ldots + \beta_p x_p$	log RR; exp(β) is RR
Binomial	Identity	π	$\alpha + \beta_1 x_1 + \ldots + \beta_p x_p$	RD

Notes: LHS stands for the left-hand side of the regression equation. RHS stands for the right-hand side of the regression equation.

$$p = \frac{\exp(\mathbf{Xb})}{1+\exp(\mathbf{Xb})} = \frac{1}{1+\exp(-\mathbf{Xb})} \tag{6.9}$$

From Equation (6.9), we can see an advantage of using the logit link function: The predicted probability is always within the range of 0 and 1, because exp(·) > 0. Figure 6.2 illustrates this with a simple logistic regression equation of

$$\ln\left(\frac{p}{1-p}\right) = 1 + 1 \times Z$$

As Z increases (decreases), the probability of a positive outcome approaches the upper (lower) asymptote of 1 (0). At Z = 0, the log odds is 1 and therefore the probability is 1/[1 + exp(−1)] = 0.73. The other link functions do not guarantee the predicted probability to stay within the 0 to 1 range. They can give strange results that some individuals are predicted to have a probability smaller than zero or larger than one in having the outcome. This is another reason for the popularity of the logistic regression model.

There are multiple approaches to the estimation of the model. A popular choice is by maximizing the log likelihood. The random component of the binary outcome has been given in Chapter 3, Equation (3.2), when we discussed maximum likelihood estimation. Substituting Equation (6.9) into the random component, the likelihood for an observation is

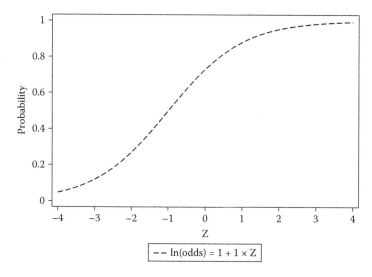

FIGURE 6.2
Predicted probability for logistic regression equation log odds = 1 + 1 × Z.

$$\left[\frac{1}{1 + \exp(-\mathbf{Xb})}\right]^{Y}\left[\frac{\exp(-\mathbf{Xb})}{1 + \exp(-\mathbf{Xb})}\right]^{1-Y} \tag{6.10}$$

The model log-likelihood is the sum of the n individuals' log-likelihood values. Iterative algorithms are used to find the set of regression coefficients that maximizes the model log-likelihood (Hardin and Hilbe 2001). The logistic regression model has a log-likelihood function that is globally concave. The iterative search algorithms always give a solution provided that there are observations with y = 0 and y = 1 and there is overlapping in the range of exposure values in the two groups of outcomes.

Table 6.6 shows the Stata commands for estimating logistic regression and other GLMs for binary outcomes. Rows 1 and 2 show commands that give identical results, but the "glm" command in row 2 has a syntax that is generic to other GLMs. Row 3 shows a command that presents the results in terms of odds ratio instead of log OR; otherwise it is the same as the commands in rows 1 and 2.

6.3.2 Statistical Inference

The Wald method is commonly used to obtain the standard errors (SEs) of the logistic regression coefficients. It first calculates the SE for the ln(OR). When there is only one binary exposure variable the SE is

TABLE 6.6

Stata Commands for the Estimation of Three GLMs and Alternatives[*]

Number	Model	Stata Command	Coefficients
1	logistic	logit y x	log OR
2		glm y x, family(binomial) link(logit)	log OR
3		logistic y x	OR
4	log-binomial	glm y x, family(binomial) link(log)	log RR
5	binomial	glm y x, family(binomial) link(identity)	RD
6	PA log-binomial[†]	glm y x, family(poisson) link(log) robust	log RR
7	PA binomial[§]	glm y x, family(poisson) link(identity) robust	RD
8	MLS binomial[**]	regress y x, hc3	RD

[*] x and y are the names of the independent and dependent variable, respectively.

[†] Poisson approximation (PA) to the log-binomial model. From Zou, G., 2004, A modified Poisson regression approach to prospective studies with binary data, *American Journal of Epidemiology* 159:702–706.

[§] Poisson approximation (PA) to the binomial model. From Spiegelman, D., and Hertzmark, E., 2005, Easy SAS calculations for risk or prevalence ratios and differences, *American Journal of Epidemiology* 162:199–200.

[**] Modified least-squares (MLS) estimation of binomial model. From Cheung, Y. B., 2007, A modified least-squares approach to the estimation of risk difference, *American Journal of Epidemiology* 166:1337–1344.

$$SE_{\ln(OR)} = \sqrt{\frac{1}{a} + \frac{1}{b} + \frac{1}{c} + \frac{1}{d}} \qquad (6.11)$$

The confidence interval of OR can be obtained by taking exponentiation of the CI of the $\ln(OR)$

$$\exp\left[\ln(OR) - z_{1-\alpha/2} \times SE_{\ln(OR)}\right] \quad \text{to} \quad \exp\left[\ln(OR) + z_{1-\alpha/2} \times SE_{\ln(OR)}\right] \quad (6.12)$$

Although the Wald SE can be used to test a null hypothesis of no association, the likelihood ratio (LR) test introduced in Chapter 3 is more accurate and preferred. The LR test can also be used to test the joint null hypothesis that more than one regression coefficients are zero. The LR test involves three steps:

1. Estimate the model with the parameter(s) concerned and obtain the maximized log-likelihood value (arbitrarily named here as L1).
2. Estimate the model without the parameter(s) concerned and obtain the maximized log-likelihood value (L0).
3. Compare the deviance, $-2 \times (L0 - L1)$, against the chi-square distribution with df equal to the difference in the number of parameters between the two models in steps 1 and 2.

Example 6.3

The Fourth Wave of the Indonesia Family Life Survey (IFLS4) (Strauss et al. 2009) collected blood pressure data of adults aged 40 years and older. This illustration defined hypertension as systolic blood pressure ≥140. As discussed in Chapter 3, it is more sensible to compare the regression coefficients if the exposure variables are z-scores. Display 6.1 regressed the log odds of this outcome upon three anthropometric variables in z-score: waist-to-hip ratio (ZWHR), body mass index (ZBMI), and height (ZHEI), controlling for demographic covariates age in years and sex. Having adjusted for the covariates, the coefficient for height was -0.04197. Therefore, the OR per one SD increase in height was $\exp(-0.04197) = 0.959$. For every one SD taller, the odds of hypertension was $(1 - 0.959)$ or 4.1% lower. According to the Wald method, height was not associated with hypertension ($P = 0.114$). The log-likelihood of the model was -5905.30. The "estimates" command stored the log-likelihood as L1. The name was arbitrary. Then a model without the height variable was fitted, and the log-likelihood value -5906.55 was stored as L0. The likelihood ratio test statistics is

$$-2 \times [-5906.55 - (-5905.30)] = 2.50$$

```
. logit hyper ZWHR ZBMI ZHEI ageyr sex

Iteration 0:  log likelihood = -6569.4839
Iteration 1:  log likelihood = -5906.9693
Iteration 2:  log likelihood = -5905.3026
Iteration 3:  log likelihood = -5905.3025
```

```
Logistic regression                        Number of obs  =    9702
                                           LR chi2(5)     = 1328.36
                                           Prob > chi2    =  0.0000
Log likelihood = -5905.3025                Pseudo R2      =  0.1011
```

hyper	Coef.	Std. Err.	z	P>|z|	[95% Conf. Interval]	
ZWHR	.1421879	.0240038	5.92	0.000	.0951414	.1892344
ZBMI	.3900568	.0251147	15.53	0.000	.340833	.4392807
ZHEI	-.0419722	.0265782	-1.58	0.114	-.0940646	.0101202
ageyr	.0655699	.0023538	27.86	0.000	.0609565	.0701833
sex	.1899392	.0454516	4.18	0.000	.1008557	.2790227
_cons	-4.340286	.143946	-30.15	0.000	-4.622415	-4.058157

```
. estimates store L1

. logit hyper ZWHR ZBMI ageyr sex

(output omitted)
```

```
Log likelihood = -5906.5505                Pseudo R2      =  0.1009
```

hyper	Coef.	Std. Err.	z	P>|z|	[95% Conf. Interval]	
ZWHR	.141892	.0239902	5.91	0.000	.0948721	.1889119
ZBMI	.3885079	.0250824	15.49	0.000	.3393472	.4376685
ageyr	.0665723	.002269	29.34	0.000	.0621251	.0710194
sex	.186207	.045381	4.10	0.000	.0972619	.2751522
_cons	-4.294727	.1408337	-30.50	0.000	-4.570756	-4.018698

```
. estimates store L0

. lrtest L1 L0
```

```
Likelihood-ratio test                      LR chi2(1)    =     2.50
(Assumption: L0 nested in L1)               Prob > chi2  =   0.1141
```

```
. logit hyper ZWHR ageyr sex

(output omitted)
```

DISPLAY 6.1
Stata codes and outputs: Logistic regression analysis of SBP > = 140 in IFLS4.

(continued)

```
Log likelihood = -6031.3419                          Pseudo R2   =  0.0819
------------------------------------------------------------------------------
  hyper |      Coef.   Std. Err.      z    P>|z|     [95% Conf. Interval]
--------+---------------------------------------------------------------------
   ZWHR |   .2689142   .0224351    11.99   0.000     .2249423    .3128862
  ageyr |   .0547516    .002079    26.34   0.000     .0506769    .0588264
    sex |   .3309052   .0438803     7.54   0.000     .2449013     .416909
  _cons |  -3.862072   .1350697   -28.59   0.000    -4.126804    -3.59734
------------------------------------------------------------------------------

. estimates store L2

. lrtest L1 L2

Likelihood-ratio test                          LR chi2(2)   =    252.08
(Assumption: L2 nested in L1)                  Prob > chi2 =    0.0000
```

DISPLAY 6.1 (CONTINUED)
Stata codes and outputs: Logistic regression analysis of SBP > = 140 in IFLS4.

The LR test statistics is used to compare the deviance against the chi-square distribution with df = 1, as the two models differed by one parameter. The P-value 0.1141 again showed that the null hypothesis of no association cannot be rejected. In this particular case, the LR test and the Wald test gave practically the same P-values. But this is not generally true.

For illustration purposes, we further test the joint null hypothesis that both height and BMI were not associated with hypertension. We fitted a model without the height and BMI variables, and saved the log-likelihood as L2. The LR test now compared the test statistics (based on L1 and L2) against the chi-square distribution with df = 2, because the models differed by two parameters. The log-likelihood of the reduced model was –6031.34. The test statistic was

$$-2 \times [-6031.34 - (-5905.30)] = 252.08$$

The joint null hypothesis was clearly rejected (P < 0.0001). At least one of the two regression coefficients were statistically significant.

6.3.3 Model Diagnostics

6.3.3.1 Hosmer-Lemeshow Test

The *Hosmer-Lemeshow goodness-of-fit test* examines whether the observed frequency matches the predicted frequency (Hosmer and Lemeshow 2000). The procedure divides the observations into G groups according to their predicted probability of a positive outcome, $P(Y = 1)$. Hosmer and Lemeshow recommended G = 10 groups for large datasets. Similar to the Chi-square test for association between categorical variables, the Hosmer-Lemeshow test statistics is based on the discrepancy between the number of observed and number of expected outcomes

$$H-L = \sum_{g=1}^{G} \frac{\left(O_g - n_g p_g\right)^2}{n_g p_g \left(1 - p_g\right)} \tag{6.13}$$

where O_g and n_g are the number of observed outcomes and number of observations, respectively, and p_g is the model predicted risk for the gth risk group. The product $n_g p_g$ is the number of expected outcomes based on the regression analysis. The test statistics in Equation (6.13) is compared against the chi-square distribution with $(G-2)$ degrees of freedom. The larger the level of discrepancy, the poorer the model fits the data and the smaller the P-value. A small P-value leads to the rejection of the null hypothesis of a good fit.

Example 6.4

In a study of early child development in developing countries, Grantham-McGregor et al. (2007) used a logistic regression model to estimate the odds ratio for school dropout before grade 11 in relation to IQ at age 7 years (exposure) in Jamaican children, adjusting for covariates. The Hosmer-Lemeshow goodness-of-fit test gave a P-value of 0.570. The null hypothesis of good fit was not rejected.

6.3.3.2 Pseudo R-Squared

In ordinary least-squares (OLS) regression, the R-squared indicates the proportion of variance explained by the regression model. A large R-squared indicates a good fit. However, there is no equivalent statistics to R-squared in logistic regression. Indeed, the principles in the model estimation are different between OLS and logistic regression models. The former aims to minimize the variance, whereas the latter aims to maximize the likelihood. The last few decades have seen substantial efforts to develop some form of pseudo R-squared. Long and Freese (2001) describe several of them in detail. These measures are called pseudo R-squared because they do not exactly have the R-squared interpretation in least-squares regression, yet they look like R-squared in the sense that they range from 0 (worst) to 1 (best). Pseudo R-squared indices concern relative, not absolute, goodness-of-fit. They are valid only in comparing models fitted to the same dataset.

As an example, consider the McFadden's pseudo R-squared, which is defined as

$$R^2_{\text{McFadden}} = 1 - \frac{\ln L(\text{full})}{\ln L(\text{intercept})} \tag{6.14}$$

where lnL(full) denotes the maximized log-likelihood value of the full model and lnL(intercept) denotes that of an intercept only model. A log-likelihood value must be a negative value. The absolute value of lnL(intercept) must be larger than that of lnL(full). The more predictive the exposure variables are, the smaller the absolute value of lnL(full), because a smaller absolute value preceded by a negative sign means a higher likelihood. So, Equation (6.14) gives a pseudo R-squared value within the 0 to 1 range. However, this is based on the log-likelihood and does not offer the proportion of variance explained interpretation equivalent to the R-squared in least-squares regression. This or other pseudo R-squared indices are relatively useful in comparing different models fitted to the same data, but they are difficult to interpret in terms of whether a model itself is fitting well.

Example 6.3 (Continued)

By default, Stata's "logit" command calculates the McFadden's R-squared. In Display 6.1, the value was 0.1011 for the first model. Note that Stata begins with calculating the log-likelihood of an intercept only model. Hence, the log likelihood at iteration 0 is lnL(intercept). In Display 6.1 it was −6569.4839. The maximized lnL(full) was −5905.3025. Therefore, the McFadden's pseudo R-squared is

$$1 - (-5905.3025 / -6569.4839) = 0.1011$$

6.3.3.3 Link Test

If the independent variables do predict the outcome, the predicted value **Xb** should also correlate with the outcome. The *link test* uses the estimated regression coefficients to calculate new variables **Xb** and $(\mathbf{Xb})^2$ for each observation, and then fit the logistic regression model to the outcome variable (Pregibon 1980)

$$\ln\left(\frac{p}{1-p}\right) = c_0 + c_1(\mathbf{Xb}) + c_2(\mathbf{Xb})^2 \tag{6.15}$$

If the model is properly specified, the variable $(\mathbf{Xb})^2$ should not be associated with the outcome. A test of the significance of the coefficient c_2 is a test of misspecification error. If the P-value for c_2 is small, misspecification is suspected. That suggests the linear form of the exposure variables on the right-hand side of the regression equation is probably incorrect.

6.3.3.4 Residuals

Examination of residuals allows us to identify influential cases or possible data errors. A trend in residuals in relation to exposure variables may also suggest ways to modify the models. Again, the locally weighted regression smooth introduced in Chapter 5 is a useful technique to go with residual analysis.

The *Pearson residual* for the ith subject is defined as

$$r_i = \frac{Y_i - M_i p_i}{\sqrt{M_i p_i (1 - p_i)}} \tag{6.16}$$

where M_i, Y_i, and p_i are, respectively, the number of observations whose covariate pattern is the same as the ith subject's, the number of positive responses among the observations whose covariate pattern is the same as the ith subject's, and the model predicted probability of a positive response among observations with this covariate pattern (Hardin and Hilbe 2001). The Pearson residuals have approximately mean 0 and SD 1. Observations with an absolute value of Pearson residual larger than 2 to 3 are worth further examination.

Example 6.3 (Continued)

Table 6.7 shows the regression coefficients and diagnostic results for the logistic regression models with WHR and BMI (both in SD), age, and sex as exposure variables (model I). All four variables were significantly related to hypertension (each P < 0.01). The Hosmer-Lemeshow test and the Link test were calculated using Stata's "estat gof, group(10)" and "linktest" commands, after estimating the logistic regression model. The Hosmer-Lemeshow test gave a P-value of 0.018. The link test showed that

TABLE 6.7

Regression Coefficients and Model Diagnostic of Logistic Regression Analysis of Hypertension in IFLS4

Regression Coefficients*	Model I[†]	Model II[§]
WHR (SD)	0.141	0.132
BMI (SD)	0.389	0.429
BMI (SD) squared		−0.049
Age (year)	0.067	0.149
Age (year) squared		−0.001
Female	0.186	0.194
Diagnostic Tests		
Hosmer-Lemeshow, P =	0.018	0.347
Link test, P =	<0.001	0.953
McFadden's R-squared	0.101	0.103

[*] All P-values < 0.01 for the regression coefficients in models I and II.
[†] Model I: right-hand side variables include WHR in SD, BMI in SD, age in years, and sex.
[§] Model II, right-hand side variables additionally include BMI^2 and age^2.

the squared predicted value was significantly associated with the outcome variable (P < 0.001). Both tests suggest that the model was not optimal.

Figure 6.3 plots the Pearson residuals against each of the three continuous variables in the model, together with a locally weighted regression smooth. The two tails of the lowess smooth in the plot against WHR showed some nonlinearity (Figure 6.3a). However, there were only a few observations at the tails, especially the left-hand side tail. The plot against BMI also showed some curvature at one end. There was a group of small residuals clustered at a very high BMI. It should raise some concerns. In the plot against age, the averaged residuals were very close to zero up to about age 80 years. But beyond that age, there was a systematic overestimation of the probability of hypertension. Unlike the plots for the anthropometric variables, the number of observations above age 80 years was substantial. Furthermore, three observations had Pearson residuals smaller than –3. Listing the outcome revealed that all three of them were nonhypertensive. Two of them aged over 90; one of them had very high BMI (+5 SD). As such, second-order polynomial terms, age-squared, and BMI-squared, were added to the model (model II).

All the regression coefficients in model II were statistically significant (each P < 0.01). The negative regression coefficient on BMI-squared and age-squared suggested that the log odds of hypertension did not increase linearly with these two variables. Instead, the increase decelerated. Both the Hosmer-Lemeshow test (P = 0.347) and the link test (P = 0.953) showed that model II was satisfactory. The pseudo R-squared also indicated better fit on the part of model II as compared to model I.

Note that the regression coefficients on WHR and sex were fairly unaffected by the addition of the polynomial terms. If the purpose of the analysis was to identify the BMI or age pattern of hypertension, the addition of the squared terms was important. However, if the purpose was to estimate the relation between hypertension and WHR or gender, with the other variables adjusted as potential confounders, the two models provided similar answers to the research question. This is to say, a model that does not fit the data well is not necessarily a disaster. It depends on whether the reason for the misfit is related to the specific research questions in hand. The issue is that if we do not investigate the reasons of the misfit, we may not know whether there is a disaster or not.

6.4 Log-Binomial and Binomial Regression Models

The generalized linear models for binary outcomes that use a log link or an identify link are sometimes referred to as the *log-binomial regression model* and *binomial regression model*. The former estimates the ln(RR) and the latter estimates the RD. Their implementation within a statistical package is in a sense straightforward, as one only needs to modify the link function component of the GLM command. Table 6.6 shows the Stata commands for

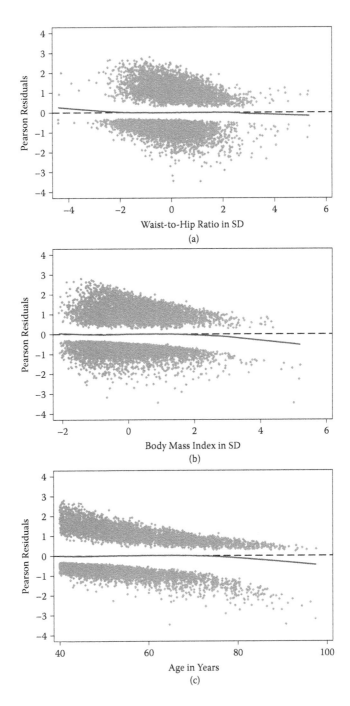

FIGURE 6.3
Scatterplots of Pearson residuals versus waist-to-hip ratio in SD (a), body mass index in SD (b), and age in years (c). A locally weighted regression smooth (solid line) is imposed on each panel.

estimating several GLMs. The syntax for estimating logistic (row 2), log-binomial (row 4), and binomial (row 5) regression models only differ in the specification of the link function.

The log-binomial model and the binomial model are more prone to estimation problems than the logistic model. The iterative search algorithms may not be able to provide a solution. This tends to happen when there are continuous exposure variables that are associated with the outcome. In this situation, during the iteration process the algorithms may find predicted probability outside the admissible range of 0 to 1 and get stuck. Several alternative methods have been proposed to perform regression estimation of risk ratios and risk differences. One approach is to use the Poisson model to approximate the (log-)binomial model (Spiegelman and Hertzmark 2005; Zou 2004). But the SE calculated based on the Poisson distribution is not correct for the binomial model. Therefore a robust, distribution-free, form of standard error estimation called *Huber-White robust standard error* is used (see Chapter 8). Since the Poisson model is also within the GLM family, the practice is simply by modifying the GLM estimation from using the binomial to the Poisson distribution and to add the "robust" option to call the robust standard error. Compare the Stata commands in row 4 against row 6 in Table 6.6 for estimation of the log-binomial model. Also compare row 5 against row 7 for estimation of the binomial model.

Another alternative to estimate risk difference is by using the least-squares method with a specific form of robust SE known as *HC3* (Cheung 2007). See row 8 of Table 6.6. Essentially, the Poisson approximation and the least-squares method give up caring about the inadmissible predicted values for individual observations and focus on estimating the regression coefficients and SEs right. If the research purpose is to develop a prognostic index so that the risk of adverse outcomes can be estimated for each individual, these approaches are not suitable. However, when the research purpose is to estimate the degree of association, the research concern is not at the individual level. The alternative methods are useful in this situation.

7

Regression Analysis of Censored Outcomes

In the analysis of survival time data, it is common that the survival times of some subjects are not exactly observed. For example, at the time of the data analysis, a subject may be 50 years old. The analyst knows that this subject's survival time is at least 50 years, but the exact value is not known. The data is *censored*. It may be tempting to exclude participants whose survival time is censored from the analysis. But this would cause a bias because the exclusion is not random. The term *survival analysis* refers to statistical methods that handle the analysis of censored data that contains only positive values (i.e., >0). It is now more common to use the term *time-to-event analysis*. This chapter will discuss regression analysis methods for such data.

7.1 Fundamentals

7.1.1 Censoring

If it is known that the underlying value must be larger (or smaller) than a certain boundary, it is called *right censoring* (or *left censoring*). If it is known that the value must be within an interval, it is *interval censoring*.

In the Early Child Health in Lahore, Pakistan (ECHLP) study (see Chapter 2), the researchers visited the participating families every month from the birth of the newborns until they were 24 months old. The infants were measured for growth status, among other things, at these home visits. If the infant had died, the date of death was recorded and the age at death in days calculated. Appendix B shows the codes that simulate a dataset that resembles this study design and data structure, the Simulated Longitudinal Study (SLS) dataset. We will use this dataset to illustrate some of the methods. Display 7.1 shows some of the variables in three selected cases in this dataset.

Example 7.1

In Display 7.1, infant ID 2 was alive and measured for height-for-age (HAZ) at the 8-month visit at the exact age of 8.21 months. After that the family was not contactable, perhaps migrated. The age of which the infant was last known to be alive was 8.21 months. The minimum of the survival time (time-to-death) of this infant was known (at least 8.21 months), but the exact survival time was not. This is right censoring. This was recorded by the pair of variables (ttd, death) = (8.21, 0). The latter variable took on the value 0 if the endpoint was censored (alive) or 1 if the endpoint was observed (dead). Infant ID 9 was alive and measured for anthropometry when visited at age 2.01 months. When his family was visited for the 4-month visit, he was reported to have died at the age of 2.29 month. The timing of event was known. Hence the pair of (ttd, death) variables took on values (2.29, 1). The value was observed, or uncensored.

The researchers may also be interested in studying the time-to-stunting, as defined by height-for-age z-score (HAZ) dropping below –2. Infant ID 3 had an HAZ value above –2 at and before the measurement at 21.95 months. But the HAZ value became –2.22 at age 23.96 months. Although the exact age at the timing of her becoming stunted was not known, it was known to be within the interval of 21.95 to 23.96 months. This is interval censoring. It needs a triplet of values (21.95, 23.96, 1) to record this interval-censored observation.

Furthermore, suppose an infant was first measured for length at 3 days after birth and was found to be already stunted. We know that his age at the timing of becoming stunted was below 3 days but we do not know the

```
. list id visit visitage haz ttd death stunt_ever if id==2 | id==3 |
id==9, noobs
```

id	visit	visitage	haz	ttd	death	stunt_~r
2	0	.019	.127	8.213	0	0
2	2	1.925	-.19	8.213	0	0
2	4	3.965	-.412	8.213	0	0
2	6	5.947	-.552	8.213	0	0
2	8	8.213	-.848	8.213	0	0
3	0	.132	.113	23.96	0	1
3	2	1.994	-.08	23.96	0	1
3	4	4.099	-.515	23.96	0	1
3	6	5.9	-.706	23.96	0	1
3	8	8.111	-.943	23.96	0	1
3	10	9.795	-1.221	23.96	0	1
3	12	11.967	-1.47	23.96	0	1
3	14	13.869	-1.68	23.96	0	1
3	16	16.125	-1.694	23.96	0	1
3	18	18.148	-1.825	23.96	0	1
3	20	19.877	-1.914	23.96	0	1
3	22	21.947	-1.999	23.96	0	1
3	24	23.96	-2.22	23.96	0	1
9	0	.048	.167	2.293	1	0
9	2	2.005	-.098	2.293	1	0

DISPLAY 7.1
Partial data of the SLS dataset.

exact timing. This is left censoring. Left censoring may be seen as interval censoring with a lower bound at 0; right censoring may be seen as interval censoring with an upper bound at infinity. While a lot of statistical methods are oriented toward the problem of right censoring, some handle data that contain the other two types of censoring or a mixture of all three.

In some studies, observation of an individual participant ends when the planned follow-up duration for each participant is reached, for example, when a child completes the final visit. In some other studies, the observations for all participants end at the same calendar date when the field operation stops. These are *administrative censoring*. The important thing is that the censoring is controlled by the researchers and is independent from the outcomes of the individuals.

Another reason for censoring is loss of follow-up. Participants may migrate to another country, withdraw their participation, become noncontactable for unknown reasons, and so on. It is known that at the last date of record, the event had not occurred. In Display 7.1, one child was loss of follow-up after about 8 months. Usually, such censoring is assumed to be independent of the time-to-event. This is an important assumption that underpins most analytic methods for censored data, including all methods described in this chapter. The validity of this assumption needs careful consideration. The analysis becomes much more complex when this assumption is not valid, and is beyond the scope of this chapter.

7.1.2 Hazard and Hazard Ratio

The *hazard* is the event rate within a very small time interval. The hazard function, h(t), describes how the hazard changes over time (or age). The hazard function; probability density function, f(t); and survival function, S(t), are related to each other as

$$h(t) = \frac{f(t)}{S(t)} = -\frac{d}{dt} \ln[S(t)] \tag{7.1}$$

where dx/dt is the first derivative of x with respect to t. The three measures represent technically equivalent ways to study the distribution of an outcome variable. But h(t) and S(t) are more convenient when the variable is censored.

Example 7.2

Mortality rate changes quickly in the first month of life. Using data from the Swedish Medical Birth Registry, a very large-scale database with over a million births and over 2000 neonatal deaths, Cheung et al. (2001c) estimated and showed a typical pattern of neonatal mortality hazard in infants with an average gestational duration and size at birth. As shown in Figure 7.1, the hazard was high at birth. But it declined quickly during the first week of life and then stabilized at a low level.

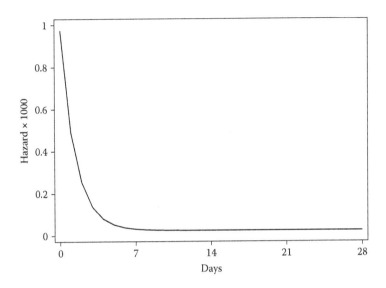

FIGURE 7.1
Estimated neonatal mortality hazard in Swedish boys born alive with average gestational duration and size at birth. (Data from Cheung, Y. B., Yip, P. S. F., and Karlberg, J. P. E., 2001c, Parametric modeling of neonatal mortality in relation to size at birth, *Statistics in Medicine* 20:2455–2466, Table 1.)

When sample size is modest, it is difficult to estimate and graph the hazard. For instance, in the SLS dataset, there were only a few deaths in the high socioeconomic status (SES) group and they happened early during the follow-up. Therefore, for a large part of the study period there is no data to estimate the hazard. The use of the *cumulative hazard*, H(t), can visualize data more clearly when the sample size is modest. The relation between the cumulative hazard function and the survival function can be derived from Equation (7.1):

$$H(t) = -\ln\left[S(t)\right] \qquad (7.2)$$

Suppose D events are observed at k distinct event times $t_1 < t_2 < \ldots < t_{k-1} < t_k$; k = D if all event times are distinct; k < D if some events occurred at exactly the same times. Let d_j be the number of individuals with the outcome event observed at time t_j and n_j be the number of individuals "at risk" of the event at time t_j. "At risk" means not yet censored and not yet experienced the outcome. Nelson (1972) and Aalen (1978) independently developed what is now called the Nelson-Aalen estimator of the cumulative hazard function:

$$\hat{H}(t) = \sum_{j=1}^{k} \frac{d_j}{n_j} \qquad (7.3)$$

Example 7.1 (Continued)

Figure 7.2 shows the cumulative hazard function for death in the SLS dataset. There were 3 deaths in the high SES group and 22 in the low SES group. Figure 7.2a uses the original scale for the y-axis, whereas Figure 7.2b uses the log-scale. The low SES group had a cumulative

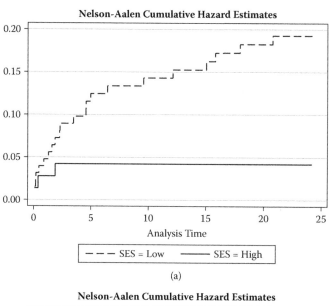

(a)

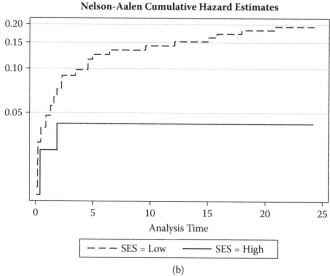

(b)

FIGURE 7.2

Plot of cumulative hazard functions for time-to-death in the SLS dataset: Original (a) and log scale (b).

hazard function that climbed faster and higher, indicating a higher level of mortality hazard. The high SES group only had a few deaths in the early part of the follow-up. So the estimated H(t) remained horizontal in a large part of the time scale.

The *hazard ratio* (HR) is a useful summary measure of association between an exposure and a censored outcome. It assumes that the relative hazard is constant over time. If this assumption is correct, the ln[H(t)] functions should be approximately parallel (Machin et al. 2006). Figure 7.2b can be used to examine the hypothesized parallelism. The curves were not exactly parallel to each other, but the number of events in the high SES was too small for us to draw a firm conclusion. In fact, the parameters in the simulation in Appendix B intended to generate the data in such a way that the curves are parallel.

If the relative hazard changes over time, it is possible to study the HR as a function of time, that is, HR(t). However, the additional complexity in the computation and interpretation may or may not be justified. Simplicity is sometimes preferred.

Example 7.3

Cheung (2000) discussed an analysis of the effect of widowhood on survival in middle-aged and older women in a longitudinal study with about 12 years follow-up. It appeared that widowhood was associated with a higher mortality (HR > 1) in the first few years. But in the last few years of the follow-up, widowhood was associated with a lower mortality (HR < 1). The author reasoned that this might be an artifact caused by a flaw in the study design.

If an exposure truly has a dramatic change in its relative hazard as in this example, the use of HR as a summary measure will blur the changes. However, let's say the relative hazard changes from 2.0 to 1.5 over the study period. The use of HR as a summary measure of association would mean averaging out the changes and giving an estimate of HR approximately in the middle between 2.0 and 1.5. In such a case, whether the use of the HR as a summary statistics is appropriate is a judgment call.

7.2 Regression Analysis of Right-Censored Data

7.2.1 Cox Regression Model

7.2.1.1 Model

Sir David Cox's (1972) seminal paper proposed a regression method that estimates associations between censored outcomes and exposure variables.

The *Cox model* assumes constant relative hazard, which is synonymous with *proportional hazard* (PH). Denote the hazard function of the outcome event among some "average" or "reference" subjects by $h_0(t)$. The study participants are measured for covariates $x_1, x_2, \ldots, x_p$. The average or reference subjects are characterized by zeros on all covariates. The Cox model describes the hazard function given covariate patterns as

$$h(t) = h_0(t) \exp(b_1 x_1 + b_2 x_2 + \ldots + b_p x_p) \tag{7.4}$$

In this formulation, b_j's are the log(HR), whereas $\exp(b_j)$, $j = 1, 2, \ldots, p$, is the HR that quantifies the proportional change in the hazard per one unit change in the value of x_j. For a binary exposure variable, x_j is either 0 or 1. Therefore, the HR compares the exposed ($x_j = 1$) versus the unexposed group ($x_j = 0$). The Cox model does not attempt to capture the average, or baseline, hazard function $h_0(t)$. Unlike the regression models described in previous chapters, the Cox model has no intercept. The Cox model is often referred to as a semiparametric model. That is because it makes no distributional assumption as it gives up capturing the baseline hazard function.

Example 7.2 (Continued)

In the study of neonatal mortality of Swedish infants born from 1982 to 1995, Cheung et al. (2001c) examined neonatal mortality related to year of birth (x_1), gender (x_2), gestational duration (x_3), ponderal index at birth (x_4), and length at birth in Z-score (x_5). Year of birth was centered at 1990 (e.g., for those born in 1989, 1990, 1991; $x_1 = -1, 0, 1$). Males and females were coded as 0 and 1 in variable x_2. Gestational duration was centered at 37 weeks and ponderal index was centered at the mean. As such, the "average" persons were boys born in 1990 after 37 weeks of gestation, with a mean ponderal index and a length Z-score of 0. The baseline hazard represented the mortality hazard in this group of persons.

Among other findings, the Cox model for term newborns gave $b_1 = -0.037$ ($P < 0.001$) and $b_2 = -0.192$ ($P = 0.001$). Therefore, the HR per one year later in year of birth was $\exp(-0.037) = 0.964$. In other words, the neonatal mortality hazard was declining at a rate of about $(1 - 0.964)$ or 3.6% per year. The ratio of hazards between neonates with a difference of 2 years in year of birth, for example, born in 1990 and 1992, can be calculated by $\exp(-0.037 \times 2) = 0.929$. Furthermore, $HR_2 = \exp(-0.192) = 0.825$, meaning girls ($x_2 = 1$) had a mortality of about $(1 - 0.825)$ or 17.5% lower than boys.

7.2.1.2 *Partial Likelihood*

The key invention in the Cox regression model is its estimation method based on the maximization of the *partial likelihood* (PL). Let $t_1 < t_2 < \ldots < t_n$ represent n observed or censored event times, n is the total sample size, and $\delta_i = 0$ if the event time is censored and $\delta_i = 1$ otherwise, $i = 1, 2, \ldots, n$. The first event time in the SLS death data occurred at $t_1 = 0.073$ and it is an observed death, giving $\delta_i = 1$. This event contributed the following element to the PL:

$$PL_1 = \left[\frac{y(1) \times h(0.073) \times e^{bx_1}}{y(1) \times h(0.073) \times e^{bx_1} + y(2) \times h(0.073) \times e^{bx_2} + \ldots + y(n) \times h(0.073) \times e^{bx_n}} \right]^{\delta_1} \quad (7.5)$$

where $y(i) = 1$ if subject i is at risk of the outcome event at time t_1, $y(i) = 0$ if subject i is no longer at risk (meaning either censored or experienced the outcome event before that time point), x_i is the value of the exposure variable x of subject i, and b is the log HR to be estimated. $h(0.073)$ is the baseline hazard at 0.073. Since this appears in every term in the numerator and denominator, this $h(t)$ term divides itself out of the equation. The likelihood is partial in the sense that it does not involve the baseline hazard. Since all n subjects were at risk at this point, the denominator of PL_1 has all $y(i) = 1$ and therefore consists of n elements. Note that $y(i)$ for each subject may change over time. For example, when the fourth events occurred (at $t_4 = 0.175$), the first three subjects were no longer at risk (already died) and so $y(1) = 0$ at t_4 despite $y(1) = 1$ at t_1. As such, PL_4 can be written as

$$PL_4 = \left[\frac{y(4) \times e^{bx_4}}{y(4) \times e^{bx_2} + y(5) \times e^{bx_5} + \ldots + y(n) \times e^{bx_n}} \right]^{\delta_4}$$

Here, we have not bothered to write the redundant baseline hazard term $h(0.175)$. Similarly, censored cases have $y(i) = 0$ after death/censoring and therefore can be omitted from the denominator. If an observation is censored, $\delta_i = 0$ and $PL_i = 1$ because anything raised to the power 0 is 1. The Cox model estimates the regression coefficient, b, by maximizing the log of the partial likelihood $PL = PL_1 \times PL_2 \times \ldots \times PL_n$:

$$PL = \prod_{i=1}^{n} \left[\frac{y(i) \times e^{bx_i}}{y(1) \times e^{bx_1} + y(2) \times e^{bx_2} + \ldots + y(n) \times e^{bx_n}} \right]^{\delta_i} \quad (7.6)$$

with respect to b. It is straightforward to extend Equation (7.6) to include p covariates by changing $\exp(bx_i)$ to $\exp(bx_{i1} + bx_{i2} + \ldots + bx_{ip})$.

The PL has two very useful characteristics. First, as already noted, the term representing the baseline hazard at time t, $h(t)$, is present in both the numerator and denominator and therefore divides itself out of the expression. Thus, unlike parametric models (see next section), one does not need to specify the form of the hazard function. Second, the "at-risk" indicator y(i) makes it easy to handle complex situations such as "time-varying" covariate values (Machin et al. 2006).

The likelihood ratio test and the Wald test are applicable to the PL estimation for testing the null hypothesis of b = 0 (or HR = 1).

Example 7.1 (Continued)

Display 7.2 shows the results of analysis of the death data in relation to SES. Note that Stata requires the "stset" procedure to define the time variable and censoring indicator variable before using the "stcox" or other commands for time-to-event analysis. We only needed one record per child. Hence, the "if" option in the stset.

The Cox model gave HR = exp(–1.458) = 0.233 for high SES as compared to low SES. The stcox command does not normally estimate a model without any exposure variable. To perform a likelihood ratio test of the SES, we can trick stcox to do this by using a variable that really is a constant. The likelihood ratio test (on one degree of freedom) gave P-value 0.0049. In fact, the stcox command also produced this result above the regression table. I only took this opportunity to demonstrate the process.

The Wald test statistics divided the ln(HR) by its SE estimates, ln(0.233)/0.616 = –2.365. Referring this to the cumulative standard normal distribution gave a P-value of 0.018. The likelihood ratio test, if available, is generally preferred to the Wald test.

7.2.1.3 Tied Data

The partial likelihood model described in Equation (7.6) is straightforward if there are no *tied event times*, meaning no more than one event occurs at exactly the same time. In the SLS dataset, t_2 and t_3 were tied. If possible, the data collection procedures should aim to have a high level of precision to avoid ties, for example, recording time of death in addition to date of death. When there are tied data, it is problematic to determine which y(i) should be set to 0 or 1. A range of methods is available to handle tied data in the estimation of the Cox model. While exact methods are theoretically superior, approximations are faster. The *Breslow method* is a commonly used approximation in computer software, but its performance decreases as the amount of tied data increases (Hsieh 1995; Hertz-Picciotto and Rockhill 1997). In a nutshell, the Breslow method includes the tied observations in the at risk population if it cannot differentiate which event occurs earlier. So, if two events occur at the same time t, both observations have y(i) = 1 in the two components of the partial likelihood.

The *Efron method* gives a more accurate approximation than the Breslow method. It is more complex in computation than the Breslow method but the computation time is still substantially smaller than the exact methods.

```
. * time to death analysis
. stset ttd, failure(death), if visit==0

(output omitted)

. * estimate cox model
. stcox SES, nolog nohr

(output omitted)
                                           LR chi2(1)    =      7.92
Log likelihood = -126.33147                Prob > chi2   =    0.0049
------------------------------------------------------------------------
      _t |    Coef.   Std. Err.     z    P>|z|   [95% Conf. Interval]
---------+--------------------------------------------------------------
     SES | -1.457937   .6155061   -2.37  0.018   -2.664307   -.2515675
------------------------------------------------------------------------

. estimates store L1

. gen ONE = 1

. stcox ONE

(output omitted)

. estimates store L2

. lrtest L2 L1

Likelihood-ratio test                      LR chi2(1)    =      7.92
(Assumption: L2 nested in L1)              Prob > chi2   =    0.0049

. * test PH assumption using Schoenfeld residuals
. stcox SES, nolog

(output omitted)

. estat phtest,detail

      Test of proportional-hazards assumption
      Time: Time
      ---------------------------------------------------------------
                  |     rho       chi2     df    Prob>chi2
      ------------+--------------------------------------------------
      SES         |  -0.27355     1.88     1      0.1703
      ------------+--------------------------------------------------
      global test |               1.88     1      0.1703
      ---------------------------------------------------------------
```

DISPLAY 7.2
Stata codes and outputs: Cox model on time-to-death and proportional hazard assumption test.

In a simulation study, Hertz-Picciotto and Rockhill (1997) concluded that "Although the Breslow approximation is the default in many standard software packages, the Efron method for handling ties is to be preferred, particularly when the sample size is small either from the outset or due to heavy censoring" (p. 1151).

7.2.1.4 Model Checking

The plot of cumulative hazard functions can be used to assess the assumption of constant HR as described in the previous section. If there are quantitative exposure variables, a useful though approximate approach is to dichotomize the variables and then examine the cumulative hazard functions across the two groups.

A more sophisticated way to test the PH assumption is to use a test based on the *Schoenfeld residuals* (Schoenfeld 1982; Therneau and Grambsch 2000). If the PH assumption holds, the Schoenfeld residuals would show no trend in relation to the analysis time. A test of nonzero slope in the residuals is therefore a test of the PH assumption.

> **Example 7.1 (Continued)**
>
> In Stata, the Schoenfeld residuals test is implemented by the "estat phtest" command. As shown in Display 7.2, the P-value was 0.170. Therefore, the null hypothesis of proportional hazard could not be rejected.
>
> The "detail" option of the test shows the test of the PH assumption for each exposure variable in addition to the global test of whether the PH assumption holds in the regression model. In the present case, they were the same because there was only one exposure variable.

If it is clinically important to capture exposure effects that change over time, the Cox model can be extended to include *time-varying covariates*. The details are beyond this chapter. Readers are referred to Machin et al. (2006) for the technical details and Cheung et al. (2001c) for an example of the non-proportional hazard effect of size at birth on neonatal mortality.

7.2.2 Parametric Models

7.2.2.1 Accelerated Failure Time Models

The Cox model has been popular for the analysis of time-to-event data in the last few decades. However, the HR may not be intuitive to interpret. In contrast, the *accelerated failure time* (AFT) *models* estimate the impact of an exposure variable in terms of difference in log time-to-event. Simply put, the AFT estimates the relation between the mean of time-to-event, had all events been observed without censoring, and exposure variable x's in the form of the following regression equation:

$$\ln(\bar{T}) = \beta_0 + \beta_1 x_1 + \beta_2 x_2 + \ldots + \beta_p x_p \tag{7.7}$$

where $\bar{T}$ is the predicted *geometric mean* of the time-to-event. The exponentiation of the regression coefficient has an interpretation of *time ratio* (TR). A

TR of 1.5 means that for one unit increase in the exposure the time-to-event is 50% longer. This is probably a metric more intuitive for lay people.

Example 7.4

In the ECHLP study, time to independent walking and time to achieving the fine motor milestone "build a tower of 3 cubes" were assessed in relation to a range of exposure variables (Cheung et al. 2001b). When the follow-up ended at age 24 months, the two milestones were not yet achieved by 12% and 21% of the infants, respectively. In the context of achieving developmental milestones, it is counterintuitive to present the "hazard" and the HR: the higher the "hazard" the earlier the achievement, which is actually better. The researchers opted to use the AFT approach and quantify the associations as time ratios. A time ratio smaller than 1 indicated an achievement of the milestones at a younger age.

The generalized gamma model was used (to be discussed shortly). Having adjusted for covariates, one z-score increase in weight-for-length at birth was associated with a 4% earlier age at independent walking (TR = 0.96; 95% CI 0.94 to 0.98). Since the average age at achieving this milestone was about 14 months, the TR suggested that infants who were one unit higher in weight-for-length at birth tended to achieve the milestone about $14 \times 0.04 = 0.56$ months earlier. Despite the statistical significance, the clinical impact appeared to be limited.

Another difference between the AFT models and the Cox model is that, unlike the Cox model that uses the partial likelihood, the AFT models use the full likelihood. The functional form of the baseline hazard needs to be specified. This is equivalent to specifying the distribution of the event time $f(t)$ or the survival function $S(t)$. There are pros and cons in using the full likelihood. One disadvantage is that misspecifying the distribution may invalidate the analysis results, similar to assuming normal distribution when the data is not normal in the analysis of quantitative data. One advantage is that parametric models are easily adaptable to the analysis of interval-censored data, as will be discussed in the next section.

7.2.2.2 Likelihood

Given a choice of the distribution for $f(t)$ and sample size n, if all n event times are observed, the likelihood of the model has the general form

$$L = \prod_{i=1}^{n} f_i(t_i) \tag{7.8}$$

In Equation 7.8, f_i refers to subject-specific distribution defined by the ith subject's covariate values. If an observation is censored, the probability of the event time larger than t_i is given by the survival function, S(t). Therefore, if m out of n observations are censored and they are indexed as i = 1 to m, the likelihood is

$$L = \prod_{i=1}^{m} S_i(t_i) \prod_{i=m+1}^{n} f_i(t_i) \tag{7.9}$$

Using the censoring indicator $\delta = 1$ for observed and $\delta = 0$ for censored values, Equation (7.9) can be written as

$$L = \prod_{i=1}^{n} \left[f_i(t_i) \right]^{\delta_i} \left[S_i(t_i) \right]^{1-\delta_i} \tag{7.10}$$

The estimation begins by choosing a probability distribution. Table 7.1 describes some of the commonly used distributions for the analysis of censored data.

7.2.2.3 Exponential Model

A single parameter characterizes the *exponential model*. It has a constant hazard function

$$h(t) = \lambda \tag{7.11}$$

The exponential model is interesting in that it can be characterized both as an AFT or a PH model. (The Cox model is not unique in assuming proportional hazard.) As an AFT model, it estimates the relationship in the form of Equation (7.7). As a PH model, it estimates the relationship

$$\ln(\lambda) = \eta_0 + \eta_1 x_1 + \eta_2 x_2 + \ldots + \eta_p x_p \tag{7.12}$$

TABLE 7.1

Some Accelerated Failure Time Models

Distribution of T	Number of Parameters	Hazard Function	Proportional Hazard	Special Case
Exponential	1 (λ)	Constant	Yes	
Weibull	2 (λ, γ)	Monotonic	Yes	Exponential ($\gamma = 1$)
Log-normal	2 (μ, σ)	Nonmonotonic	No	
Generalized gamma	3 (λ, σ, κ)	Nonmonotonic	No	Exponential ($\kappa = \sigma = 1$), Weibull ($\kappa = 1$), Log-normal ($\kappa = 0$)

where λ is the hazard and η_j is log HR and equals the negative of β_j in Equation (7.7), $j = 1, 2, \ldots, p$. That is,

$$\eta_j = -\beta_j \tag{7.13}$$

In the AFT metrics, a positive regression coefficient (β) means the exposure variable is associated with a longer time-to-event. In the PH metrics, a negative regression coefficient (η) means a lower level of the hazard of the event (and therefore longer time-to-event).

Example 7.1 (Continued)

Display 7.3 shows the Stata codes for fitting the exponential model in both AFT and PH metrics using the SLS death data. Table 7.2 shows the results. In the AFT formulation, the exposure status had a regression coefficient 1.499, giving a time ratio exp(1.499) = 4.476. In Display 7.3 we fitted the model twice, with and without using the "tr" option to request the time ratio formulation of the model. That was actually unnecessary. Using Equation (7.13), we know that in the PH formulation the regression coefficient is –1.499 and the HR was exp(–1.499) = 0.223. For models that can be expressed as either AFT or PH, the two formulations were the same thing. They were just expressed in different ways. As such, the likelihood ratio test gave the same P-value (0.0037) regardless of which formulation was used.

```
** Parametric models
stset ttd, failure(death), if visit==0
streg SES,distribution(exp) tr
streg SES,distribution(exp)
streg SES, distribution(weibull) tr
streg SES, distribution(weibull)
streg SES, distribution(lognormal) tr
streg SES, distribution(gamma) tr

* Wald test comparing gamma vs weibull distribution
test [kappa]_cons = 1

* Likelihood ratio test and AIC comparing gamma vs weibull
streg SES, distribution(weibull)
estimates store weibull
streg SES, distribution(gamma)
estimates store gamma
lrtest gamma weibull,force stat
```

DISPLAY 7.3
Stata codes for estimating and comparing AFT models.

TABLE 7.2

Analysis of Mortality in Relation to SES in the SLS Dataset

Model	Metrics	Regression Coefficient (Exponentiated)	95% CI	P-Value (LR Test)	Ancillary Parameters	Log-Likelihood	AIC
Exponential	HR	0.223	(0.067 to 0.746)	0.0037			
	TR	4.476	(1.340 to 14.955)	0.0037		−132.916	269.832
Weibull	HR	0.230	(0.0689 to 0.769)	0.0045			
	TR	33.722	(1.413 to 804.609)	0.0045	$\gamma = 0.4175$	−118.935	243.870
Log-normal	TR	23.585	(1.464 to 379.925)	0.0102	$\sigma = 4.6926$	−118.584	243.168
Generalized gamma	TR	22.303	(1.095 to 454.356)		$\kappa = -0.1230;$ $\sigma = 4.9756$	−118.580	245.160

7.2.2.4 Weibull Model

The hazard function of the *Weibull model* is

$$h(t) = \gamma\lambda(\lambda t)^{\gamma-1} \qquad (7.14)$$

The ancillary parameter γ describes how the h(t) changes. Figure 7.3 gives some examples of the Weibull hazard functions. If $\gamma = 1$, Equation (7.14) simplifies to Equation (7.11). Hence, the exponential model is a special case of the Weibull model. If $\gamma > 1$, the hazard function monotonically increases over time. If $\gamma < 1$, the hazard function monotonically decreases over time. This captures the neonatal mortality hazard reasonably well. Note that some books and software may prefer using $p = 1/\gamma$ instead of γ itself in writing the model.

Like the exponential model, the Weibull model can also be expressed in AFT and PH formulations, with β and η to denote the regression coefficients in the two formulations, respectively. The coefficients can be converted to each other by

$$\eta_j = -\beta_j\gamma \qquad (7.15)$$

Example 7.1 (Continued)

Table 7.2 also shows the results of the analysis of the SLS death data using the Weibull model. In the AFT formulation, the exposure status had a regression coefficient of 3.518, giving a time ratio exp(3.518) = 33.722,

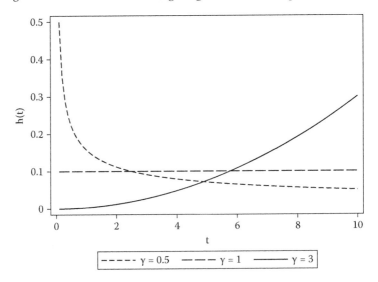

FIGURE 7.3
Hazard functions of the Weibull model with $\lambda = 0.1$ and $\kappa = 0.5, 1,$ and 3.

and the ancillary parameter 0.4175. Using Equation (7.15), the log HR is $-3.518 \times 0.4175 = -1.469$, giving HR = $\exp(-1.469) = 0.230$.

The SLS death data was indeed simulated with an underlying ancillary parameter $\gamma = 0.5$. Stata gives the 95% confidence interval for the estimate 0.4175 as 0.2853 to 0.6110, covering the true value and excluding the value 1 that indicates an exponential model. Since the exponential model is a special case of the Weibull, we can also use the likelihood ratio test to examine whether the sample data is consistent with a constant hazard, by calculating likelihood ratio statistics $-2 \times [(-132.916) - (-118.935)] = 27.962$ and comparing this with the chi-square distribution on 1 degree of freedom. It gives $P < 0.0001$, leading to the rejection of the exponential model. The hazard was not constant.

7.2.2.5 Log-Normal Model

The *log-normal model* assumes that the log of event times follows a normal distribution with mean μ and standard deviation σ. The hazard function is given by

$$h(t) = \frac{\dfrac{1}{t\sqrt{2}}\exp\left[-\dfrac{1}{2}\left(\dfrac{\log(t)-\mu}{\sigma}\right)^2\right]}{1 - \Phi\left[\left(\dfrac{\log(t)-\mu}{\sigma}\right)\right]} \tag{7.16}$$

where Φ is the cumulative standard normal distribution function. This hazard function is not monotonic, meaning it can increase and decrease in different ranges of the time scale. Figure 7.4 shows some examples of the log-normal hazard functions. They go upward initially and then go downward. Unlike the exponential and Weibull models, the log-normal model is only an AFT model. It cannot be formulated as a PH model. The results are always expressed as time ratios or their logarithmic values.

7.2.2.6 Generalized Gamma Model

The *generalized gamma model* is flexible in shape (Stacy 1962). The flexibility is achieved by having three parameters $(\lambda, \sigma, \kappa)$ to fully specify the distribution. Its algebraic form is omitted here as it is mathematically complex. An important feature of the model is that it includes the exponential, Weibull, and log-normal models as special cases (Table 7.1). In other words, they are nested models of the generalized gamma model. A natural question at this point is that, if it is so flexible, why not always use this and forget the exponential, Weibull, and log-normal models? The major reason is that the model is difficult to fit. The maximum likelihood procedure may not always find a solution. If external knowledge suggests that the simpler models can work, they may be preferred. The likelihood ratio test and Wald test can be

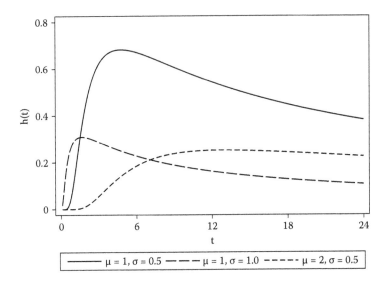

FIGURE 7.4
Hazard functions of the log-normal model with $\mu = 1$ and 2, and $\sigma = 0.5$ and 1.0.

used to compare the nested and parent models. The Akaike information criterion (AIC),

$$AIC = -2LL + 2(c + p) \tag{7.17}$$

where LL is the log-likelihood, c is the number of regression coefficients estimated, and p is the number of model parameters in Table 7.1, can be used to compare models. The smaller the AIC, the better the model fits the data.

Example 7.1 (Continued)

Continuing with Table 7.2, the log-normal and the generalized gamma models gave time ratios 23.585 and 22.303, respectively. The log-likelihood of the generalized gamma model was clearly larger than that of the exponential model. Recall that the data was actually simulated based on an underlying Weibull model. If the ancillary parameter κ of the generalized gamma model is close to 1, it would suggest that the generalized gamma model was taking the shape similar to the Weibull distribution. Despite the ancillary parameter estimated to be −0.1230 (95% CI −2.881 to 2.635), which looked different from 1, the log-likelihood values of the two models concerned were not very different. The estimate could be imprecise when based on such a small number of events. By calculating likelihood ratio statistics −2 ×[(−118.935) − (−118.580)] = 0.71 and comparing this with the Chi-square distribution on 1 degree of freedom, P = 0.399 was obtained. The test was on 1 degree of freedom because the difference in the number of parameters between the two models was 1. There was no evidence to reject the hypothesis that the Weibull model fitted the data sufficiently. Alternatively, the Wald test can also be used to assess the null hypothesis of $\kappa = 1$ ($P = 0.425$). The Stata codes for comparing the two models by the Wald test, LR test, and AIC are included in Display 7.3. Note that the "lrtest" command normally refuses to compare the likelihoods from two different estimation commands (as opposed to the same command implemented twice with different variables). The "force" option is needed to force lrtest to perform the test. The "stat" option provides the AIC.

The AIC of the Weibull model was $-2 \times (-118.935) + 2 \times (1 + 2) = 243.87$. This was slightly better (smaller) than the AIC of the generalized gamma model $-2 \times (-118.580) + 2 \times (1 + 3) = 245.16$. Again, there was no evidence to say the latter model fitted the data better.

Nevertheless, the log-normal model had the smallest AIC (243.168), despite our knowledge that the underlying hazard was simulated according to a Weibull hazard. This provides a cautionary note that we should not read too much from a small difference in AIC and that with a small number of events we may not be able to understand the shape of the baseline hazard function well.

Table 7.2 also demonstrates the impact of model misspecification. The exponential model gave a time ratio (4.476) that was much smaller than the other models. A serious misspecification can lead to serious error. However, among the two- and three-parameter models, the time ratios were fairly similar and their 95% confidence intervals did not exclude each other.

7.2.2.7 Model Diagnostics

It is easier to know what is "better" than to know what is "good." The likelihood ratio test and the AIC can be used to compare different models. They tell us which model is better. The worse model is usually not the choice unless the external knowledge strongly supports the worse model. The difficulty is that we still do not know whether the model that fits better is good enough.

If a model fits, the *Cox-Snell residuals* follow an exponential distribution (Machin et al. 2006). The Cox-Snell residuals are censored if the event times are censored. Once the Cox-Snell residuals are calculated, we can treat them like survival time data, fit the residuals to the exponential model and more complex models that include the exponential as a special case, and then use the likelihood ratio test to compare whether the exponential model holds for the Cox-Snell residuals. Graphical assessments of the residuals can be found in Machin et al. (2006).

Example 7.1 (Continued)

The first part of Display 7.4 shows the Stata codes for the survival time analysis using a log-normal model and saving the Cox-Snell residuals as a new variable named "csln." The second part of the display treats the Cox-Snell residuals as censored data. It fitted the residuals to an exponential model and a Weibull model (without covariates). Then the

```
* illustrate cox-snell residuals for model diagnostic

* fit log-normal model to time-to-death data
* and calculate C-S residuals
stset ttd, failure(death), if visit==0
streg SES,distribution(lognormal)
predict csln,csnell

* treat C-S residuals from log-normal model as censored data
* test whether the C-S residuals is exponential distribution
stset csln,failure(death), if visit==0
streg,distribution(exp)
estimates store A
streg,distribution(wei)
estimates store B
lrtest B A,force
streg,distribution(gamma)
estimates store C
lrtest C A,force
```

DISPLAY 7.4
Stata codes for the assessment of Cox-Snell residuals.

likelihood ratio test was performed to check the plausibility of the Cox-Snell residuals following an exponential distribution versus a more complex Weibull model. The P-value was 0.9995. It also fitted the residuals to a three-parameter gamma model. Then the likelihood ratio test was performed to compare the exponential model versus the three-parameter Gamma model. The P-value was 0.9990. In either case we could not reject the null hypothesis that the Cox-Snell residuals followed an exponential distribution. Therefore, we could not reject the hypothesis that the log-normal model was suitable for the death data.

The same procedure could be repeated to assess the other models for the death data. For example, the Cox-Snell residuals from an exponential model for the death data certainly did not follow an exponential model (P < 0.0001 in LR test against Weibull).

7.3 Analysis of Interval-Censored Data

Interval censoring occurs when we know that an event took place between two time points α and β, $\alpha < \beta$, but we do not know the exact timing. Left-censored time-to-event data can be seen as interval-censored data with $\alpha = 0$. Right-censored data can be seen as interval censored with $\beta = $ infinity. One approach to handle interval censoring is by imputation, which will be deferred until Chapter 13. In this chapter, we discuss the use of the midpoint of intervals to substitute the interval-censored value and the extension of parametric models to the analysis of interval-censored data.

7.3.1 Midpoint Approximation

The midpoint of an interval can be used to approximate a value censored within that interval if the following criteria are met (Gómez et al. 2003; Nauta 2010):

- The timing of assessment is predefined, not triggered by the events.
- The width of intervals is small in comparison to the range of data.
- The true distribution (possibly after transformation) is not very skewed.

After the substitution, the values are analyzed as if they are uncensored.

> **Example 7.4 (Continued)**
>
> In the ECHLP study, infants were assessed on a monthly basis. The first two criteria above were considered satisfied. The third criterion depended on what outcome variables it was. From Chapter 1, Figure 1.3, we can see that there is some but not a lot of skewness in gross motor milestones.
>
> Ages at achievement of motor milestones were assumed to be at the midpoint of the previous visit and the visit when performance of the milestones was observed (Cheung et al. 2001b). Otherwise it was right censored at the last visit. So the analysis dataset only had right censoring. The three-parameter gamma model was used to analyze the right-censored age at achievement of motor milestones in relation to early growth variables.

As shown by the World Health Organization (WHO) Multicentre Growth Reference Study Group (2006b), the distributions of age at achieving motor milestones demonstrate some positive skewness, as indicated by a larger distance between the 75th and 90th percentiles than between the 10th and 25th percentiles (Chapter 1, Figure 1.3). A log-normal distribution $\exp[N(2.5, 0.0225)]$ gives percentiles that approximate the percentiles of age at starting to walk alone in Figure 1.3. We used simulation to visualize the accuracy of replacing interval-censored values by midpoints. Suppose the true values in Group A and Group B follow $\exp[N(2.5, 0.0225)]$ and $\exp[N(2.6, 0.0225)]$ respectively. The true arithmetic means for the two distributions are 12.32 and 13.62, respectively.

Data were simulated for the two distributions, with n = 500 per group. The exact sample values were analyzed first. Then we assumed that the children were measured at regular intervals of either 1, 2, 3, or 4 months, starting from 4 months of age, and assigned the midpoint of the intervals as the observed value. As shown in Table 7.3, even if data was collected up to 4-month intervals and the interval-censored values replaced by midpoints, the group means and difference between two means were little affected. However, the association between group membership and the age values (in terms of odds ratio) was slightly diluted by the use of midpoint replacement.

TABLE 7.3

Comparison of Statistics of True Values and Mid-Values for Simulated Exact and Interval-Censored Data (n = 500 per group)

			Mean			
			Midpoint Approximation			
Statistics	Distributions	Exact	1 Month	2 Months	3 Months	4 Months
Mean, Group A	exp[N(2.5,0.0225)]	12.46	12.46	12.43	12.48	12.42
Mean, Group B	exp[N(2.6,0.0225)]	13.51	13.51	13.53	13.53	13.51
Odds ratio*		1.34	1.33	1.32	1.28	1.24

* Logistic regression of Group B membership in relation to true age and midpoint substituted ages at event (Y = 1 for Group B and Y = 0 for Group A).

Frequency of outcome monitoring and acceptability of midpoint replacement is a matter to be determined according to the accuracy required by the researchers. For situations similar to this simulation and the ECLHP study, the practice of using midpoints is quite accurate as long as the intervals of assessment are not too wide. Lindsey (1998) examined a range of scenarios and made similar conclusions about the accurate of this approximation.

7.3.2 Parametric Modeling

7.3.2.1 Model and Estimation

Parametric models are the major armamentarium for analysis of such data. They are "remarkably robust to changing distributional assumptions and generally more informative than the corresponding non-parametric models" (Lindsey 1998, p. 329). The maximum likelihood estimation method described in Section 7.2.2 is readily extendable for the analysis of interval-censored data. Recall that a right-censored event time contributes $S_i(t_i)$ to the likelihood of a parametric model. That is the basis of Equation 7.10. Given two time points $\alpha_i < \beta_i$, the probability of event time larger than α_i is $S_i(\alpha_i)$ and the probability of event time larger than β_i is $S_i(\beta_i)$. The probability of an event occurring between the two time points is therefore $S_i(\alpha_i) - S_i(\beta_i)$. The subscript i here emphasizes that the censoring intervals may vary from one person to another. The likelihood in Equation (7.10) can be modified to the following to incorporate interval censoring:

$$L = \prod_{i=1}^{n} \left[f_i(t_i) \right]^{\delta_i} \left[S_i(\alpha_i) - S_i(\beta_i) \right]^{1-\delta_i} \tag{7.18}$$

where $\delta = 0$ if interval censored and $\delta = 1$ otherwise. The model estimation is then practically the same to fitting parametric models for right-censored data.

The user-written Stata macro "intcens" (Griffins et al. 2006) implements a wide range of parametric models for regression analysis of interval-censored data.

In reality, quantitative variables are often interval-censored due to limited measurement precision, and rounded to the nearest unit. If the measurement is precise enough such that the interval is very small (e.g., days not weeks), one can feel comfortable to treat the interval-censored data as if they are the exact observations.

Example 7.5

Kwok et al. (2011) examined the association between early life infections and puberty onset. The data collection procedures included extracting puberty status data from records of annual checkups from age 6 to 12 years in school health services. Age at puberty onset was interval censored. A parametric model in the general form of Equation (7.18) was applied to analyze the interval-censored puberty onset data, using the log-normal model.

Example 7.1 (Continued)

Age became stunted was interval censored in the SLS dataset. Display 7.5 analyzed the data twice using the three-parameter generalized gamma models. The first analysis used the midpoints of intervals where stunting first occurred as the proxy of the timing of the event. The age at last visit was used to indicate right-censored value if stunting was not observed during the study. The analysis followed Equation (7.10). The time ratio estimated for the exposure was 2.099 (95% CI 1.697 to 2.596).

The second analysis formally recognized the interval censoring. It follows Equation (7.18). The macro "intcens" (Griffins et al. 2006) gave a time ratio of 2.095 (95% CI 1.691 to 2.594). The results were practically identical. This was expected if the intervals were narrow relative to the spread of the true event times.

The syntax of the Stata macro is somewhat different from the official Stata software. The official Stata command "streg" requires the "stset" and uses the option "distribution(gam)" to request the three-parameter generalized gamma model. The "intcens" macro does not require the "stset" and uses "distribution(gen)".

7.3.2.2 Model Diagnostics

Model diagnostics for analysis of interval-censored data is substantially more difficult than that for right-censored data. Goodness-of-fit tests are often based on a comparison of nested and parent models (Lawless and Babineau 2006), as we have done in Section 7.2. A simple though approximate approach for examining distributional assumption is to substitute some values for the censored values before performing a graphical assessment or statistical tests

```
. ** time to stunting as interval censoring
. stset tts_midage, failure(stunt_ever), if visit==0

(output omitted)

. * using midpoint as time of events
. streg SES, distribution(gamma) nolog tr

(output omitted)
```

```
-------------------------------------------------------------------
       _t |  Tm. Ratio   Std. Err.     z    P>|z|   [95% Conf. Interval]
----------+--------------------------------------------------------
      SES |   2.098723    .2276424   6.83   0.000    1.696789    2.595867
    _cons |   14.11704    1.47775   25.29   0.000     11.4985     17.3319
----------+--------------------------------------------------------
  /ln_sig |  -.5203802    .0815212  -6.38   0.000   -.6801588   -.3606016
   /kappa |  -.7878368    .4010579  -1.96   0.049   -1.573896   -.0017778
----------+--------------------------------------------------------
    sigma |   .5942946    .0484476                    .5065366    .6972568
-------------------------------------------------------------------
```

```
. * interval-censoring
. intcens tts_lbage tts_ubage SES, distribution(gen) nolog eform, ///
> if visit==0

(output omitted)
```

```
-------------------------------------------------------------------
          |    exp(b)    Std. Err.     z    P>|z|   [95% Conf. Interval]
----------+--------------------------------------------------------
      SES |   2.09451    .2287262    6.77   0.000    1.690943    2.594394
    _cons |  14.01136    1.509827   24.50   0.000    11.34376    17.30627
-------------------------------------------------------------------
```

DISPLAY 7.5
Stata codes and outputs: Analysis using midpoint of interval as event time versus analysis using interval-censored regress, both using the three-parameter generalized gamma model.

(Royston 2007), such as using the Q-normal plot introduced in Chapter 5. For interval-censored data, the midpoint is a likely choice for the substitution. For right-censored data, the substitution needs deliberation and probably requires some knowledge on the subject matter. This will be discussed in more detail in Chapter 13.

8

Analysis of Repeated Measurements and Clustered Data

8.1 Introduction

Correlated data may arise as a result of *repeated measurements* in longitudinal studies. They may also arise from *cluster sampling,* such as studies of samples of families, within each there are multiple persons. The statistical issues are similar in these two situations, although the former has an additional issue of a time dimension. This chapter will mostly use the word *panel* to refer to a group of correlated observations, no matter if it is about repeated measurements or cluster sampling. However, in the terminology of the *mixed models,* it is typical to use the term "subject" to refer to a panel of observations because the context is often about repeated measurements of each person. So these words are used interchangeably sometimes.

Example 8.1

A clinical trial randomized children to one of three dosages of atropine to study their impact on the prevention of myopia progression (Chia et al. 2012). Both eyes of each participant received the same dosage. The children were the panels. Each panel consisted of two observations (eyes). The degree of myopia progression in the two eyes was likely to be correlated, because the two eyes shared the same environment, genetics, or other factors.

Example 8.2

In the Simulated Longitudinal Study (SLS) dataset, each child was measured once every 2 months up to the age of 24 months. Each child was a panel. Each panel contained up to 13 repeated measurements of anthropometry. Such repeated measurements were unlikely to be independent. Figure 8.1 plots the height-for-age z-score (HAZ) against age for the first 10 children. Child 10 consistently had lower HAZ than the others throughout the follow-up period. Child 6 consistently had higher HAZ. The variation between persons generated similarity (correlation) among the observations within panels.

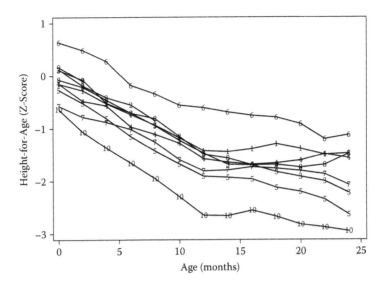

FIGURE 8.1
Height-for-age z-score (HAZ) versus age for the first 10 children in the SLS dataset.

8.1.1 Why Analyze Repeated Measures?

The studies of growth and development may involve questions like: (1) Have people acquired or lost a characteristic by a certain age? (2) When did they acquire that characteristic? (3) How fast does that characteristic change over time? (4) What interventions or exposure history affect the acquisition, level, and rate of change? Longitudinal studies with repeated measurements of individuals can help answer some of these questions. Furthermore, if used appropriately, analysis of repeated measures may increase the statistical power and improve precision.

The research questions should decide the methods. Because one can repeatedly measure the subjects does not necessarily mean one should do it. Because there are repeated measurement data does not necessarily mean one should analyze them. For example, question (1) requires analysis of a binary outcome at a single age, even if the outcome was also measured at other ages. Sometimes a clinical trial may have the primary endpoint measured at, say, two time points after initiation of intervention. The primary aim of the trial may be to assess the "long-term" effect of the intervention, and it is the final measure that is really of interest. The secondary aim of the trial may be to assess the "short-term" effect, and it is the first measure that is of interest. If some of the methods described later are used indiscriminately to pool the two measures in a single analysis, it could potentially mix up the short- and long-term effects and fail to address the specific aims.

Example 8.1 (Continued)

The myopia trial defined the change in spherical equivalent from baseline to 2 years as the primary endpoint. Each child has two data points for the analysis. The study actually measured the children every 4 months for 2 years. Despite the availability of repeated measures of spherical equivalent that served secondary purposes, the main analysis only used the change from baseline to 2 years.

8.1.2 Naïve Analysis

Suppose a researcher wanted to estimate the difference in a cognitive function test between two groups of older people. He aimed to recruit a total of N participants. He (mis)reasoned that if he tests each participant twice, on two consecutive days, the sample size would become 2N and therefore the analysis would become more powerful. This is clearly wrong because, apart from measurement error and possibly a "practice effect," the second measure is just a duplication of the first measure. It does not provide additional information. Most commonly used statistical methods assume that the observations are independent. In this hypothetical example, but for measurement errors and practice effect, the two data points should have a correlation close to 1. So, despite the number of observations being 2N, the *effective sample size* is still approximately N. Naïve analysis of these data as if there are 2N independent observations will underestimate the standard error (SE), inflate the type I error rate (P-values become too small), and provide confidence intervals that are too narrow. However, the estimate of the difference in the cognition function test scores between groups would basically be unaffected, because the second measure did not provide new information. This is a common result of naïve analysis of correlated data: the point estimates (of mean, difference, association, etc.) are largely unaffected, but the standard errors are too small and the statistical inference is therefore wrong.

Intuitively, in the aforementioned hypothetical example, if the second test was conducted 1 year later instead of on the next day, the two scores would not be so strongly correlated and the second score should provide some additional information. If it was taken 5 years later, the correlation should be even weaker and the amount of additional information would be even larger. The degree of within-person correlation between observations can be quantified by the *intraclass correlation coefficient* (ICC), which we will discuss in Chapter 12. In the meantime, we note that the increase in variance, or *design effect*, due to correlated data is given by (Lohr 1999)

$$1+(M-1)\times\text{ICC} \tag{8.1}$$

where M is the number of observations per panel. In the hypothetical example of cognitive function test conducted on two consecutive days, $M = 2$ and ICC ~ 1, and the design effect is therefore about 2. Despite the apparent sample size being 2N, dividing this number by the design effect would give $2N/2 = N$. So the effective sample size is still about N despite the 2N observations.

8.1.3 Population- versus Subject-Level Impact

A population consists of heterogeneous individuals. Some degree of heterogeneity always exists (Aalen 1988). With heterogeneity, an estimate of odds ratio (OR) at the individual level can differ from the estimate at the population level. Table 8.1 illustrates it with a hypothetical population. There are 1000 males and 1000 females in a control group. Suppose the odds of disease is 2.0 among males and 0.5 among females in the control group, showing substantial heterogeneity between genders. For the control group as a whole, 667 males and 333 females are diseased, and the odds in the control group as a whole is therefore $(667 + 333)/(333 + 667) = 1.0$. Suppose there is an intervention that can reduce each individual's odds by half, that is, $OR = 0.5$. So the odds are 1.0 and 0.25 among males and females in the intervention group. Thus, among 1000 males and 1000 females in the intervention group, there are 500 diseased males and 200 diseased females. For the intervention group as a whole, the odds are $(500 + 200)/(500 + 800) = 0.54$. So, at the population level, the OR for intervention versus control is $0.54/1.0 = 0.54$, even though we know that at the subject level the OR is 0.50! It is general that, with heterogeneity, the odds ratio at the population level is closer to the null value of 1 than the odds ratio at the subject level (e.g., Hu et al. 1998; Zeger et al. 1988). The larger the degree of heterogeneity, the bigger the difference between the two ORs. It is not a discrepancy. It is just looking at the data from different angles. Furthermore, the SEs of the two log ORs estimated from the data are also different, such that the P-values are similar and no one method is "preferred" in terms of rejecting or accepting the null hypothesis of no association. The method in Section 8.2 estimates the population-level impact. The methods in Sections 8.3 and 8.4 estimate the subject-level impact.

Most measures of association are the same for the two levels of analysis. Following the steps in Table 8.1, the readers can verify that risk ratio and mean difference are the same at the population and individual levels.

8.1.4 Panel- and Observation-Level Variables

Panel-level variables have a fixed value for all observations within a panel. For example, both eyes of a participant have either male or female for gender. Observation-level variables may vary within a panel. For example, in repeated measurements of participants, age or time-on-study vary across

TABLE 8.1

Hypothetical Population with 1000 Males and 1000 Females in Each Group

Gender	Control Group			Intervention Group		
	Diseased	Nondiseased	Odds	Diseased*	Odds	Nondiseased
Males	667	333	667/333 = 2.0	$1000 \times (1/2) = 500$	$2.0 \times 0.5 = 1.0$	500
Females	333	667	333/667 = 0.5	$1000 \times (0.25/1.25) = 200$	$0.5 \times 0.5 = 0.25$	800
Total	1000	1000	1000/1000 = 1.0	700	700/1300 = 0.54	1300

Notes: The intervention reduces individual odds by 50% (individual OR = 0.5), but giving population-level OR = 0.54/1.00 = 0.54.

* Calculated by Risk = Odds/(1 + Odds).

observations within the same subject. Since many studies concern correlated data arising from repeated measurements of individuals, it is common to use the terms *time-constant* or *time-varying covariates* to mean the panel-level and observation-level variables, respectively. Some of the methods discussed in this chapter can handle both panel- and observation-level variables in the same analysis. For example, a regression analysis may simultaneously include gender (panel level) and time-on-study (observation level) as independent variables.

8.1.5 Weighting

Statistical methods implicitly or explicitly weigh the observations. In terms of the point estimation of the parameter of interest, the naïve analysis and the use of the *robust variance estimator* (Section 8.2) give equal importance to each observation. So, a panel with a larger number of observations is more influential in the analysis than a panel with a smaller number of observations. In contrast, the *summary statistics* approach (Section 8.3) gives equal importance to each panel. So, an observation in a small panel is more influential than an observation in a large panel. The *mixed models* (Section 8.4) assign weights to observations differentially, depending on the degree of intraclass correlation and pattern of within-panel correlation. These methods tend to give similar regression coefficients if the number of observations per panel does not vary substantially.

8.2 Robust Variance Estimator

The robust variance estimator has many names, such as the sandwich estimator; empirical variance estimator; and Huber, or White, or Huber-White variance estimator, after the two pioneers who independently proposed the method (see review by Binder 1983). The robust variance estimator was originally developed for the analysis of uncorrelated data. Inference using the robust variance estimator is robust to model misspecification. It was subsequently extended to provide inference for correlated data, assuming that the panels are independent but the observations within a panel may be correlated (Gould and Sribney 1999; Williams 2000). The point estimate of the parameter is not affected by the robust variance estimator.

We look at least-squares regression in more details. Recall that Equation (5.15) (Chapter 5) uses a single estimate of the variance of Y given X. It assumes equal variance for all observations and therefore makes the equation simple. In contrast, the robust variance estimator (for uncorrelated data) does not make the assumption of equal variance:

$$\text{Var}(\hat{\beta}) = \theta(\mathbf{X'X})^{-1}\mathbf{X'}\begin{pmatrix} \hat{e}_1^2 & & \\ & \ddots & \\ & & \hat{e}_n^2 \end{pmatrix}\mathbf{X}(\mathbf{X'X})^{-1} \quad (8.2)$$

$$= \theta(\mathbf{X'X})^{-1}(\mathbf{X'eeX})(\mathbf{X'X})^{-1}$$

$$= \theta(\mathbf{X'X})^{-1}(\mathbf{u'u})(\mathbf{X'X})^{-1}$$

where $\hat{e}_i$, $i = 1, 2, \ldots, n$, is the regression residual for the ith observation, $\mathbf{e}$ is a diagonal matrix with $\hat{e}_i$ on the diagonal, and $\theta = n/(n - p - 1)$ is a finite-sample adjustment factor and p is the number of exposure variables. This is sometimes called the sandwich estimator because, ignoring the adjustment factor, Equation (8.2) looks like two pieces of "bread" with some "filling" in the middle.

For panel data, the filling is calculated differently. Define

$$u_{G_k} = \sum_{j \in G_k} u_j,$$

where $G_1, G_2, \ldots, G_M$ are the panels and u_j, $j = 1, 2, \ldots, n$, is the jth row of $\mathbf{eX}$. That is, the products of the regressors (including the intercept term) and residuals are summed within a panel. The robust variance estimator for cluster data is

$$\text{Var}(\hat{\beta}) = \theta(\mathbf{X'X})^{-1}(\mathbf{u_G'u_G})(\mathbf{X'X})^{-1} \quad (8.3)$$

where $\mathbf{u_G}$ is a $M \times (p + 1)$ matrix and the finite-sample adjustment factor is $\theta = [(n - 1)/(n - p - 1)] \times [M/(M - 1)]$. Different types of regression use different formulas, but the principle of empirically estimating the filling for each observation and then summing within a panel is the same.

If there is a within-panel correlation in the sense of within a panel the outcomes tend to be similar, one would expect the residuals within a panel to be mostly of the same sign, for example, subject 6 and subject 10 in Figure 8.1 have mostly positive and negative residuals, respectively. Summing before squaring them would produce large values. If there is no within-panel correlation, each residual within a panel is equally likely to be positive or negative. Summing before squaring would produce small values because the positive values and negative values cancel each other out. So the robust variance for cluster data reflects the within-panel correlation.

The robust variance estimator is asymptotically correct, meaning it requires at least a moderate number of panels. Simulation studies have found that it tends to give confidence intervals that are too narrow if the number of panels is smaller than 20 (Feng et al. 1996; Wang and Long 2011).

It performs well when the number of panels is 50 or more. Between 20 and 50 is a gray area.

Using robust variance for the analysis of correlated data usually but not necessarily gives a larger estimate of SE than naïve analysis does. If there is a large amount of within-panel variation, meaning some very large positive and very large negative residuals within the same panel, the robust variance estimate may turn out to be smaller than the naïve variance estimate. The within-panel variation may sometimes be reduced by including important observation-level covariates, for example, time or age in the context of repeated measurements, and therefore makes the robust standard error more sensibly larger than the naïve standard error. The omission of important observation-level covariates may be considered a form of model misspecification. Despite its name, the robust variance estimator is not absolutely robust and still needs to be used with care. The number of clusters and the selection of covariates are important considerations.

Example 8.1 (Continued)

The myopia progression trial used the ordinary least-squares method together with the robust SE for statistical inference in the analysis of correlated data, that is, two eyes per child.

Example 8.2 (Continued)

The analysis requires the data in the long format. In the SLS data, which was in the long format, each child was measured for height-for-age (HAZ) every 2 months from birth up to age 24 months. We might hypothesize that the children grew slower than the reference population during infancy and therefore the HAZ would drop over time. We conducted a naïve analysis and an analysis using the robust standard error. The Stata codes and outputs are shown in Display 8.1. The variable "visit" took on the values 0, 2, 4, ..., or 24, indicating the target age (in months) of the visits. The naïve analysis showed a decline of 0.086 z-score per month; the SE was 0.00188. In Stata, the use of the robust standard error in regression analysis is very simple, just add the option "cluster(*varname*)", where *varname* is the variable name that identifies the panels. In this case, the variable name is "id". The outputs showed that there were 200 clusters defined by the id variable. The regression coefficient was the same as the naïve analysis, showing a decline of 0.083 z-score per month. However, the SE was about 0.0020, about 6% larger than the naïve SE.

The intercept was −0.179, meaning at visit 0 (at birth) the infants were 0.179 z-score shorter than the reference population. The naïve SE is 0.0257 whereas the robust SE was 0.0396, about 54% larger than the naïve SE. This showed again the typical impact of the within-panel correlation on the precision of the estimate.

```
. regress haz visit
    Source |      SS        df        MS              Number of obs =    2226
-----------+-------------------------------            F( 1, 2224)   = 2100.72
     Model | 931.880237     1    931.880237            Prob > F      =  0.0000
  Residual | 986.566029   2224   .443599833            R-squared     =  0.4857
-----------+-------------------------------            Adj R-squared =  0.4855
     Total | 1918.44627   2225   .862223041            Root MSE      =  .66603


-------------------------------------------------------------------------------
       haz |      Coef.   Std. Err.      t     P>|t|     [95% Conf. Interval]
-----------+-------------------------------------------------------------------
     visit | -.0859507   .0018753    -45.83   0.000    -.0896282   -.0822732
     _cons | -.1788908   .025728      -6.95   0.000    -.2293442   -.1284375
-------------------------------------------------------------------------------


. regress haz visit,cluster(id)

Linear regression                                      Number of obs =    2226
                                                       F( 1, 199)    = 1850.98
                                                       Prob > F      =  0.0000
                                                       R-squared     =  0.4857
                                                       Root MSE      =  .66603
                             (Std. Err. adjusted for 200 clusters in id)
-------------------------------------------------------------------------------
           |              Robust
       haz |      Coef.   Std. Err.      t     P>|t|     [95% Conf. Interval]
-----------+-------------------------------------------------------------------
     visit | -.0859507   .0019978    -43.02   0.000    -.0898902   -.0820112
     _cons | -.1788908   .0396288     -4.51   0.000    -.2570371   -.1007446
-------------------------------------------------------------------------------
```

DISPLAY 8.1
Stata codes and outputs: Analysis of repeated height-for-age measurements in SLS data using naïve analysis and robust standard errors.

8.3 Analysis of Subject-Level Summary Statistics

8.3.1 Mean and Rate of Change

Correlated data within a panel can be summarized by some summary statistical measures. Mean response and rate of change are two major choices for summarizing data within a panel (Frison and Pocock 1992; Laird and Wang 1990). In cluster randomized trials, the mean response of the individuals is used to summarize the response in the cluster (Hayes and Moulton 2009). Ophthalmology studies like the aforementioned myopia progression trial are also cluster randomized trials: a group of two eyes is randomly allocated to receive one of the interventions. One possible and valid approach to analyze the data was to calculate the mean myopia progression of the two eyes, use this summary statistics (mean) for the outcome data such that there is only one outcome value per subject, and analyze the data using standard statistical methods for independent observations.

For repeated measurement data, it is common to use rate of change over time as the summary statistics. A regression model in the form of

$$y = b_0 + b_1 \times time$$

is fitted to the data for each panel, with time or age as the explanatory variable. The regression coefficient b_1 represents the rate of change per unit of the time variable. This is used as the summary statistics in the analysis. The analysis then reduces to one data point per panel. Standard statistical methods that assume independency across data points are now applicable.

Example 8.3

A clinical trial examined the impact of inhaled corticosteroids on growth in children (Bensch et al. 2011). Height was measured at each of the eight visits within the 1-year follow-up period. The primary endpoint was growth velocity assessed by linear regression analysis estimated for each child with at least three height measures using height as the dependent variable and time as the independent variable. The rate of change or regression coefficients was expressed as centimeters per 52 weeks. The rate of change was then analyzed using standard statistical methods.

This approach of analyzing summary statistics treats each panel equally regardless of how many observations there are within the panel. Panels that have too few data points may not be precise enough for meaningful analysis. It is useful to have an inclusion criterion that a panel has to have a minimum number of observations for it to be included in the analysis. Panels that do not have at least two data points are not analyzable for a rate of change at all and have to be treated as missing values. However, if many panels are excluded due to a small number of observations, the exclusion would raise doubts about selection bias.

Example 8.2 (Continued)

Display 8.2 demonstrates how to use linear regression to estimate the rate of change in HAZ in the first 12 months for each participant in Stata. First, it saved the largest panel (ID) number as a local macro called N. Then it counted the number of nonmissing HAZ within each panel. Then, using the "forvalues" loop, the procedures within the curly brackets were automated to run from the first to the last panel number indexed in the forvalues statement.

Suppose the panel (ID) numbers in the dataset were 1, 2, 4, 5, and so on. Without the "capture" command the loop would stop and issue an error

```
keep if visit<=12

** Save the largest id number in local macro N
sum id
local N=r(max)

** Count number of non-missing values
by id: egen total=count(haz)

** Estimate rate of change if at least 3 measures; save rate in var
** slope
** loop over id 1 to `N'
gen slope = .
quietly forvalues i=1(1) `N' {
capture regress haz visit if total>=3 & id==`i'
capture replace slope =_b[visit] if total>=3 & id==`i'
}

** Generate a variable for selection of one record per person for
** analysis
** Compare two groups by ANOVA
bys id (visit): gen sequence=_n
oneway slope SES, tab, if sequence==1
```

DISPLAY 8.2
Stata codes for estimating rates of change from the SLS dataset by linear regression for each individual with at least 3 HAZ measures in the first 12 months.

message when it tried to find participant ID 3 but failed. The "capture" command told Stata to ignore the error and move on to the next step. In this case, it moves on to find and operate on participant ID 4. The regression coefficients are stored in the variable "slope". The operation was performed only for participants with at least three nonmissing HAZ.

Since we only want one record (row) per participant for the rate of change, we may generate a variable called "sequence" that runs from 1 to the total number of records within each subject. In the analysis of summary statistics, we only analyze the record if the sequence is 1. Among 185 participants with at least three HAZ measures, the mean rates of change (SD) in the high and low SES groups were −0.104 (0.023) and −0.151 (0.023) per month, respectively. An ANOVA or regression analysis with a binary exposure variable gave $P < 0.001$. The low SES group declined faster in HAZ than the high SES group. Although the difference in the slope $(-0.104) - (-0.151) = 0.047$ looked small at first glance, we need to be mindful that a regression coefficient refers to a change in the outcome in relation to one unit change in the exposure variable, which is 1 month here. Over 1 year, the difference between the two groups would be about 0.047×12 or 0.564 HAZ. The difference was quite substantial.

A limitation of the summary statistics approach is that sometimes there is no ideal yet interpretable summary statistics available. Knowledge in the subject matter or graphical analysis would tell us that in developing countries HAZ does not decline linearly in the first 2 years of life, although it may behave so in the first year. That is why the example was limited to summarizing the rate of change in the first 12 months. Using a linear regression to summarize the slope in the first 24 months is not ideal. Using a polynomial may describe the relation between HAZ and age better, but the coefficients are difficult to interpret. As such, the analysis of a single slope parameter tends to be more useful for studies with a relatively short duration of follow-up because in a short duration of time a nonlinear relationship may be sufficiently approximated by a straight line.

8.3.2 Spline Models

The aforementioned rate of change estimation can be extended by the use of splines to approximate a nonlinear relationship. Yang et al. (2011) proposed a linear spline model to capture weight gain from birth to the age of 5 years in the following form:

$$y_{ij} = b_{0i} + (b_{1i} \times S_{1ij}) + (b_{2i} \times S_{2ij}) + (b_{3i} \times S_{3ij}) \tag{8.4}$$

where $i = 1, 2, \ldots, n$ indices the children; j indices the time of observations within children; and S_{1ij}, S_{2ij}, and S_{3ij} represent the amount of time the ith child spent in the period 0 to 3 months, 3 to 12 months, and 12 to 60 months, up to the jth month. For example, if y_{i50} is the ith child's HAZ measured at age 50 months, $S_{1i50} = 3 - 0 = 3$, $S_{2i50} = 12 - 3 = 9$, and $S_{3i50} = 50 - 12 = 38$. The coefficients b_{1i}, b_{2i}, and b_{3i} represent the growth velocity in the three periods, respectively, and the estimated coefficients can be used as three independent variables.

> **Example 8.2 (Continued)**
>
> Display 8.3 demonstrates how to use linear splines to estimate the rate of change in HAZ in two time periods, from 0 to 12 months and from 12 to 24 months. It is based on Display 8.2, with the modifications italicized. The "cond(a,b,c)" function says that if condition a is satisfied, the data value is b, otherwise it is c. The ANOVA showed that the mean of the change rate in the second period was −0.041 and −0.022 in the two groups (P < 0.001). In the period from 12 to 24 months, there was still a statistically significant difference in the rate of change in HAZ between the two groups, but the difference was milder during this period, at 0.019 per month. Over 12 months the difference was predicted to accumulate to 0.019 × 12 = 0.228 z-score. It still appeared to be practically significant although the difference was milder than in the first 12 months.

Chapter 11 will discuss the estimation of the characteristics of trajectories and the age at which the characteristics occurs.

```
keep if visit<=24

** Save the largest id number in local macro N
sum id
local N=r(max)

** Count number of non-missing values
by id: egen total=count(haz)

** Estimate rate of change if at least 3 measures; save rate in var
** slope
** loop over id 1 to `N'
gen slope1 =.
gen slope2 =.
gen visit1=cond(visit<=12,visit,12)
gen visit2=cond(visit>12,visit - 12,0)
quietly forvalues i=1(1) `N' {
capture regress haz visit1 visit2 if total>=3 & id==`i'
capture replace slope1=_b[visit1] if total>=3 & id==`i'
capture replace slope2=_b[visit2] if total>=3 & id==`i'
}

** Generate a variable for selection of one record per person for
** analysis
** Compare two groups by ANOVA
bys id (visit): gen sequence=_n
oneway slope1 SES, tab, if sequence==1
oneway slope2 SES, tab, if sequence==1
```

DISPLAY 8.3
Stata codes for estimating rate of change from the SLS dataset by the linear spline model for each individual with at least 3 HAZ measures in the first 24 months, with slope1 and slope2 denoting the rate of change in the periods 0–12 months and 12–24 months, respectively.

The analysis of subject-level summary statistics has the beauty of simplicity. It is easy to explain to the nonstatistical audience and may fit "the day-to-day reality of statistical reporting in medical journals" better than more advanced statistical models (Frison and Pocock 1992). A common question some readers may ask: Is this not a loss of information to reduce multiple observations to a single summary statistics for analysis? No, it is not. The larger the number of observations is used to estimate the panel-level summary statistics, the more precise the summary statistics is. This translates into a more powerful analysis of summary statistics (Hayes and Moulton 2009).

8.4 Mixed Models

Mixed models are regression models that include both fixed effects and random effects in the regression equations (Rabe-Hesketh and Skrondal 2008; Searle et al. 1992). The fixed effects are regression coefficients for exposure variables like what we have discussed in previous chapters. The random effects accommodate, for a random sample of panels, the panel-level deviations from the central tendency. The mixed models are sometimes called *subject-specific models*. Hence, in this terminology a subject means a panel. We will discuss two common types of mixed models. We will initially focus on using linear regression for quantitative outcomes to illustrate the concepts. They are called *linear mixed models* (LMMs). Later, we will discuss other outcome variables. In addition to its value in assessing associations, the mixed models play an important role in the development of longitudinal growth and development references as well as in defining trajectory characteristics (Chapter 11). That is why this chapter pays more attention to the mixed models and chooses not to include more methods that are useful for population-level analysis.

8.4.1 Random Intercept Models

When there is only one observation per participant, a multivariable regression equation describes the outcome in the ith subject in relation to p independent variables:

$$y_i = b_0 + b_1 x_{1i} + b_2 x_{2i} + \ldots + b_p x_{pi} + e_i$$

In the case of multiple observations per subject, let y_{ij} represent the jth observation of the ith subject. The *random intercept model*

$$y_{ij} = (b_0 + \beta_{0i}) + b_1 x_{1ij} + b_2 x_{2ij} + \ldots + b_p x_{pij} + e_{ij} \qquad (8.5)$$

where e_{ij} denotes residuals and β_{0i} is an unobserved variable that captures the subject-level variation by allowing each subject to have its own intercept $(b_0 + \beta_{0i})$. The $b_0 + b_1 x_{1ij} + \ldots + b_p x_{pij}$ in Equation (8.5) is the fixed effects part of the model and is already described in previous chapters. The x's may be subject-level (constant within panel) or observation-level (varying within subject) variables. We may remove the subscript j in the notation for brevity when the meaning is clear. The β_{0i} is the random effect of subjects on the intercept. It shifts the individual regression line up or down from the average regression line with intercept b_0. The model assumes that β_{0i} and e_{ij} are uncorrelated, β_{0i} and x's are uncorrelated, and, given β_{0i}, the residuals e_{ij} nested within subject i are independent from each other. The random effects are assumed to follow a normal distribution with mean zero and variance to be estimated from the data. The random intercept model can be more generally represented in matrix notation:

$$\mathbf{y}_i = \mathbf{X}_i \mathbf{b} + \mathbf{1}_i \mathbf{r}_i + \mathbf{e}_i \qquad (8.6)$$

where $i = 1, 2, \ldots, n$, $\mathbf{y_i}$ consists of n_i rows (observations), $\mathbf{1_i}$ is a $n_i \times 1$ vector of one's, and $\mathbf{r_i}$ is the value for the random intercept. X_i is a $n_i \times (p + 1)$ design matrix that may comprise subject- or observation-level variables or both.

8.4.2 Random Coefficient Models

The *random coefficient model* expands Equation (8.5) to include random effects of subjects on one or more of the regression coefficients (slopes):

$$y_{ij} = (b_0 + \beta_{0i}) + (b_1 + \beta_{1i})x_{1ij} + b_2 x_{2ij} + \ldots + b_p x_{pij} + e_{ij} \tag{8.7}$$

In Equation (8.7), β_{1i} is a random effect on the slope in relation to x_{1ij}. It is assumed to be normally distributed with mean 0 and variance to be estimated. It is possible to further include β_{2i} and so on in the equation. Again, it is assumed that the random effects are uncorrelated with x's and the residuals e_{ij}, and, given the random intercepts and slopes, the residuals e_{ij} are independent. Under this model, the independent variable(s) has a different degree of impact on the outcome of different individuals, as reflected by $(b_1 + \beta_{1i})$. If the independent variable x_{1ij} represents a treatment, β_{1i} may be interpreted as person–treatment interaction or heterogeneity of treatment effects. If the independent variable concerned is time or age, the random effect on the slope reflects different growth curves.

The random coefficient model can be more generally represented in matrix form as

$$\mathbf{y_i = X_i b + Z_i r_i + e_i} \tag{8.8}$$

where $\mathbf{Z_i}$ is a $n_i \times q$ design matrix for the random effects and $\mathbf{r_i}$ is the $q \times 1$ vector of the random intercept and coefficient(s). The random intercept model in Equation (8.6) is a special case of Equation (8.8) where $q = 1$ and $\mathbf{Z_i = 1_i}$. If the model has one random effect on the slope (in addition to the intercept), $q = 2$, and so on. For $q \geq 2$, the elements of the $q \times q$ variance–covariance matrix $\mathbf{G}$ characterize the random effects. Table 8.2 describes some common variance–covariance matrices. If the random effects can be assumed to be uncorrelated, $\mathbf{G}$ is set to a diagonal matrix. The diagonal elements represent the variances of each random effect. If the variances of all random effects are constrained to be the same, $\mathbf{G}$ is an *identity* matrix. If each variance is estimated freely from the data, $\mathbf{G}$ is an *independent* matrix. If correlation between random effects is allowed, the diagonal and off-diagonal elements are estimated from the data. The off-diagonal elements are the covariances. (A covariance divided by the two standard deviations concerned is the correlation.) An *exchangeable*, also known as *compound symmetry*, matrix assumes a common variance and a common covariance. An *unstructured* matrix freely

TABLE 8.2

Common Variance–Covariance Matrices
(an Example of Three Random Effects)

Pattern	Matrix
Identity	$\begin{pmatrix} \sigma^2 & & \\ 0 & \sigma^2 & \\ 0 & 0 & \sigma^2 \end{pmatrix}$
Independent	$\begin{pmatrix} \sigma_1^2 & & \\ 0 & \sigma_2^2 & \\ 0 & 0 & \sigma_3^2 \end{pmatrix}$
Exchangeable	$\begin{pmatrix} \sigma^2 & & \\ \alpha & \sigma^2 & \\ \alpha & \alpha & \sigma^2 \end{pmatrix}$
Unstructured	$\begin{pmatrix} \sigma_1^2 & & \\ \alpha_{12} & \sigma_2^2 & \\ \alpha_{13} & \alpha_{23} & \sigma_3^2 \end{pmatrix}$

estimates all the variances and covariances. Furthermore, the random effects and residuals have a variance–covariance matrix

$$\mathrm{Var}\begin{bmatrix} \mathbf{r}_i \\ \mathbf{e}_i \end{bmatrix} = \begin{bmatrix} \mathbf{G} & \mathbf{0} \\ \mathbf{0} & \sigma_e^2 \mathbf{R} \end{bmatrix} \tag{8.9}$$

and $\mathbf{R} = \mathbf{I}$, an identity matrix. That is, the random effects are uncorrelated with the residuals and the residuals are independent with constant variance.

Example 8.4

De Jager et al. (2012) studied the association between genetic alleles and age-related cognitive decline in middle-age and older people. The participants were administered various cognitive function tests annually, on average, for 9 years. Linear mixed models were used to allow for random effects on the intercept and the regression coefficient on age. The random coefficient allowed a different linear trend of cognitive decline in relation to age for different individuals. In addition to a SNP at the well-known APOE locus, the AA, AG, and GG groups defined by a SNP at the PDE71/MTFR1 locus differed significantly in average rate of cognitive decline, at –0.0679, –0.0596, and –0.1003 per year, respectively.

8.4.3 Growth Curves

8.4.3.1 Interaction Model

The use of interaction terms between time and categorical variable(s) in the fixed effects part of a mixed model can reveal the average growth curves of different groups of panels. If the independent variable is quantitative, the interaction term reveals the *additional* slope over time per one unit increase in the independent variable.

Example 8.2 (Continued)

Display 8.4 shows the codes and outputs of a random coefficient model fitted to the HAZ data in the first 12 months among subjects with at least three observations. The regression equation fitted was

$$HAZ_{ij} = (b_0 + \beta_{0i}) + (b_1 + \beta_{1i})visit_{ij} + b_2 SES_i + b_3(visit_{ij} \times SES_i) + e_{ij}$$

By default, Stata uses an independent covariance matrix. The estimated SDs of the random intercept β_{0i} and β_{1i} were 0.528 and 0.020, respectively.

Since SES was constant over time (subject-level variable), there was no subscript j here. Since the SES variable was 0 and 1 coded, the fixed effect parts tell us that among the average growth curve for the low SES group was (by substituting SES = 0 in the equation):

$$HAZ = b_0 + b_1 \times Visit$$

Taking the regression coefficients from Display 8.4, it can be seen that the average HAZ among the low SES group from birth (0 month) to 12 months was given by the growth curve $HAZ = 0.029 - 0.149 \times Visit$. In contrast, the average growth curve for the high SES group was (by substituting SES = 1 in the equation):

$$HAZ = (b_0 + b_2) + (b_1 + b_3) \times Visit$$

From Display 8.4, the growth curve for the high SES was $HAZ = (0.029 + 0.099) + (-0.149 + 0.046) \times Visit$, or $HAZ = 0.128 - 0.103 \times Visit$. In other words, the high SES was associated with less decline in HAZ, by 0.046 per month ($P < 0.001$). We had reached this conclusion earlier by the use of summary statistics. High SES was also associated with a higher HAZ at birth, by 0.099 z-score, but it was not statistically significant ($P = 0.220$). The results were expected to agree well with the simple solutions based on summary statistics as there was only a small variation in the number of observations per subject. In studies that have a highly variable number of observations per subject, the results may disagree more.

Using the aforementioned equation, we may calculate that at, for example, 12 months, the difference between groups were $b_2 \times SES + b_3 \times 12 \times SES$. Stata's "nlcom" (nonlinear combination) commands provided not only this estimate, $0.099 + 0.0463 \times 12 = 0.655$, but

```
. ** mixed models

.
. xtset id visit
       panel variable: id (unbalanced)
        time variable: visit, 0 to 24, but with gaps
               delta: 1 unit
. xi: xtmixed haz i.SES*visit || id: visit if total>=3 & visit<=12

(output omitted)
------------------------------------------------------------------------
         haz |    Coef.   Std. Err.     z    P>|z|  [95% Conf. Interval]
-------------+----------------------------------------------------------
      _ISES_1 |  .0989761   .0806905    1.23  0.220  -.0591744    .2571267
        visit | -.1493598   .0019984  -74.74  0.000  -.1532766   -.1454429
_ISESXvisit_1 |  .0463087   .0032681   14.17  0.000   .0399034    .052714
         _cons |  .0290297   .0492798    0.59  0.556  -.067557    .1256163
------------------------------------------------------------------------


------------------------------------------------------------------------
Random-effects Parameters | Estimate   Std. Err.   [95% Conf. Interval]
--------------------------+---------------------------------------------
id: Independent           |
               sd(visit) | .0197906    .0012194     .0175392    .0223309
               sd(_cons) | .5280338    .0277092     .4764243    .5852339
--------------------------+---------------------------------------------
            sd(Residual) | .0776788    .0018594     .0741186    .0814099
------------------------------------------------------------------------
LR test vs. linear regression:  chi2(2) = 3432.47 Prob > chi2 = 0.0000

Note: LR test is conservative and provided only for reference.

. nlcom _b[_ISES_1]+ _b[_ISESXvisit_1]*12

       _nl_1: _b[_ISES_1]+ _b[_ISESXvisit_1]*12
------------------------------------------------------------------------
    haz |    Coef.    Std. Err.     z    P>|z|     [95% Conf. Interval]
--------+---------------------------------------------------------------
  _nl_1 |  .6546804    .0886112   7.39   0.000     .4810057    .8283551
------------------------------------------------------------------------
```

DISPLAY 8.4
Stata codes and outputs: Random coefficient model analysis of HAZ in the SLS dataset.

also the standard error and confidence interval, as shown in the display. Polynomials or splines may be added to the regression model to examine the nonlinear trend when needed.

8.4.3.2 *Best Linear Unbiased Prediction*

The mixed models can also produce individual growth curves. The random effects in mixed models are characterized by their variances. Unlike the (fixed effects) regression coefficients that can be applied to make a prediction of the outcome in straightforward ways, the individual random effects values need to be predicted for each subject before the outcome can be predicted. This is based on a class of method called *best linear unbiased prediction*, or BLUP, which involves using the **G** and **R** matrices in Equation (8.9). A major characteristic of BLUP is shrinkage toward the mean. The smaller the amount of information available for a subject, the more it shrinks toward the center tendency, or vice versa. Intuitively speaking, when there is not enough information for knowing a subject well, the BLUP tends to say this subject is not very different from the average. This is in contrast to the summary statistics approach, in which the summary statistics for each subject are estimated equally despite the variable amount of information about them. Readers are referred to specialized texts such as Rabe-Hesketh and Skrondal (2008), Robinson (1991), and Searle et al. (1992) for details about BLUP.

Example 8.2 (Continued)

Our interest in this example has been about whether the low and high SES groups differed in the rate of change in HAZ during the first year of life. Earlier we used summary statistics to answer this question. In Display 8.5, we fitted a random coefficient model to the data, without the interaction between SES and month (because now we want to capture the difference in rate of change by the individual random effect values instead of by the interaction term). Then we used the "predict *varname*, reffects" function of Stata to obtain the BLUP values of the random intercept and random slope for each subject. We used the "oneway" ANOVA command to show the results. The mean (SD) of the β_{1i} was -0.016 (0.0198) in the low SES and 0.027 (0.021) in the high SES group. The means were smaller than the mean of the rate of change (slope) estimated using the summary statistics approach earlier. This is because the random slope values represent the individual *deviations* from the average slope, whereas the summary statistics approach estimated the individual slopes themselves. The difference in the random slope values between the two groups was 0.043 (P < 0.0001).

8.4.3.3 Latent Growth Model

In psychometrics and structural equation modeling, there is a method called latent growth model or latent growth curve (Bollen and Curran 2006). It is common to present the model in graphical form. Figure 8.2 shows a typical latent growth model with 12 observed variables (Y_j, j = 1, 2, ..., 12). There are two latent (unobserved) variables that determine the observed variables, one is called *intercept* and the other called *slope*. Each arrow pointing from the intercept variable represents a relationship with a regression coefficient $\beta_{0j} = 1$. Each arrow pointing from the slope variable represents a relationship with a regression coefficient $\beta_{1j} = j$. Furthermore, the observations are

```
. capture drop beta*

. xi: xtmixed haz SES visit || id: visit if total>=3 & visit<=12

(output omitted)
```

| haz | Coef.   Std. Err.   z   P>|z| [95% Conf. Interval]
-----+--
| SES | .1162704   .0807004   1.44   0.150   -.0418995   .2744403
| visit | -.1324697   .0022996 -57.61   0.000   -.1369767   -.1279626
| _cons | .0232724   .0492933   0.47   0.637   -.0733408   .1198855

Random-effects Parameters | Estimate   Std. Err.   [95% Conf. Interval]
--------------------------+---
   id: Independent |
           sd(visit) | .0300195   .0016964   .0268721   .0335356
           sd(_cons) | .528216   .0277342   .4765615   .5854694
--------------------------+---
        sd(Residual) | .0775772   .0018566   .0740223   .0813028

LR test vs. linear regression: chi2(2) = 3333.69   Prob > chi2 = 0.0000

Note: LR test is conservative and provided only for reference.

. predict beta*, reffects
(18 missing values generated)
(18 missing values generated)

. oneway beta1 SES,tab, if seq==1
```

```
Socio-Econ | Summary of BLUP r.e. for id: visit
    Status |      Mean      Std. Dev.     Freq.
-----------+------------------------------------
       Low |  -.01592077    .01976003      116
      High |   .02676534    .02097468       69
-----------+------------------------------------
     Total |   9.282e-12    .02889745      185
```

	Analysis of Variance				
Source	SS	df	MS	F	Prob > F
Between groups	.078833082	1	.078833082	192.82	0.0000
Within groups	.074818472	183	.000408844		
Total	.153651554	184	.000835063		

DISPLAY 8.5
Stata codes and outputs: Obtaining best linear unbiased predictors.

also partially determined by individual factors e_j. In psychometrics, the individual factors are often considered as representing measurement errors. The two-way arrow between the intercept and slope denotes the correlation between them. The observations are the means for understanding the unobserved variables that represent the starting point and linear trend of each
```

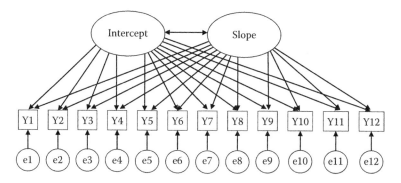

FIGURE 8.2
An example of a typical latent growth model.

subject's trajectory. Despite different terminology, if the individual factors have equal variances and are uncorrelated, this model is equivalent to the random coefficient model

$$y_{ij} = (b_0 + \beta_{0i}) + (b_1 + \beta_{1i})x_{1ij} + e_{ij}$$

with an unstructured covariance matrix **G**. The random effects β_{0i} and β_{1i} are equivalent to the latent variables intercept and slope. The covariance in **G** is equivalent to the two-way arrow between intercept and slope.

8.4.4 Generalized Linear Mixed Models

The previous sections describe the linear mixed models (LMMs) for quantitative outcomes. The concepts of the mixed models can be applied to model binary or other outcome variables, giving *generalized linear mixed models* (GLMMs) (Rabe-Hesketh and Skrondal 2008). Estimation of LMMs and GLMMs involves integrating out the random effects in the likelihood. For LMMs, the integral can be solved analytically. For GLMMs, the solution can only be approximated. With the availability of high-speed personal computers, a computationally intensive approximation method called *adaptive Gaussian quadrature* has become the main method to estimate GLMMs (Rabe-Hesketh and Skrondal 2008). The method uses a number of nodes, or quadrature points, to approximate the integral. Using a larger number of nodes improves the approximation, at the cost of increasing the computation time roughly proportionally. After estimating a GLMM, it is useful to check whether the results are sensitive to the number of nodes, by comparing the results with those estimated with a different number. A rule of thumb is that if the regression coefficients do not change by more than a relative difference of 0.0001 when a different number of nodes is used, the solution is robust (StataCorp 2011). If the coefficients do change substantially, try using a larger number of nodes to improve the approximation. If that still does not stabilize, one may need to switch to other methods such as the methods described earlier in this chapter. A reason for persistent failure in the estimation is that the within-panel correlation is too strong. In that case, a panel appears almost like an observation and therefore there is not enough information to estimate the model.

Example 8.2 (Continued)

Define stunting$_{ij}$ as HAZ$_{ij} < -2$ Z-score. Display 8.6 first fitted a random intercept logistic regression

$$\text{logit(stunting}_{ij}) = (b_0 + \beta_{0i}) + b_1 \text{visit}_{ij} + b_2 \text{SES}_i + b_3 (\text{visit}_{ij} \times \text{SES}_i)$$

using Stata's "xtlogit" command, which by default uses 12 nodes in the adaptive quadrature approximation. Then the "quadchk" (quadrature check) command was used to examine the changes in the regression coefficients if the number of nodes increased or decreased by one-third. The results showed substantial differences. The log OR for the SES variable in the default estimation (12 nodes) was -3.298 but it was -4.123 and -3.542 when the number of nodes decreased to 8 or increased to 16. The relative differences were $(4.123 - 3.298)/3.298 = 0.25$ or 25% and $(3.542 - 3.298)/3.298 = 0.07$ or 7%, respectively. Both were much larger than the 0.0001 reference value in the rule of thumb. The instability persisted even after increasing the number of nodes further (details not shown). As such, it is unsafe in this case to use the result from the GLMM. Alternative methods such as logistic regression with robust variance estimator for cluster data will need to be considered.

8.4.5 Further Remarks

8.4.5.1 Presentation

The random effects in the mixed models are characterized by their variances (or SDs). Near the bottom of the mixed models regression results in Display 8.5, the SDs of the random effects are presented. For example, Display 8.5 shows that the SD of the rate of change in HAZ per month was 0.030 (95% CI 0.027 to 0.034) and the SD of the intercept, which denotes HAZ at birth in this case, was 0.528 (95% CI 0.476 to 0.585). This information is often not reported in scientific journals. The correlation among observations within a panel arising from heterogeneity in intercept or slopes is often considered a nuisance to be removed. According to this view, the interest and the focus of presentation are on the fixed effects part of the regression. Nevertheless, it is important not to forget the variance estimates as a potentially useful findings in some circumstances. For instance, in a cluster randomized trial and ophthalmology studies, power and sample size requirements are affected by the SDs of the random effects. Knowing the SD estimates from previous

```
. xi: xtlogit stunt i.SES*visit if total>=3 & visit<=12, nolog
```

(output omitted)

```
. quadchk
```

(output omitted)

Quadrature check

	Fitted quadrature 12 points	Comparison quadrature 8 points	Comparison quadrature 16 points	
Log likelihood	-124.78878	-124.55696	-121.94531	
		.23182484	2.843474	Difference
		-.00185774	-.02278629	Relative difference
stunt: _ISES_1	-3.2977662	-4.123179	-3.5424923	
		-.8254128	-.24472602	Difference
		.25029451	.07420963	Relative difference
stunt: visit	3.1285067	3.5432406	6.110235	
		.41473386	2.9817282	Difference
		.13256607	.95308352	Relative difference
stunt: _ISESXvisi~1	-.84234849	-.63293437	-1.8688869	
		.20941412	-1.0265384	Difference
		-.24860747	1.2186624	Relative difference
stunt: _cons	-40.562054	-47.773977	-78.939544	
		-7.2119228	-38.37749	Difference
		.17779974	.94614267	Relative difference
lnsig2u: _cons	4.4349061	4.8175302	5.7741856	
		.3826241	1.3392796	Difference
		.08627558	.301986	Relative difference

DISPLAY 8.6
Stata codes and outputs: Estimation and checking of a random intercept logistic regression model.

studies would be tremendously helpful in the planning of new studies. Unfortunately, such information is often not presented, leaving future investigators to fumble in the dark. It is desirable to present the SD estimates even if they are not the focus of a study.

8.4.5.2 Population- versus Subject-Level Odds Ratio

For a random intercept model that assumes a normal distribution for the random effects, there is an approximate relationship between the population- and subject-level log OR (Zeger et al. 1988):

$$\beta_{PA} \approx \beta_{SS} / \sqrt{1 + 0.346 \times \text{Variance(intercept)}} \tag{8.10}$$

Since variance is always larger than zero, the population-level log OR tends to be smaller, or closer to the null value.

With observation-level covariates like age, the mixed model provides coefficients that are easy to interpret. They represent the change in the outcome per unit change in the covariate for a given panel. But they are difficult to interpret if the exposure variables are constant within a panel.

Example 8.5

Neuhaus et al. (1991) analyzed and discussed a study of women that involved obtaining a sample of fluid from both breasts of each participant. Each panel (woman) contributed two observations (breasts). The binary outcome was fluid availability from the breast. One of the binary covariates was parity: A woman could either be nulliparous or parous. Therefore, two observations (breasts) share the same panel-level exposure status. For this covariate, the random intercept model gave OR 3.49. It "almost invites an unjustified causal statement about the change in the odds of fluid availability for a given woman who ceases to be nulliparous" (Neuhaus et al. 1991, p. 33), but in fact there was no change in exposure status within a panel. Nevertheless, the regression coefficient of the random intercept model may be interpreted as indicating that for women who have the same underlying risk (same random effect β_{0i}), the odds of the outcome was 3.49 higher for those women who were parous.

8.4.5.3 Missing Values

Packages for structure equation modeling (latent growth model) tend to use data in the wide format. If one of the repeated measurements is missing, the data from that subject (row) is not usable at all. Packages for mixed models use data in the long format, and incomplete data from subjects with missing values are usable. The nice feature of the mixed model in the presence of missing values will be further discussed in Chapter 12.

8.4.5.4 Robustness

The random effects are usually assumed to be either normally distributed or gamma distributed. This is mainly a result of computational convenience. It is sometimes said that the mixed model is not robust if the random-effect distribution is misspecified. However, both theoretical and simulation results about the mixed models for regression analysis of different types of outcomes have suggested that it "yields inferences about the regression parameters that are quite robust to misspecification" (Neuhaus et al. 1992, p. 761; Xu et al. 2012). Furthermore, although the power can be seriously affected, the type I error rate is quite unaffected by the misspecification (Litière et al. 2007).

9

Quantifying Growth:
Development of New Tools

Existing growth references do not satisfy the needs of all studies. For instance, fetal growth and size at birth references based on singletons may not fit multiple births. For novel parameters that have only recently become measurable due to advances in technology, there will be no existing reference at all. This chapter describes how to generate new references (norms). Analysis of nonlinear relationships is a key element in the development process. In addition, we will need parametric or nonparametric methods to describe the distribution.

One question to ask before doing statistical analysis is: What is the purpose of the work? If it is for developing standards, inclusion and exclusion criteria should be set so that the sample represents an ideal population suitable for setting the bar. For studies of associations, usually the purpose is to facilitate comparisons across individuals within the same sample and make findings more easily interpretable. In this context, there is usually no attempt to make reference to an ideal population. The following sections do not discuss inclusion or exclusion criteria further.

9.1 Capturing Nonlinear Relationships

9.1.1 Fractional Polynomials

9.1.1.1 Natural Polynomials

A simple way to describe nonlinear relationships is by using polynomials. That is, by adding power terms of the x-axis variable as covariates. A mth degree polynomial regression equation is

$$\hat{y}_i = b_0 + b_1 x_i + b_2 x_i^2 + \ldots + b_m x_i^m \tag{9.1}$$

Some growth references were developed using such polynomial functions, for example, an earlier version of the Swedish size at birth-for-gestational age references used a third-degree polynomial to fit the nonlinear relationship (Niklasson et al. 1991). To easily differentiate them from another type of polynomial to be discussed later, we called them *natural polynomials* (NPs).

If the equation is fitted by the least-squares method, the coefficient of determination, R-squared, can be used to compare different natural polynomial models. Since R-squared always increases as the number of independent variables increases, it is not very informative in the comparison of models with different polynomials. An *adjusted R-squared* is given by

$$R^2_{\text{Adjusted}} = 1 - \frac{SS_{\text{Residual}} / \left[n - (p+1) \right]}{SS_{\text{Total}} / (n-1)} \qquad (9.2)$$

where n is the sample size and (p + 1) is the number of regression coefficients (including the intercept). The model with a larger adjusted R-squared is usually preferred.

Another measure of a regression model's *goodness-of-fit* is the *deviance*, D, defined as the model's maximized log-likelihood multiplied by –2. For models with normally distributed residuals, it can also be calculated with the formula (StataCorp 2011)

$$D = n \left[1 + \ln \left(\frac{2\pi SS_{\text{Residual}}}{n} \right) \right] \qquad (9.3)$$

A larger deviance indicates a poorer fit. In a sense, D may be better called a measure of "badness-of-fit." Furthermore, a simpler model "nested" within a larger model can be tested for their difference in goodness-of-fit based on the model deviances, by referring the difference in D to a chi-square distribution with the degree of freedom equal to the difference in the number of parameters of the two models. This is equivalent to performing a likelihood ratio test. For example, a second-degree natural polynomial model is a simpler model nested within a third-degree natural polynomial model. This example of a simpler model constrains the coefficient of the third-degree term to zero. Since the simpler model has one less parameter to be estimated, the degree of freedom for the test is one.

Example 9.1

Figure 9.1 plots the weight of 1968 Indonesian boys aged less than 5 years in the Fourth Wave of the Indonesia Family Life Survey (IFLS4) (Strauss et al. 2009). Table 9.1 shows the results of fitting a series of natural polynomial models to the data. The fifth-degree natural polynomial (NP5)

$$\text{Weight} = 3.654 + 10.416\text{Age} - 7.545\text{Age}^2 + 3.035\text{Age}^3 \\ - 0.559\text{Age}^4 + 0.038\text{Age}^5$$

where age was in years, had adjusted R-squared larger than that of a fourth-degree polynomial (NP4) and similar to that of a sixth-degree polynomial (NP6). Referring the difference in deviance between NP5 and NP6

$$7703.29 - 7699.86 = 3.43$$

to the chi-square distribution on one degree of freedom gave a P-value of 0.064. The NP6 did not fit the data significantly better than NP5. But NP5 did fit significantly better than NP4 (P < 0.001). So, NP5 was chosen as the best fitting model among the natural polynomial models.

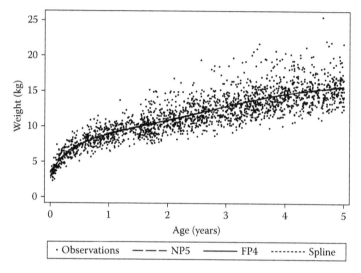

FIGURE 9.1
Weight of boys aged under 5 years in IFLS4, with fitted NP5, FP4, and spline-based regression lines.

TABLE 9.1

Fractional and Natural Polynomial Regression Analysis of Weight of 1968 Boys Aged under 5 Years in IFLS4

Model	Degree	R-Squared	Adjusted R-Squared	Deviance	P*
Natural polynomial	6	74.43%	74.35%	7699.86	0.064
	5	74.38%	74.32%	7703.29	<0.001
	4	74.20%	74.15%	7717.09	—
Fractional polynomial	5	74.48%	74.41%	7695.95	0.976
	4	74.47%	74.41%	7696.93	0.004
	3	74.33%	74.29%	7707.79	—

* Testing NP6 versus NP5, NP5 versus NP4, FP5 versus FP4, and FP4 versus FP3.

9.1.1.2 Fractional Polynomials

The natural polynomials are not very flexible in shape. An NP2 imposes a quadratic trend and an NP3 imposes a cubic trend on the data, and so on. They have a tendency to create artificially wavy curves and they tend not to fit the edges of the range of data well. An alternative that mitigates the problems is *fractional polynomial* (FP) regression models (Royston and Altman 1994; Royston and Sauerbrei 2005). A mth degree fractional polynomial regression model is

$$\hat{y}_i = b_0 + b_1 x_i^{(p_1)} + b_2 x_i^{(p_2)} + \ldots + b_m x_i^{(p_m)} \tag{9.4}$$

where

$$x^{(p_j)} = \begin{cases} x^{p_j}, & if \ p_j \neq 0 \\ \ln(x), & if \ p_j = 0 \end{cases}$$

and the power terms (p_j) are estimated from the data and not limited to integers. Furthermore, *repeated powers* of p that are defined as

$$b_1 x^p + b_2 x^p \ln(x) \tag{9.5}$$

are allowed. The estimation of an FP regression model requires a systematic search for the combination of powers that minimizes the deviance. To avoid the difficulty in searching among an infinite set of combinations, some restrictions are imposed. A restricted set (–2, –1, –0.5, 0, 0.5, 1, 2, 3) has been recommended as this limited set of powers for the first- and second-degree FP (FP1 and FP2, respectively) models already offers sufficient flexibility in many situations (Royston and Altman 1994).

Tests for difference in deviance between the mth and $(m-1)$th degree FP models are made by referring to a chi-square distribution with two degrees of freedom. It is two degrees of freedom because the mth degree model not only had one more regression coefficient b_m but also one more power term p_m to be estimated.

Example 9.1 (Continued)

Table 9.1 shows the R-squared, adjusted R-squared, and deviance of FP3 to FP5 models for the child weight data. The FP4 model with repeated powers (1, 1, 2, 2)

$$\text{Weight} = 2.451 - 0.833\text{Age} - 9.230\text{Age} \times \ln(\text{Age}) + 7.296\text{Age}^2 \\ - 2.257\text{Age}^2 \times \ln(\text{Age})$$

had adjusted R-squared equal to that of FP5 but larger than FP3. In terms of the difference in deviance, FP5 was not better than FP4 (P = 0.976), while FP4 fitted better than FP3 (P = 0.004). Figure 9.1 plots the NP5 and FP4 curves. Although they were quite close to each other for a large part of the age range, there was a visible difference between the FP4 and NP5 at the lower edge and, to a lesser extent, the upper edge. Figure 9.2 zooms into the data points below age 1 year. The FP4 curve climbed more sharply from a lower starting point than the NP5 does. Since the number of boys assessed very close to the date of birth was small in this data, a comparison of the intercepts of the FP4 and NP5 was not very meaningful. Nevertheless, among the 34 boys aged below 1 month, the mean weight observed was 3.6 kg. The fitted value based on FP4 was 3.7 kg and that based on NP5 was 4.1 kg. It appeared that FP4 fitted the data at the lower end better than NP5.

In data-driven model building, indices and tests of goodness-of-fit should be considered only approximate. Such repeated analyses do not usually recognize model uncertainty and inflation of type I error due to multiple testing. Furthermore, the adjusted R-squared and deviance of the NP and FP models are not strictly comparable. The standard way of calculating adjusted R-squared does not include adjustment for the estimation of the FP power terms. NP and FP models are not nested within each other. The indices are primarily for comparison within the same type of polynomial models. In addition, the test for difference in deviance between FP models is also approximate and it tends to be somewhat conservative. This is because one degree of freedom is given to each power term, but in fact it is only a restricted subset of candidate values that is searched for in the power term.

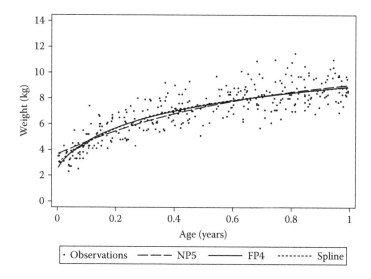

FIGURE 9.2
Zooming in the age range 0 to 1 year.

Example 9.2

Display 9.1 shows the Stata codes and outputs of analysis of the males in the Simulated Clinical Trial (SCT) dataset. Note that we know the birth weight-for-gestational age data were simulated according to a NP2 function. Fitting a NP2 model to the data gave R-squared and adjusted R-squared 0.4066 and 0.4051, respectively. The "predict" command after the "regress" command generated a new variable that holds the regression predicted values. The "fracpoly" command with the "degree(2)" and "compare" options not only fitted a second-degree FP model (m = 2) but also fitted without showing the details of the m = 1 model, and compared the goodness-of-fit between the two models, under the output column P(*). The test of difference in deviance between FP2 and FP1 models gave P < 0.0001. This rejected the FP1 model and favored the FP2. The algorithm found 2 and 3 for the FP2 power terms, shown under the column Powers. The R-squared and adjusted R-squared of the FP2 model were very similar to those of the NP2 model. In this case, the NP2 and FP2 gave a similar fit to the data. Figure 9.3 plots the data and the three fitted curves. The FP1 clearly was not sufficient in describing the central tendency of birth weight in the lower range of gestational age. The NP2 and FP2 almost totally overlapped.

9.1.2 Cubic Splines

The *cubic spline* is a convenient tool for growth studies. It ensures that the fitted curve is smooth and flexible in shape. A regression line based on a simple form of cubic spline with j internal knots at locations $x = k_1, k_2, \ldots, k_j$ can be estimated simply by fitting the least-squares regression equation

$$\hat{y} = b_0 + b_1 x + b_2 x^2 + b_3 x^3 + I(x > k_1)b_4(x - k_1)^3 + I(x > k_2)b_5(x - k_2)^3 \quad (9.6)$$
$$+ \ldots + I(x > k_j)b_{j+3}(x - k_j)^3$$

where $I(x > k_j)$ is an indicator function that has value 1 if the condition $x > k_j$ is true and 0 otherwise. In this formulation, there is a global NP3 curve that is represented by the first four terms. Then there are "local" curves that are modifications of the global curve whenever it passes through a knot. The price of this simplicity is that the terms are highly correlated. When there are many knots the estimation may become numerically unstable.

Example 9.1 (Continued)

Display 9.2 provides the Stata codes for generating the variables for three knots at 0.5, 1.0, and 3.0 (years of age), and fitting a least-squares regression model to the weight data of the boys aged under 5 in IFLS4. The R-squared and adjusted R-squared were 74.47% and 74.40%, respectively. These values were very similar to those of the FP4 model (Table 9.1).

```
** gest2, np2 and fp2 are new variables generated

. gen gest2 = gestweek^2

.

. regress bw gestweek gest2 if male ==1

      Source |       SS          df       MS              Number of obs =      769
-------------+------------------------------              F( 2,  766)   = 262.44
       Model |  64.1136211        2   32.0568106          Prob > F      = 0.0000
    Residual |  93.5649219      766   .122147418          R-squared     = 0.4066
-------------+------------------------------              Adj R-squared = 0.4051
       Total |  157.678543      768   .205310603          Root MSE      = .3495

(output omitted)

. predict np2
(option xb assumed; fitted values)


.

. fracpoly regress bw gestweek,compare degree(2) noscaling ///
> adjust(no), if male==1

(output omitted)

      Source |       SS          df       MS              Number of obs =      769
-------------+------------------------------              F( 2,  766)   = 262.53
       Model |  64.1267647        2   32.0633823          Prob > F      = 0.0000
    Residual |  93.5517784      766   .122130259          R-squared     = 0.4067
-------------+------------------------------              Adj R-squared = 0.4051
       Total |  157.678543      768   .205310603          Root MSE      = .34947


----------------------------------------------------------------------------
          bw |     Coef.   Std. Err.      t    P>|t|    [95% Conf. Interval]
-------------+--------------------------------------------------------------
    Igest__1 |   .0191847   .0024795    7.74   0.000    .0143172    .0240521
    Igest__2 |  -.0003059   .0000435   -7.04   0.000   -.0003912   -.0002205
       _cons |  -7.961382   1.187018   -6.71   0.000   -10.29158   -5.631187
----------------------------------------------------------------------------
Deviance: 562.37. Best powers of gestweek among 44 models fit: 2 3.

Fractional polynomial model comparisons:
----------------------------------------------------------------------------
gestweek        df      Deviance     Res. SD     Dev. dif.    P (*)   Powers
----------------------------------------------------------------------------
Not in model    0       963.822      .453112     401.451      0.000
Linear          1       601.342      .358206      38.971      0.000    1
m = 1           2       578.660      .352962      16.290      0.000   -2
m = 2           4       562.371      .349471       --          --     2 3
----------------------------------------------------------------------------
(*) P-value from deviance difference comparing reported model with
m = 2 model

. predict fp2
(option xb assumed; fitted values))
```

DISPLAY 9.1

Stata codes and outputs: Fitting second-degree natural and fractional polynomial models to the males in the SCT dataset on birth weight and gestational age.

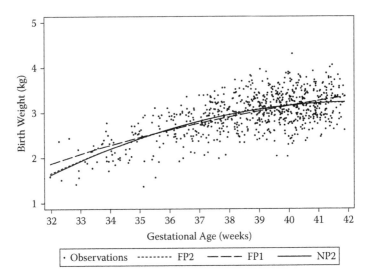

FIGURE 9.3
Birth weight by gestational age in the SCT dataset, with FP1, FP2, and NP2 regression lines.

```
** ageyr is variable name
gen ageyr2=ageyr^2
gen ageyr3=ageyr^3
gen agek1=cond(ageyr>0.5,(ageyr-0.5)^3,0)
gen agek2=cond(ageyr>1,(ageyr-1)^3,0)
gen agek3=cond(ageyr>3,(ageyr-3)^3,0)
regress wei ageyr ageyr2 ageyr3 agek1 agek2 agek3
```

DISPLAY 9.2
Stata codes for generating cubic splines with knots at 0.5, 1, and 3 years.

They were almost indistinguishable when plotted on the same figure (Figures 9.1 and 9.2).

The use of cubic splines requires specification of the number and locations of knots. More knots are needed for a more flexible shape. The knots do not need to be equally spaced. It is desirable to place knots closer to each other in the region where the y-axis variable is expected to change rapidly. Hence, the locations {0.5, 1, 3} are chosen in Example 9.1. In a study of the height of Finnish children and adolescents, Wei et al. (2006) placed the knots at {0.2, 0.5, 1.0, 1.5, 2.0, 5.0, 8.0, 10.0, 11.5, 13.0, 14.5, 16.0}. The spacing was decided according to the need for more flexibility in describing rapid changes in infancy and puberty.

For a large number of knots, such as that in Wei et al. (2006), it is desirable to put constraints on the cubic spline so as to increase stability. One way of constraining is to fix the spline to be linear beyond two boundary knots. The boundary knots can be the minimum and maximum observed values, or they can be within the range of the observations. It is the researcher's choice. This constrained version of cubic spline is called a *natural cubic spline*. See Royston and Sauerbrei (2007) for a suite of Stata macros for generating and analyzing natural cubic splines. Another form of constrained spline model is called the *penalized spline model*. That plays a particularly useful role in the studies of longitudinal trajectories and will be discussed in Chapter 11.

9.1.3 Transformation of X-Axis Variable

Transformation of the x-axis variable can help to linearize a relationship and thus make the modeling easier. In growth studies that have age as the x-axis variable, two transformations have been proposed (Cole et al. 1998; Royston and Wright 1998; WHO 2006). One of them aims to linearize the early part of a growth curve, while the other targets the end part.

The rapid growth rate and rapid deceleration in growth rate produces a non-linear relationship during infancy. A power transformation of the form

$$f(T) = T^\lambda \tag{9.7}$$

attempts to linearize the early part of the curve. For the growth variables in the World Health Organization (WHO) Multicentre Growth Reference Study (MGRS) group, the power term λ for the x-axis transformation ranged from 0.05 for length-based body mass index (BMI) ($\leq$24 months of age) to 1 (i.e., no transformation) for weight-for-length/height and height-based BMI (24–60 months). But most of the λ's were in the range 0.15 to 0.65 (WHO 2006, 2007). The value can be found by searching over a grid between 0 and 1 at preset intervals. The value that gives the best goodness-of-fit index such as deviance or R-squared is adopted.

On the other hand, a growth variable may reach a stable level as age increases, for example, height stabilizing after puberty. The transformation

$$x = \exp\left[\frac{t - T_{min}}{T_{max} - T_{min}} \ln(\rho)\right] \tag{9.8}$$

where T_{min} and T_{max} are the minimum (often 0) and maximum of the age range, respectively, and ρ is a small positive value below 1, helps to linearize the later part of such a growth curve. Royston and Wright (1998) suggested that a value of 0.01 for ρ is suitable, but other small positive values can also be considered.

The use of transformation of the x-axis requires caution. There may not be a single transformation that works for the whole range of the x-axis data. Furthermore, the process is data driven and therefore hypothesis testing about goodness-of-fit is only approximate.

9.2 Modeling

9.2.1 Quantile Regression

Quantile regression is a convenient way to identify specific quantiles and therefore identify individuals who are "abnormal," statistically speaking. It does not assume normality in the raw data. Chitty and Altman (2003) used quantile regression to estimate the 10th, 50th, and 90th percentiles of fetal renal pelvis. In their application, the percentiles are straight lines. Growth data tend to show nonlinear trends. Kulich et al. (2005) used splines to go with quantile regression to develop a lung function percentile chart. Wei et al. (2006) estimated height-for-age percentiles similarly.

Example 9.1 (Continued)

Display 9.3 shows the Stata codes for fitting the 5th, 50th, and 95th percentile curves for the weight of boys under age 5 years, using the cubic splines with three knots created in Display 9.2. It then obtained the predicted values (percentile curves). Two new variables for under- or overweight according to the 5th and 95th percentiles were then generated accordingly. Among the 1968 boys, 95 (4.83%) and 99 (5.03%) were classified by the quantile regression as under- and overweight, respectively (Table 9.2). Figure 9.4 plots the percentile curves together with the data points. Even though the lower and upper ends of the range of the x-axis tended to be more difficult to fit, among the 175 infants below 0.5 year, 9 (5.14%) fell below the 5th percentile and 8 (4.57%) fell above the 95th percentile. The quantile regression with splines still worked fairly well at this end in this sample. There was some oddness near age 5 years that the 95th percentile appeared to drop although the 50th percentile still increased almost linearly. This was the characteristic in this sample. Tabulating the overweight variable for 195 children aged over 4.5 years showed that 8 (4.10%) of them were below the 5th percentile curve and 9 (4.62%) are above the 95th percentile curve. In a 10-by-3 table representing 10 half-year age intervals and 3 weight-for-age groups (<5th, 5th to 95th, and >95th percentiles), a chi-square test gave a P-value much larger than 0.05. There was no evidence of gross misfit.

```
** wei is dependent variable; age* are independent variables
** q5, q50 and q95 are new variables generated
qreg wei ageyr ageyr2 ageyr3 agek1 agek2 agek3,q(0.05)
predict q5
qreg wei ageyr ageyr2 ageyr3 agek1 agek2 agek3,q(0.5)
predict q50
qreg wei ageyr ageyr2 ageyr3 agek1 agek2 agek3,q(0.95)
predict q95
```

DISPLAY 9.3
Stata codes for quantile regression with cubic splines, q = 0.05, 0.50, and 0.95.

TABLE 9.2

Characteristics of Various Growth Reference Models, 1968 Boys Aged under 5 Years in IFLS4

	Quantile Regression	FP-N	FP-EN	LMS Age$^{0.25}$	LMS Original Age
<5th percentile	95 (4.83)%	42 (2.13%)	103 (5.23%)	103 (5.23%)	106 (5.39%)
>95th percentile	99 (5.03)%	128 (6.50%)	106 (5.39%)	103 (5.23%)	104 (5.28%)
Chi-square (P-value)*	9.27 (0.953)	17.55 (0.486)	13.77 (0.744)	16.84 (0.534)	15.62 (0.619)
Mean	—	0.000	0.000	0.002	0.003
SD	—	1.001	1.000	1.001	1.001
Skewness	—	0.776	0.006	−0.005	−0.006
Kurtosis	—	4.074	2.932	3.044	3.121
Deviance	—	7439.9	7254.2	7261.2	7284.2

* 10-by-3 table, representing 10 half-year age groups and below 5th, above 95th, and between 5th and 95th percentiles.

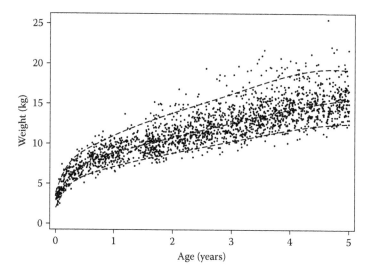

FIGURE 9.4

Weight of boys aged under 5 years in IFLS4, 5th, 50th, and 95th percentile curves estimated by quantile regression.

In quantile regression, each percentile curve is estimated from the data individually, without using information from other percentile curves. This may result in biologically doubtful curves, especially for percentiles near the extremes of the distribution where a small number of observations near the extremes can be influential. Nevertheless, the quantile regression is a very powerful tool for conditioning on covariates or previous growth history.

9.2.2 Parametric Method

Royston and Wright (1998) proposed a parametric method that is based on the *exponential transformation* and fractional polynomials. They also developed a Stata macro "xriml" to implement this method (Wright and Royston 1997; available from http://fmwww.bc.edu/RePEc/bocode/x). The simplest, two-parameter version of the method is suitable if the growth variable Y conditional on the x-axis variable T is normally distributed (without any transformation). Let M(t) and S(t) be the median and SD of Y as a function of T. The standardized variable

$$Z_N = \frac{Y - M(t)}{S(t)} \tag{9.9}$$

follows N(0,1) if the model assumption of normality is correct. The subscript N indicates the normal distribution. The steps to estimate the model are

- Fit a fractional polynomial for the mean value to determine the power terms for the M(t) equation.
- Calculate the *scaled absolute residual* (SAR), which was introduced in Chapter 5 and has an expected value equal to the SD (Altman 1993):

$$SAR = \sqrt{\pi/2}\left|Y - M(t)\right| \tag{9.10}$$

- Fit a FP for the SAR to determine the power terms for the S(t) equation.
- Based on the power terms selected for M(t) and S(t), estimate the regression coefficients for both equations simultaneously by maximum likelihood and calculate the z-scores accordingly.

Example 9.1 (Continued)

Display 9.4 illustrates the Stata codes for applying this method to the weight of the boys aged under 5 years in the IFLS4. We did not expect the weight data to be normal, but that helps to illustrate the weakness of the simple version of the method. In the previous section we have determined that a four-degree FP with repeated powers (1, 1, 2, 2) captured the mean well. Based on deviance, a two degree FP did not fit the SARs better than a one degree FP (P = 0.118). A one degree FP with the power term 0.5 fitted the SARs better than a linear regression (P = 0.026). Using the power terms (1, 1, 2, 2) for M(t) and 0.5 for S(t), the full maximum likelihood model was estimated by using the "xriml" macro, which estimated the regression coefficients for both curves and calculates the z-scores for each observation using Equation (9.9). The "dist(n)" option specified that the normal distribution was used.

Table 9.2, model FP-N, clearly shows that the z-scores are not normal: too few children (2.13%) are below the 5th percentile, too many children

```
*** wei and ageyr are variable names

** fit FP degree 4 for the mean and obtain predicted values
fracpoly regress wei ageyr, degree(4) center(no) noscaling compare
predict pred

** calculate the scaled absolute residuals
gen absres = sqrt(_pi/2)*abs(wei-pred)

** fit FP degree 1 for the scaled absolute residuals
fracpoly regress absres ageyr, degree(1) adjust(no) noscaling compare

** Normal model
xriml wei ageyr,fp(m: 1 1 2 2, s:.5) dist(n) centile(5 50 95)

** EN model with G(t) being a linear function of ageyr
xriml wei ageyr,fp(m: 1 1 2 2, s:.5, g: 1) dist(en) centile(5 50 95)

** MEN model with D(t) being a constant
xriml wei ageyr,fp(m: 1 1 2 2, s:.5, g: 1) dist(men) centile(5 50 95)
```

DISPLAY 9.4
Stata codes for applying Royston's parametric method.

(6.50%) are above the 95th percentile, skewness >0, and kurtosis >3. More in-depth model diagnostics will be discussed in the last section.

A three-parameter version of the method is to further transform Y such that

$$Z_{EN} = \frac{\exp\left[G(t)Z_N\right] - 1}{G(t)} \tag{9.11}$$

where $G(t)$ is a parameter related to skewness as a function of T. Royston and Wright (1998) called this an *exponential normal* (EN) distribution, hence the subscript EN for the z-scores. There is no preselection of the FP term for $G(t)$. Instead, models with G being a constant (intercept), a linear function, or quadratic function of T can be estimated, and the difference in deviance can be compared with the chi-square distribution with one degree of freedom.

Example 9.1 (Continued)

For the IFLS4 data, on top of the aforementioned FPs for $M(t)$ and $S(t)$, a quadratic function, linear function, and intercept term only for $G(t)$ gave model deviance 7251.7, 7254.2, and 7259.6, respectively. Referring the difference in deviance to a chi-square distribution on one degree of freedom, the difference between quadratic and linear function was not significant ($P = 0.114$), but the difference between linear function and constant was ($P = 0.020$). Therefore, the linear function $G(t) = b_0 + b_1 t$ was chosen. As shown in Display 9.4, the "xriml" macro could estimate this by using the option "dist(en)" to specify that the exponential normal distribution was required and the "g: 1" within the "fp" option to request G(t) being linear, that is, power term = 1. The resultant z-scores were practically normally distributed, with mean, SD, skewness, and kurtosis all being close to the expected normal distribution (Table 9.2, model FP-EN). The percentages of children above and below the 5th and 95th percentile curves were also close to their nominal levels.

A four-parameter version of the method is to further allow for kurtosis to change in relation to T

$$Z_{MEN} = \text{sign}(Z_{EN})\frac{1}{D(t)}\left[(1 + |Z_{EN}|)^{D(t)} - 1\right] \tag{9.12}$$

This is called a *modulus-exponential normal* (MEN) distribution. If $D(t) = 1$, Z_{MEN} is standard normal and equals Z_{EN}. In that case, the fourth parameter is redundant and one may prefer the EN model. Again, there is no preselection of the functional form of the D(t) curve. Models with an intercept-only model, a function model, and a quadratic function are fitted and compared.

Example 9.1 (Continued)

In the IFLS example, allowing D(t) to change as a linear or quadratic function did not give significantly smaller deviance than an intercept-only model for D(t). An intercept-only model for D(t) estimated that $D(t) = 1.002$ (95% confidence interval (CI) 0.875 to 1.128), suggesting no significant improvement over the EN model. The model deviance was 7254.2, identical to the aforementioned EN model. There was no advantage in using the MEN model here. As shown in Display 9.4, the MEN model could be requested by using the "dist(men)" option. That D(t) was not specified in the "fp" option meant fitting D(t) as a constant.

9.2.3 LMS Method

The *LMS method* was introduced in Chapter 4, Section 4.1.2.2. To recap, it describes the distribution of a growth variable Y at $T = t$ by its median, M(t); coefficient of variation, S(t); and a measure of skewness, L(t) (Cole 1988; Cole et al. 1995; Cole and Green 1992). While the parametric method in the previous section models the standard deviation, the LMS method models the coefficient of variation, meaning the standard deviation divided by the mean. The transformation aims to create a standardized variable

$$Z = \begin{cases} \dfrac{\left[Y/M(t)\right]^{L(t)} - 1}{L(t)S(t)} & \textit{if } L(t) \neq 0 \\[20pt] \dfrac{\ln\left[Y/M(t)\right]}{S(t)} & \textit{if } L(t) = 0 \end{cases} \tag{9.13}$$

that is normally distributed. Having estimated the values of the three parameters at each t, the 100α percentile is given by

$$P_{100\alpha}(t) = \begin{cases} M(t)\left[1 + L(t)S(t)Z_\alpha\right]^{1/L(t)} & \textit{if } L(t) \neq 0 \\[14pt] M(t)exp\left[S(t)Z_\alpha\right] & \textit{if } L(t) = 0 \end{cases} \tag{9.14}$$

where Z_α is the critical value that gives 100α% probability under the cumulative standard normal distribution, for example, $Z_{0.95} = 1.645$ for the 95th percentile.

The M(t), S(t), and L(t) as functions of T are estimated by a smoothing method based on the *maximum penalized likelihood*, which is equivalent to natural cubic splines with knots at every distinct value of t (Cole and Green 1992). Because of the smoothing procedure, the LMS method is considered a semiparametric method. The degree of smoothing is controlled by the *equivalent degrees of freedom* (EDF) (Hastie and Tibshirani 1987). The LMS method does not produce a regression equation to relate the LMS parameters to T explicitly. Instead, the estimated parameter values are presented for each small time interval. The three functions do not need to have the same EDF. The smaller the EDF, the smoother (less wavy) the curve. A large EDF will allow a curve to fit the data more closely, at the price of risking the curve fluctuating erratically and becoming biologically unbelievable. Usually, the EDF for M(t) is larger than the EDF for S(t), which in turn is larger than the EDF for L(t).

The *generalized Akaike information criterion* is defined as

$$GAIC(\#) = D + \#p \tag{9.15}$$

where D denotes the fitted deviance, p is the total number of EDF, and # is a penalty constant to be decided by the researcher (Rigby and Stasinopoulos 2004). The Akaike information criterion (AIC) is a special case of # = 2, that is, AIC = GAIC(2). The heavier the penalty, #, the smoother the fitted curve. The smaller the GAIC(#), the better the relative fit of the model. The WHO MGRS used the AIC to select the EDF for the M(t) and the GAIC(3) for the S(t) and

L(t) (WHO 2006). This implied that the WHO MGRS wanted the S(t) and L(t) curves to be smoother than the M(t).

The LMS method is implemented in specialized software packages such as LMSChartMaker Light® (Medical Research Council, United Kingdom). Stasinopoulos and Rigby (2007) included this method in an R package called GAMLSS, which stands for *Generalized Additive Models for Location, Scale, and Shape*, and is available from the Comprehensive R Archive Network (www.CRAN.R-project.org). The modeling process may begin with a grid search of λ for age transformation (age$^\lambda$) at some small intervals (see Section 9.1.3), for example, from 0.05 to 1.00 at 0.05 inter-val, based on an initial set of EDFs, for example, 2, 6, and 3 for L(t), M(t) and S(t), respectively. The transformation that gives the lowest deviance is preferred. Then change the EDFs, first for M(t), then S(t) and L(t), using the deviance, AIC, or GAIC(3) as a guide (Pan and Cole 2005; WHO 2006). Graphical examinations and goodness-of-fit tests are used to fine-tune and confirm the choices. In terms of a significance test, if allowing one more EDF to one of the LMS parameters reduces the deviance by about 4, this would be approximately statistically significant at the 0.05 level (Pan and Cole 2005). However, this may not be practically significant. Pan and Cole (2005) suggested that a reduction in deviance of less than 8 is likely to be practically unimportant. If sample size is very large, say >10,000, even a reduction of 20 may be unimportant. In practice, the determination of the EDFs is somewhat subjective (Royston and Wright 1998).

Example 9.1 (Continued)

For the IFLS4 data, $\lambda = 0.25$ and EDFs 2, 7, and 3 for L(t), M(t), and S(t), respectively, gave deviance 7261.2 and

$$\text{GAIC}(3) = 7261.2 - [3 \times (2 + 7 + 3)] = 7225.2$$

This model was selected as the best fit. For demonstration purposes, a model without age transformation, that is, $\lambda = 1$, was also fitted. This required EDFs 2, 11, and 3 to give deviance 7284.2 and GAIC(3) 7336.2. Both indicators suggested that the model with age transformation fitted better. Looking at Table 9.2, the 5th and 95th percentiles from the LMS analyses with and without age transformation classified approximately the right percentages of children in the appropriate category. A descrip-tive summary of the z-scores also showed that they were approximately normally distributed.

The resultant 5th, 50th, and 95th percentile curves of these two models were very close to each other over a large part of the age range. They were also very similar to those of the previously described FP model using the exponential normal (without age transformation). But the three sets of percentile curves were visibly different in the first half year of the age range. Figure 9.5 zooms into this age range. The 5th and 50th percentiles from the LMS method with age transformation (dash lines) were almost

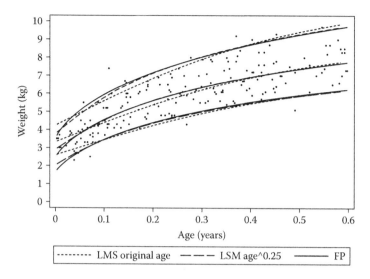

FIGURE 9.5
Weight of boys in IFLS4, zooming in the age range 0 to 0.6 years, 5th, 50th, and 95th percentile curves estimated by LMS and FP-based models.

indistinguishable from those of the FP method except for the first 0.1 year. In contrast, the (dotted) curves based on the LMS method without age transformation clearly deviated from the others, and showed a less steep slope in the first 0.1 year. The poorer fit of the LMS model based on original age was largely due to its relatively poor fit at the lower end of the age range.

9.2.4 Model Diagnostics

Goodness-of-fit tests and graphical diagnostics can be applied to the z-scores derived, in addition to the intuitive examinations discussed earlier.

9.2.4.1 Goodness-of-Fit Tests

If z-scores are generated from an appropriate model, their mean, variance, skewness, and kurtosis should be independent of the x-axis variable and approximately equal to 0, 1, 0, and 3, respectively. Royston and Wright (2000) developed five hypothesis tests to test each of the four aspects of goodness-of-fit as well as general departures from normality. They are called *Q tests* and are implemented by the Stata macro "xriqtest" (Wright and Royston 1999; www.homepages.ucl.ac.uk/~ucakjpr/stata).

The Q tests partition the z-scores into G contiguous groups of approximately equal sizes $n_1, n_2, \ldots, n_G$ defined by the x-axis variable. The closer the group means are to their expected value of zero, the smaller the test statistics

$$Q_1 = \sum_{g=1}^{G} n_g \bar{z}_g^2 \tag{9.16}$$

where $\bar{z}_g$ is the mean of the z-scores in the gth group. The test statistics follows a chi-square distribution with G − #(mean), where #(mean) is the number of regression coefficients estimated in the regression model for the mean. The number of FP or NP power terms is not counted here. In the FP4 model above, #(mean) = 5 (including the intercept).

The test statistics

$$Q_2 = \sum_{g=1}^{G} \left[v_g^{1/3} - (1 - V_g) \right]^2 / V_g \tag{9.17}$$

where $V_g = 2/\left[9(n_g - 1) \right]$, compares the sample variances v_g against their expected values $(1 - V_g)$. Q_2 approximately follows a chi-square distribution with

$$G - \frac{1}{2}\#(\text{SD}) + \frac{1}{2}$$

degrees of freedom, where #(SD) is the number of parameters estimated in the modeling of the standard deviation.

The P-values in the test of zero skewness test, p(s), and test of normal kurtosis, p(k), form the basis of the Q_3 and Q_4 tests:

$$Q_3 = \sum_{g=1}^{G} \left\{ \Phi^{-1} \left[\frac{1}{2} p(s)_g \right] \right\}^2 \tag{9.18}$$

$$Q_4 = \sum_{g=1}^{G} \left\{ \Phi^{-1} \left[\frac{1}{2} p(k)_g \right] \right\}^2 \tag{9.19}$$

where Φ^{-1} is the inverse cumulative standard normal distribution function and the two statistics follow chi-square distributions with G − #(skewness) and G degrees of freedom, respectively. The overall test of departures from normality is based on the P-values from the Shapiro-Wilk W test of nonnormality:

$$Q_5 = \sum_{g=1}^{G} -2\ln(p_g) \tag{9.20}$$

The test statistics follows a chi-square distribution with 2G − #(skewness) degrees of freedom.

Example 9.1 (Continued)

The Q tests were applied to the normal and exponential normal models fitted to the IFLS4 data. Display 9.5 shows the codes and results. The z-scores from the two models were named nZ and enZ, respectively. The normal model used 4 df for the fractional polynomial terms plus an intercept for the mean, and 1 df for the slope plus an intercept for the SD. So #(mean) = 5 and #(SD) = 2. Using 10 groups, Q3 to Q5 showed statistically significant departures from zero skewness, normal kurtosis, and overall normality (each P < 0.001).

The exponential normal model additionally added 2 df for the modeling of the slope and intercept of the skewness equation. All five Q tests led to acceptance of the null hypothesis of no departures from normality. The EN model looked appropriate.

```
* Test FP-Normal model

. xriqtest nZ ageyr,par(m: 5,s: 2) groups(10)

Q-tests on Z-scores in nZ with 10 groups by ageyr (N = 1968):
```

Null hypothesis	Test	Chi-sq	DF	P-value
Means = 0	Q_1	.83336	5	0.975
Variances = 1	Q_2	10.388	9	0.320
Skewness coeffs = 0	Q_3	171.53	10	0.000
Kurtosis coeffs = 3	Q_4	60.965	10	0.000
Normality	Q_5	194.58	20	0.000
N(0,1) in each group	Q1..4	243.71	34	0.000

```
Warning: Tests validated only for 50< = n< = 1000

* Test FP-Exponential Normal model

. xriqtest enZ ageyr,par(m: 5,s: 2, g: 2) groups(10)

Q-tests on Z-scores in enZ with 10 groups by ageyr (N = 1968):
```

Null hypothesis	Test	Chi-sq	DF	P-value
Means = 0	Q_1	1.4574	5	0.918
Variances = 1	Q_2	4.9175	9	0.841
Skewness coeffs = 0	Q_3	5.3503	8	0.720
Kurtosis coeffs = 3	Q_4	2.9375	10	0.983
Normality	Q_5	15.063	18	0.658
N(0,1) in each group	Q1..4	14.663	32	0.996

DISPLAY 9.5
Stata codes and outputs for Q tests on the FP-normal and FP-exponential normal models.

9.2.4.2 Detrended Q-Q Plot

Section 5.1.2 discussed the Q-Q plot. A *detrended Q-Q plot* replaces the y-axis values (quantiles of the empirical distribution) by the difference between the y- and x-axis values, that is, empirical quantile minus normal distribution quantile. The detrended Q-Q plot magnifies the deviations from normality. Van Buuren and Fredriks (2001) suggested applying the detrended Q-Q plot to the z-scores by contiguous groups of observations defined according to the x-axis variable in the growth reference, usually age. They also showed that the 95% confidence interval for a given quantile z, whose cumulative standard normal distribution is p, can be obtained by

$$\pm 1.96 \times f(z)^{-1} \sqrt{p(1-p)/n} \qquad (9.21)$$

where $f(\cdot)$ is the standard normal density function and n is the sample size for the age group. Van Buuren and Fredriks (2001) called the plot the *worm plot*, because the data points look like a worm. A flat worm indicates a good fit. The patterns of worm plots have nice interpretations (van Buuren and Fredriks 2001); they are summarized in Table 9.3. A Stata macro to generate the detrended Q-Q plot is given in Appendix C.

Example 9.1 (Continued)

Figure 9.6 shows the worm plots for nine contiguous age groups for the weight z-scores generated by the FP exponential normal model. The data points in all nine age groups formed approximately flat worms, lying around the horizontal reference line zero and mostly between the 95% CI. This suggested a good fit across all age groups.

For illustration purposes, Figure 9.7 showed the worm plot for the z-scores derived from the FP normal model. Other than in the youngest

TABLE 9.3

Interpretation of Patterns in the Worm Plot

Worm	Interpretation
Passes above the origin	Observed mean larger than fitted mean
Passes below the origin	Observed mean smaller than fitted mean
Has a positive slope	Observed variance larger than fitted variance
Has a negative slope	Observed variance smaller than fitted variance
Has a U-shape	Observed distribution more positively skewed than fitted
Has an inverted U-shape	Observed distribution more negatively skewed than fitted
Has an S-shape rotated clockwise (∿)	Observed distribution less pointed (lower kurtosis) than fitted
Has an inverted S-shape rotated clockwise (~)	Observed distribution more pointed (higher kurtosis) than fitted

Source: Based on van Buuren, S., and Fredriks, M., 2001, Worm plot: A simple diagnostic device for modeling growth reference curves, *Statistics in Medicine* 20:1259–1277, Table II.

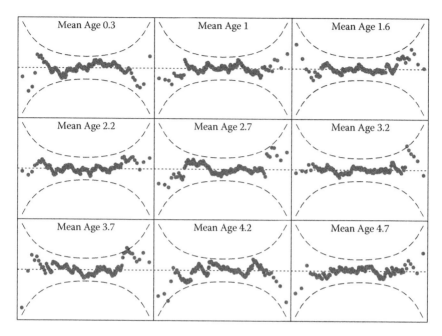

FIGURE 9.6
Worm plot for the weight of boys in IFLS4: FP-exponential normal model.

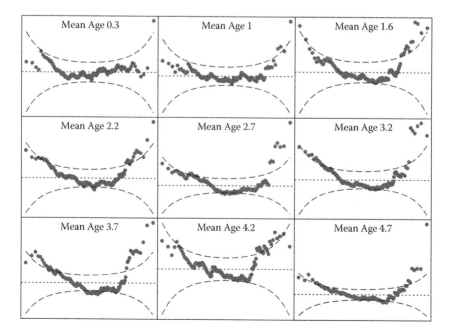

FIGURE 9.7
Worm plot for the weight of boys in ILFS4: FP-normal model.

age group, there was a distinct U-shape pattern in all the age groups. Many data points fell outside the 95% CI, suggesting a poor fit. The U-shape demonstrated positive skewness unaccounted for.

10

Quantifying Development: Development of New Tools

This chapter begins with introducing the use of *item characteristic curves* and *item-response theory* to quantify development based on binary items. Then it will discuss the development of references (norms) and the use of factor analysis for quantitative variables.

10.1 Summary Index Based on Binary Items

10.1.1 Estimation of Ability Age Using Item Characteristic Curves

10.1.1.1 Item Characteristic Curves

The probability (P) of passing a test item can be described by a logistic regression equation

$$\ln\left(\frac{P}{1-P}\right) = \alpha + \beta x \tag{10.1}$$

where x is an independent variable such as age or ln(age). The intercept coefficient α relates to the probability of success when $x = 0$, whereas the slope coefficient β relates to the rate the log odds increases as x increases. The ratio $-\alpha/\beta$ indicates the value of x at which the probability of success is 0.5. If x is age, this ratio gives the median age at passing the test item. If x is ln(age), $\exp(-\alpha/\beta)$ gives the median age. Rearranging Equation (10.1) gives equivalently the *item characteristic curve* (ICC) that describes the probability of success in relation to x

$$P = \frac{\exp(\alpha + \beta x)}{1 + \exp(\alpha + \beta x)} \tag{10.2}$$

In a study of Brazilian children, Drachler et al. (2007) estimated and showed the logistic regression coefficients for over 100 items in the Denver Developmental Screening Test-Revised (DDST-R), using ln(age in months)

TABLE 10.1

Estimates of Intercept (α) and Slope (β) Coefficients in Logistic Regression Models Using ln(age) as an Independent Variable, and Illustration for Calculating the Likelihood for a Child at Age 20 Months

Items	α^*	β^*	$\exp[\alpha + \beta \times \ln(20)]$	$P_{j,20}$
Drinks from cup (y_1)	−17.97	7.03	21.98	$21.98/(1 + 21.98) = 0.956$
Removes garment (y_2)	−18.70	6.66	3.50	$3.50/(1 + 3.50) = 0.778$
Separates from mother easily (y_3)	−14.01	4.30	0.33	$0.33/(1 + 0.33) = 0.248$

* α and β coefficients from Drachler, M. L., Marshall, T., and de Carvalho Leite, J. C., 2007, A continuous-scale measure of child development for population-based epidemiological surveys: A preliminary study using item response theory for the Denver Test, *Paediatric and Perinatal Epidemiology* 21:138–153, Table 2.

as the independent variable. They proposed to use the ICCs for estimating the underlying developmental status. For illustration, Table 10.1 shows the coefficient estimates of 3 example items from the 22 items in the DDST-R Personal-Social domain. The items will be referred to as y_j, $j = 1, 2,$ or 3, in the illustration. Using Equation (10.2) and the coefficients from Table 10.1, the item characteristics curves can be plotted. Figure 10.1 shows the ICCs for "Drinks from cup" and "Removes garment" in the reference population (the Brazilian children). They show that older children had a higher probability of being able to perform these two tasks. We can also calculate $1 - P$ instead of P. That gives the curve for the probability of failure. Figure 10.1 includes the probability of failing the item "Separates from mother easily" in relation to age.

In the reference population, the joint probability, or likelihood, of observing the successes and failures in test items y_j's, $j = 1, 2, ... J$, for a child at age x can be calculated as

$$\text{likelihood} = \prod_{j=1}^{J} P_{j,x}^{\delta_j} (1 - P_{j,x})^{(1-\delta_j)} \tag{10.3}$$

where $P_{j,x}$ is the probability of a child at age x succeeding at item y_j and is calculated using Equation (10.2) and the known coefficients (α and β) from the reference population, $\delta_j = 1$ if the test result in y_j is a success and $\delta_j = 0$ if it is a failure. Suppose a child passed the first two items but failed the third. The likelihood of passing the first two items ($y_1 = y_2 = 1$) but failing the third ($y_3 = 0$) for a child at age 20 months, or ln(age) = 2.996, can be calculated as

$$\text{likelihood} = \prod_{j=1}^{3} P_{j,20}^{\delta_j} (1 - P_{j,20})^{(1-\delta_j)} \tag{10.4}$$

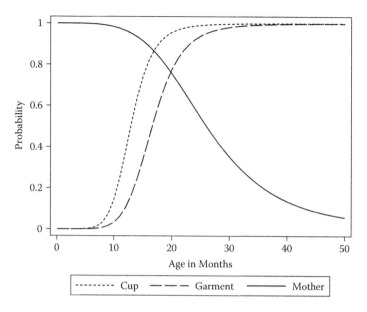

FIGURE 10.1
Item characteristic curves for "Drinks from cup" and "Removes garment" and the characteristic curve for failing the item "Separates from mother easily." (ICC coefficients from Drachler, M. L., Marshall, T., and de Carvalho Leite, J. C., 2007, A continuous-scale measure of child development for population-based epidemiological surveys: A preliminary study using item response theory for the Denver Test, *Paediatric and Perinatal Epidemiology* 21:138–153, Table 2.)

The calculation of $P_{j,20}$, j = 1, 2, and 3, using Equation (10.2) is illustrated in Table 14.1. The likelihood of a child at age 20 months having this pattern of test result is

$$(0.956^1 \times 0.044^0) \times (0.778^1 \times 0.222^0) \times (0.248^0 \times 0.752^1) = 0.559$$

The likelihood of having this response pattern at other ages can be calculated in the same way by brute force. Figure 10.2 shows the likelihood as a function of age for the response pattern ($y_1 = 1$, $y_2 = 1$, $y_3 = 0$). It peaks at 0.5742 at age 21.2488 months. On the ln(age) scale the peak is at 3.0563. This estimate is the "most likely" value, or maximum likelihood estimate (MLE). This is an estimate of the *ability age* of a child who had this response pattern as compared to the reference population.

Display 10.1 shows a computational shortcut to find the MLE with the generalized linear model (Chapter 6), using the earlier example of one child tested on three items. The data is arranged in the long format as one record per item. The estimation turns Equation (10.1) around. Given that we know the coefficients α and β from the reference population, we can solve the logistic regression equation using β as an independent variable, x as the

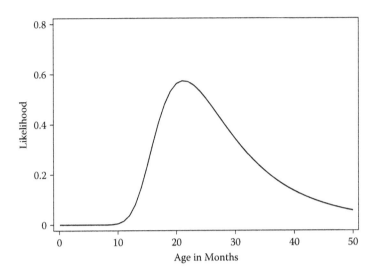

FIGURE 10.2
Likelihood function for the pattern ($y_1 = y_2 = 1$) and ($y_3 = 0$).

```
. ** Example data of one subject with y1=1, y2=1 and y3=0

. list,clean noobs

    child_id        item    y     alpha      beta
           1         cup    1    -17.97      7.03
           1     garment    1     -18.7      6.66
           1      mother    0    -14.01       4.3

.
. ** Obtain ability age by logistic regression
. glm y beta, fam(bin) link(logit) offset(alpha) nocons

(output omitted)
---------------------------------------------------------------------
             |                 OIM
          y  |    Coef.    Std. Err.      z    P>|z|   [95% Conf. Interval]
---------+------------------------------------------------------------
       beta  |   3.0563   .2986637    10.23   0.000    2.47093   3.64167
      alpha  |  (offset)
---------------------------------------------------------------------
```

DISPLAY 10.1
Stata codes and outputs: MLE estimation of ability age using logistic regression.

regression coefficient to be estimated, and α as another independent variable whose regression coefficient is fixed at one. The "offset(*varname*)" option of Stata's "glm" command fixes the regression coefficient for α at 1, whereas the "nocons" (no constant) option prevents further inclusion of an intercept to the equation (which is not needed because the known α is itself the intercept term). The regression coefficient in this estimation is therefore the x in Equation (10.1). That is, ln(age) in the present example. The estimate 3.0563 agrees with the previous brute force estimate. Furthermore, it provides the standard error (SE) for the estimate of the ability age. The larger the number of items, the higher the level of precision is. This simplistic example of three items is only meant for illustration of the concept.

Once the ability age of a child is estimated, we can further calculate an ability-to-chronological age ratio or the natural logarithm of it. Multiplying the ratio by 100 gives an index that is similar to the developmental/intelligence quotient in terms of interpretation.

10.1.1.2 Local Independence and Chained Items

Note that this approach to scoring is optimal if, conditional on the ability age, the responses on the test items are mutually independent. In other words, the only reason the items are correlated is the ability of the examinees. This is known as the *local independence* condition and the converse is *local dependence* (LD). If this condition is not met, the estimation of the ability age would be heavily weighted toward the locally dependent items. Some *chained items* are correlated. An example of chained items is "building a tower of two cubes" and "building a tower of four cubes" in some inventories of developmental milestones. Passing the previous item is a prerequisite for passing the subsequent item. Although the α and β coefficients are estimated for all items, the estimation of the ability age does not have to use all of the chained items for every individual. Drachler et al. (2007) recommended that in a chain of two items if the first (easier) item is a success and the second (more difficult) item is a failure, both items are included in the estimation of ability age of that individual. Otherwise, if the first is a failure, the second is excluded. If the second item is a success, the first is excluded. The same logic extends to chains of over two items. This reduces the influence of locally dependent items. Note that the conventional way of establishing age-related references for developmental status implicitly also gravitates toward the locally dependent items. The ICC approach is not inferior to the conventional approach in the presence of LD; it only makes the condition explicit.

10.1.1.3 Reference for Item Subset

The aforementioned treatment of chained items implies that the scoring of ability age does not require every subject to have complete data on all items administered in the reference study. The conventional references for development described in Chapter 4 tabulate the distribution of scores by age, without giving the details about individual items. In contrast, the ICC approach provides the probability of success at each item in relation to age. As such, even if a new study does not administer all the items used in the reference study, it is still possible to use the outcomes on the subset of items and the ICCs from the reference study to estimate the subjects' ability age. This is a convenient feature because, as previously discussed, researchers may want to exclude culturally inappropriate items from a scale developed and normed in another culture. It is also possible that some studies need to balance between spending resources on a larger sample size versus a larger set of test items, like the Early Child Health in Lahore, Pakistan (ECHLP) (Yaqoob et al. 1993).

10.1.1.4 Weighted Maximum Likelihood Estimation

An important limitation of the maximum likelihood estimation is that if an individual passes or fails all test items, the estimation using Equation (10.3) will fail because there will not be a unique solution that maximizes the likelihood. The *weighted maximum likelihood estimation* (WMLE) method can provide an ability age estimate for all response patterns, including all successes and all failures.

   Intuitively, an item that a 24-month-old has a 50% chance of passing and a 48-month-old has a 99% chance of passing is more informative for assessing infants age about 24 months than about 48 months. This is to say that the degree of informativeness of each item is variable across ability age and is at its maximum when the probability of success is 0.5. The *Fisher's information*, $I(\theta)$, is a statistical measure of the informativeness of the item. A general form of the Fisher's information for the jth binary item is

$$I_j(\theta) = \frac{\left[P_j^*(\theta)\right]^2}{P_j(\theta)\left[1 - P_j(\theta)\right]} \tag{10.5}$$

where $P_j(\theta)$ is the probability of passing the jth item at ability age θ, and $P_j^*(\theta)$ is its first derivative with respect to θ. In the special case of the two-parameter logistic model of Drachler et al.

$$I_j(\theta) = \frac{\beta^2 \exp(\alpha + \beta\theta)}{\left[1 + \exp(\alpha + \beta\theta)\right]^2} \tag{10.6}$$

where α and β are the ICC coefficients for the jth item. Figure 10.3 shows the information function of the three example items. The items are most informative at different ages. Furthermore, the information of a test consisting of J items is the sum of the item information

$$I(\theta) = \sum_{j=1}^{J} I_j(\theta) \tag{10.7}$$

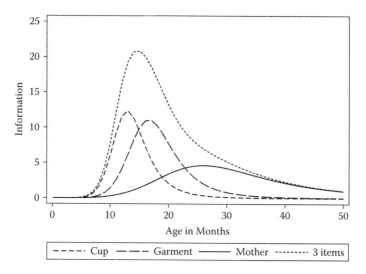

FIGURE 10.3
Information functions of three example items and the three-item test.

The test information is also shown in Figure 10.3. The weighted likelihood uses the information to weight the likelihood and maximizes the quantity

$$\text{weighted likelihood} = \sqrt{I(\theta)} \prod_{j=1}^{J} P_{j,x}^{\delta_j} (1 - P_{j,x})^{(1-\delta_j)} \tag{10.8}$$

Using brute force estimation, we can find that $\theta = 32.6$ months maximizes the weighted likelihood for the hypothetical example of a participant passing all three items in the three-item test.

The SE of the WMLE is given by

$$\text{SE}(\hat{\theta}) = 1 / \sqrt{I(\hat{\theta})} \tag{10.9}$$

where $I(\hat{\theta})$ is the information evaluated at the estimated value of ability age.

A Stata program "wmle" developed by Xu implements the WMLE method (personal communication). It is included in Appendix D (do file) and Appendices D1 and D2 (ado files). Display 10.2 shows a hypothetical example using the three items. Two input data files are required. The first data file provides the α and β parameters of the ICC for each of the items. The second data file contains the outcomes of each subject. Note that one of the subjects (ID 2) passed all three items and another failed all three (ID 3). The MLE approach would fail. The wmle program performs the estimation based on Equations (10.6) to (10.9) and keeps the ability age estimates for each subject

```
. use item_par,clear

. list

      +------------------------+
      |  item    parA    parB  |
      |------------------------|
  1.  |    1   -17.97    7.03  |
  2.  |    2    -18.7    6.66  |
  3.  |    3   -14.01     4.3  |
      +------------------------+

. use item_response,clear

. list

      +------------------------------+
      |  ID   score1   score2  score3  |
      |------------------------------|
  1.  |   1      1       1       0   |
  2.  |   2      1       1       1   |
  3.  |   3      0       0       0   |
      +------------------------------+

. quietly do WMLE
. list

   +-------------------------------------------------------------------+
   | ID  score1  score2  score3  Estimate    SE      LB      UB |
   |-------------------------------------------------------------------|
 1.|  3     0       0       0    2.3860  0.2330  1.9293  2.8426 |
 2.|  1     1       1       0    2.9597  0.2444  2.4807  3.4387 |
 3.|  2     1       1       1    3.4828  0.3990  2.7007  4.2648 |
   +-------------------------------------------------------------------+
```

DISPLAY 10.2
Stata codes and outputs: Weighted maximum likelihood estimation using the programs in
Appendix D.

in the current data file. The WMLE estimate for the subject with all passes
was 3.483, or exp(3.483) = 32.55 months.

10.1.2 Estimation of Latent Trait Using Item-Response Theory

The estimation of ability age in the previous section is based on the observed
age and observed test outcomes in a reference population. A different way
of thinking is that there is an unobserved (or unobservable) variable that
represents ability. This is called a *latent trait*. The *item-response theory* (IRT)
is a method for making an inference about a latent trait by using multiple
observed indicators that measure it.

10.1.2.1 Rasch Model

Georg Rasch (1901 to 1980), a Danish mathematician, developed a probabilistic model for intelligence and achievement tests that are based on binary items (Rasch 1960). The *Rasch model*, which is sometimes called a *one-parameter logistic* (1PL) *model*, describes the probability of subject i passing test item j as

$$P(y_{ij} = 1 | \theta_i, \beta_j) = \frac{\exp(\theta_i - \beta_j)}{1 + \exp(\theta_i - \beta_j)} \tag{10.10}$$

where θ_i is the *ability parameter* of subject i and β_j is the *difficulty parameter* of item j. The ability level of an individual is called a *latent trait*, meaning it is not directly observable. Since each item is characterized by only one parameter, β_j, the item characteristic curves are parallel to each other on the log odds scale. The difficulty parameters determine the vertical distance between the ICCs. Equation (10.10) can be equivalently presented in the log odds form as

$$\ln \left[\frac{P(y_{ij} = 1 | \theta_i, \beta_j)}{1 - P(y_{ij} = 1 | \theta_i, \beta_j)} \right] = \theta_i - \beta_j \tag{10.11}$$

This model puts the ability level and difficulty level on a common metric: the subject's ability increases by x or the item difficulty decreases by x will lead to the same change in the log odds. For example, if $\theta_i = \beta_2$, the log odds of subject i in passing item 2 is zero and therefore the probability is 0.5. Similarly, if θ_i is larger than β_1 but smaller than β_3, the probability of subject i passing item 1 is larger than 0.5 but the probability of passing item 3 is smaller than 0.5. Figure 10.4 illustrates with three hypothetical items that have difficulty parameters −1, 0, and 1, respectively. Suppose Peter has an ability level 0. The probability that Peter passes item 2 ($\beta = 0$) is 0.5, and the probability of him passing item 1 ($\beta = -1$) and item 3 ($\beta = 1$) is

$$\exp(0+1) / \left[1 + \exp(0+1) \right] = 0.73$$

and

$$\exp(0-1) / \left[1 + \exp(0-1) \right] = 0.27$$

respectively. Similarly, the probabilities of John, whose ability level is −2, passing items 1, 2, and 3 are 0.27, 0.12, and 0.05, respectively.

Figure 10.5 shows the same probabilistic patterns but using the log odds scale for the y-axis. It highlights the parallelism and the common metrics for the ability and difficulty parameters. A difference of 2 in ability between

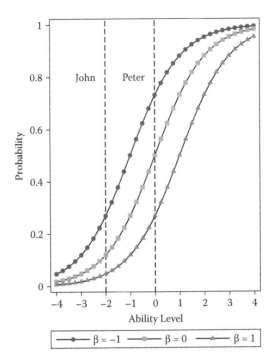

FIGURE 10.4
Probability of passing three hypothetical items with difficulty parameters –1, 0, and 1.

Peter and John translates to a difference of 2 in the log odds of passing each item. For instance, Peter has a log odds of 1 in passing item 1 while John only has a log odds of –1.

The Rasch model assumes local independence. The likelihood of observing a set of J outcomes equals the product of the probabilities of observing each outcome. If the item difficulty parameters are known, the ability level may be estimated by substituting the probability term in Equation (10.11) into the likelihood in Equation (10.3) and applying the MLE method. Figure 10.6 shows the likelihood values of a hypothetical study of three individuals tested on the three items with difficulty parameters –1, 0, and 1 as aforementioned. The first individual succeeded at only the easiest item, that is, the response pattern was Y = (1,0,0). The most likely ability level of this individual is about –0.78. The second and third individuals both passed two items, but the response patterns were different: Y = (1,0,1) and Y = (1,1,0), respectively. Despite the difference in response patterns, the likelihoods for both individuals are maximized at ability level 0.78. This illustrates an important point that the ability estimate in a Rasch model only depends on the total number of successes. In other words, all subjects with the same total score will have the same estimate for the ability levels. Statistically speaking, the total number of successes is the *sufficient statistics* on the ability parameter in

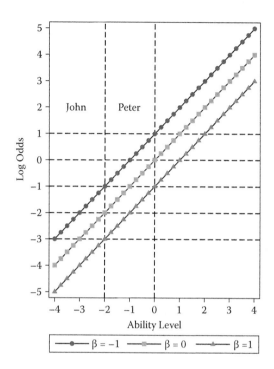

FIGURE 10.5
Log odds of passing three hypothetical items with difficulty parameters –1, 0, and 1.

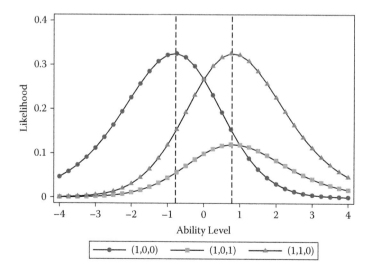

FIGURE 10.6
Likelihood of ability level of three individuals with different response patterns on the three hypothetical items.

a Rasch model. The response pattern (1,0,1) is intuitively an unlikely result because it means the individual fails an easier item ($\beta = 0$) but passes the most difficult item ($\beta = 1$). In contrast, it is more likely that an individual passes two easier items but fails the most difficult one, that is, $Y = (1,1,0)$. This intuition is reflected by the likelihood curve for (1,0,1) in Figure 10.6 being lower than that for (1,1,0).

10.1.2.2 Estimation

We do not directly observe the subjects' ability level or the items' difficulty level. Instead, we observed the outcomes y_{ij}'s and use these data to estimate the parameters. The *conditional maximum likelihood* (CML) method can be used to estimate the item difficulty parameters. This involves using the fact that the total score (total number of successes), denoted by S, is a sufficient statistic. It can be shown that the probability of observing a response pattern y_i given the subject's total score $S = s_i$, $s_i = 0, 1, 2, \ldots, J$, depends on the difficulty parameters but not the ability parameters (e.g., Embretson and Reise 2000), and that the probability can be expressed as

$$P(y_i|\theta_i, \beta, S = s_i) = P(y_i|\beta, S = s_i) = \frac{\exp\left(\sum_{j=1}^{J} y_i \beta_j\right)}{\sum_{z \in \Omega} \exp\left(\sum_{j=1}^{J} z_i \beta_j\right)} \quad (10.12)$$

where $\beta = (\beta_1, \beta_2, \ldots, \beta_J)$ and Ω is the set of response patterns that give the total score s_i. For instance, in the three-item example, if the total score is 2, Ω includes $z = (1,1,0)$, $z = (1,0,1)$, and $z = (0,1,1)$, and the denominator on the right-hand-side of equation (10.12) involves a sum of the three corresponding terms. The conditional likelihood for the whole dataset of n individuals is

$$L_{\text{conditional}}(\beta|Y, S) = \prod_{i=1}^{n} P(Y = y_i|\beta, S = s_i) \quad (10.13)$$

The difficulty parameters are estimated by maximizing the conditional likelihood with respect to the β's.

The maximum likelihood estimation of the ability parameters based on the estimated difficulty levels $\hat{\beta}$ are known to be biased. Furthermore, it fails to work if a subject passes all items ($s_i = J$) or fails all items ($s_i = 0$). The WMLE method gives unbiased estimates and can be used even if the subject passes or failed all items (Hoijtink and Boomsma 1995). They are obtained by maximizing the weighted likelihood function

$$\sqrt{I(\theta)}\left\{\frac{\exp(s_i\theta)}{\prod_{j=1}^{J}\left[1+\exp(\theta-\hat{\beta}_j)\right]}\right\} \qquad (10.14)$$

with respect to θ, where $I(\theta)$ is the information function

$$I(\theta) = \sum_{j=1}^{J}\frac{\exp(\theta-\hat{\beta}_j)}{\left[1+\exp(\theta-\hat{\beta}_j)\right]^2} \qquad (10.15)$$

10.1.2.3 Metrics

The Rasch model differs from the ability age method of Drachler et al. (2007) in an important way. Drachler et al. benchmarked the item characteristic in relation to the observed age of a sample of participants. The ability age is not arbitrary. It indicates the most likely age in the reference population a person would be observed to have the specific test result. In contrast, the Rasch model conceptualizes the ability level as a latent trait. The latent trait and the item difficulty levels are defined only in relation to each other. An ability level of 1 means being expected to have log odds 0 in passing an item with difficulty level 1, log odds 1 in passing an item with difficulty level 0, and so on, as indicated by Equation (10.11).

One way of fixing the metric that is commonly available in Rasch analysis software packages is to center the difficulty parameters at zero. Another is to fix one of the difficult parameters at zero. Although the estimated parameter values look different, they are in essence identical.

Example 10.1

The Simulated Clinical Trial (SCT) dataset include 12 binary items assessed at about age 36 months. Item 10 was the most difficult item, with a 37% success rate in this data. Item 12 was the easiest, with a 79% success rate.

Display 10.3 shows part of the estimation results using the Stata macro "raschtest" (Hardouin 2007). The option "meandif" specifies fixing the mean of the difficulty parameters to 0. The macro gave a footnote in the outputs about this specification. The difficulty level of item 12 was smallest ($\beta = -1.308$), reflecting the highest success rate among the items. Item 10 was the most difficult ($\beta = 0.630$), reflecting the lowest success rate. The outputs included the estimated ability levels. Recall that in the Rasch model, people with the same number of passes have the same ability level estimates. The "genlt(*varname*)" option in the command saved the estimated ability level for each participant as a new variable. The values ranged from -3.405 to 3.344.

```
. raschtest m1-m12,meandif dif(male) genlt(lt)

(output omitted)
```

| | Difficulty | Std. | | | | Standardized | | |
Items	parameters	Err.	R1c	df	P-value	Outfit	Infit	U
m1	-0.25364	0.05111	8.123	10	0.6168	-0.597	-0.746	-0.388
m2	-0.09907	0.05073	7.633	10	0.6646	0.105	0.463	0.001
m3	-0.05449	0.05066	6.932	10	0.7319	0.639	0.342	0.735
m4	-0.03780	0.05064	11.817	10	0.2975	-1.596	-1.578	-1.456
m5	0.14820	0.05063	5.155	10	0.8806	0.437	0.762	0.293
m6	0.28772	0.05087	7.102	10	0.7158	0.868	1.023	0.970
m7	0.36945	0.05110	9.196	10	0.5136	-0.100	-0.305	-0.365
m8	0.46646	0.05146	23.527	10	0.0090	-0.033	-0.558	0.140
m9	0.56223	0.05192	3.774	10	0.9570	-0.095	0.343	-0.026
m10	0.63016	0.05231	11.818	10	0.2975	0.977	1.071	1.497
m11	-0.71089	0.05375	8.404	10	0.5894	-1.682	-1.713	-2.070
m12	-1.30833	0.06091	9.099	10	0.5227	0.408	0.679	0.163

```
R1c test                    R1c= 120.215   110    0.2379
Andersen LR test            Z= 124.766    110    0.1590
```

```
The mean of the difficulty parameters is fixed to 0
You have groups of scores with less than 30 individuals. The tests can
be invalid.
```

| | | Ability | | | Expected | |
Group	Score	parameters	Std. Err.	Freq.	Score	ll
0	0	-3.405	2.102	1	0.45	
1	1	-2.177	0.509	6	1.36	-9.3643
2	2	-1.536	0.271	32	2.28	-116.4692

```
(output omitted)
```

| | | Ability | | | Expected | |
Group	Score	parameters	Std. Err.	Freq.	Score	ll
8	8	0.693	0.132	204	7.86	-1184.1578
9	9	1.074	0.175	117	8.79	-590.6407
10	10	1.528	0.263	47	9.71	-185.2122
11	11	2.147	0.495	15	10.63	-26.5051
12	12	3.344	2.029	1	11.54	

```
-------------------------------------------------
Test of Differential Item Functioning (DIF)
-------------------------------------------------
Variable: male    Number of groups: 2
LR Test Z= 367.441    ddl=   11    P-value=0.0000
```

DISPLAY 10.3
Stata codes and outputs: Rasch model analysis of the SCT dataset.

The readers may verify that, without using the "meandif" option, the Rasch model estimation would fix the difficulty level of the last item (item 12) to 0, but the difference in difficulty levels between the items would remain the same as those estimated using the meandif option. They are the same results presented in different metrics.

Figure 10.7 plots the item difficulty levels together with the distribution of total scores, each of which correspond to a latent trait value. Most of the items had fairly similar levels of difficulty, clustering at about 0 to 1. Item 12 stood out to be relatively easy. In the development of a new assessment, the desire for degree of items spreading out or concentrating depends on the specific aim for which the test is developed (van der Linden 1998). If the aim is to decide on whether a subject's latent trait reaches a particular level, it would be desirable to have items clustering at the target level of difficulty so that the accept-or-reject decision has a high level of accuracy. On the other hand, if the purpose is to describe the population distribution of the latent trait, it would be more appropriate to have items that are spread out. The numeric scores in the figure represent the distribution of the latent trial/number of passes. For instance, 2.28% of the subjects passed only two items and had a latent trait level near the difficulty level of item 12. That means those who passed two items were estimated to have very roughly a 50% chance of passing item 12.

As discussed earlier, the numeric values of the parameters of the Rasch model are arbitrary. It is possible to transform them so that they become more convenient to interpret. Van Buuren (2013) made an analogy to the Celsius

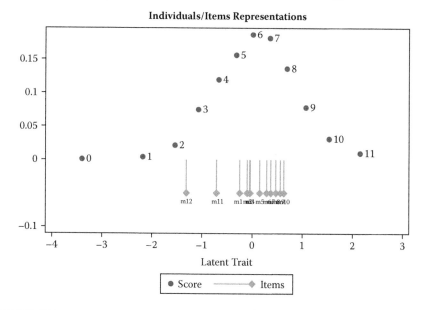

FIGURE 10.7
Plot of item difficulty and distribution of latent trait.

temperature scale. By anchoring 0 degree and 100 degrees to the freezing and boiling points of water at a pressure of one standard atmosphere, respectively, the Celsius scale becomes easy to understand. Similarly, van Buuren (2013) proposed to find two items that are commonly understood to anchor to. Let M be a multiplication factor and L be a location shift factor, the transformed item difficulty parameter β^* and ability parameter θ^* and the transformed log odds are given by the equation system

$$
\left.
\begin{aligned}
\beta_j^* &= L + M \times \beta_j \\
\theta_i^* &= L + M \times \theta_i \\
\ln\left(\frac{p_{ij}}{1-p_{ij}}\right) &= \frac{(\theta_i^* - \beta_j^*)}{M}
\end{aligned}
\right\}
\tag{10.16}
$$

Note that after the transformation of the item and ability parameters, the calculation of the log odds of a subject passing an item also needs to be rescaled by division by M.

Example 10.1 (Continued)

Suppose that items 3 and 6 in Example 10.1 are "sit without support" and "walk independently," two commonly used gross motor milestones psychologists and pediatricians are familiar with. We may scale the parameters such that the latent trait values 30 and 60 represent a 50% chance of passing the sitting and walking milestones, respectively. Here, $\beta_3^* = 30$ and $\beta_6^* = 60$ as per our intended result; $\beta_3 = -0.05449$ and $\beta_6 = 0.28772$ as per Display 10.3. Solving for the simultaneous equations

$$L + M \times (-0.05449) = 30$$

$$L + M \times (0.28772) = 60$$

gives L = 34.777 and M = 87.665. Then the scaled latent trait value is calculated using the second line in Equation (10.16). From Display 10.3, the latent trait value for a subject who passed eight items was 0.693. The scaled latent trait value is 34.777 + (87.665 × 0.693) = 95.529. The log odds of this subject passing, for example, item 6 was (95.529 − 60)/87.665 = 0.405. Back-transforming the log odds to odds and then the odds to risk, this means about a 60% chance of the subject being able to pass item 6.

10.1.2.4 Model Diagnostics

There is a huge number of model diagnostic procedures for the Rasch model. Some of them are reviewed here. The importance of these diagnostic results is somewhat debatable. Similar to the earlier discussion about statistical significance versus clinical significance, the model diagnostics may suggest that the model does not fit the data well but that does not necessarily mean the model is useless. In the context of measuring patient-reported outcomes, it has been argued that the use of IRT does not require strict *unidimensionality*, instead it requires the data to be unidimensional "enough" (PROMIS 2005). How to define enough is a call of judgment.

10.1.2.4.1 Item and Person Fit Indices

The *OUTFIT* and *INFIT* indices are summaries of residuals (Linacre and Wright 1994). Based on the estimated ability level for subject i and difficulty level for item j, an expected probability of success p_{ij} can be calculated. For each cell (i,j) of the data, the residual is $e_{ij} = y_{ij} - p_{ij}$, where y is the observed binary response. The squared standardized residuals for the cell is

$$\frac{e_{ij}^2}{p_{ij}(1-p_{ij})} \tag{10.17}$$

The mean of the squared standardized residuals for item j across n subjects is

$$\text{OUTFIT}_j = \frac{1}{n}\sum_{i=1}^{n}\frac{e_{ij}^2}{p_{ij}(1-p_{ij})} \tag{10.18}$$

which indicates whether the response to the *j*th item is well predicted by the model. Similarly, the mean for subject i across J items gives OUTFIT_i, which indicates whether the responses of the *i*th subject is well predicted by the model.

The INFIT is a weighted average of the squared standardized residuals, using $p_{ij}(1-p_{ij})$ for the weight.

$$\text{INFIT}_j = \frac{1}{\sum_{i=1}^{n} p_{ij}(1-p_{ij})}\sum_{i=1}^{n}\frac{e_{ij}^2}{p_{ij}(1-p_{ij})}\left[p_{ij}(1-p_{ij})\right] = \frac{\sum_{i=1}^{n} e_{ij}^2}{\sum_{i=1}^{n} p_{ij}(1-p_{ij})} \tag{10.19}$$

Similarly, INFIT_i is calculated by taking the weighted average for subject i across items. Note that if $p = 0.5$, the quantity $p(1-p)$ is at its maximum, that is, 0.25. Therefore, the weighting makes the INFIT more sensitive to the residuals from the cells where the predicted probability is close to 0.5. In other words, the cases whose estimated ability levels are close to the estimated item difficulty levels are more influential in the INFIT. These fit indices can be standardized so that their distributions are approximately standard normal. Values larger than 2 or 3 suggest too much unpredictability, while values smaller than –2 or –3 suggest too much predictability (Linacre 2002).

Example 10.1 (Continued)

The raschtest command shows standardized $OUTFIT_j$ and $INFIT_j$ for the items (Display 10.3). Item 11 had standardized OUTFIT and INFIT values somewhat close to but not below –2. Notice that the data was indeed simulated to demonstrate a poor fit for item 11: the slopes of the ICCs for items 1 to 10 were identical but that for item 11 was about twofold (Appendix A). Thus, this item was more predictable than the others. It did not satisfy the parallelism as shown in Figure 10.5. Nevertheless, the deviation from the parallelism was not serious enough to be picked up by the OUTFIT and INFIT. As shown in Display 10.3, in practice the OUTFIT and INFIT usually give roughly the same answer about item fitness.

10.1.2.4.2 Differential Item Functioning

Differential item functioning (DIF) refers to a situation that the true underlying ICC is different in different subpopulations. This is an important issue in studies across cultures. For instance, the milestone "turn doorknob" may be meaningful in some societies but not in others where houses do not usually have doors or doorknobs. DIF can be tested by the likelihood ratio (LR) test. The LR test compares the likelihood of a null model that all item parameters are the same across a test variable versus an alternative model that the item parameters can be different. Another aspect of DIF concerns whether the items behave differently among people at different levels of the latent trait. This is called the *Andersen likelihood ratio test* (Andersen 1973).

Example 10.1 (Continued)

The "dif(*varname*)" option in the raschtest assesses for DIF across the variable specified. The set of items was tested for DIF in relation to gender (*varname* = male). The outputs showed P-value <0.0001. This rejected the null hypothesis of no DIF across gender. Note again that, for demonstration purposes, the simulation program indeed generated item 12 differently for males and females (Appendix A). However, the dif(*varname*) option did not tell which item was showing DIF. The user may rerun the estimation, each time excluding one item, and find out which excluded item contributes to the DIF test result.

Recall that in a Rasch model the number of successes in the raw data is sufficient for determining the latent trait value. The Andersen LR test tested for differences in item parameters across groups defined by the number of successes, as shown in the panel in the middle of Display 10.3 with the first column titled Group. In this case, the null hypothesis of no DIF in relation to the latent trait was not rejected (P = 0.159).

10.1.2.4.3 Local Dependency

To assess whether local dependency (LD) is present, correlation analysis of residuals can be used. Plugging the estimated parameters into Equation

(10.10) provides the expected probability of a person passing an item. The item residual is the observed 0 or 1 value minus the expected probability. Then a Pearson's correlation coefficient can be estimated for each pair of item residuals. In the absence of LD, the pairwise correlation coefficients (r) should be close to zero. Theoretically the expected r is $-2/(n-1)$, where n is the number of subjects, and is slightly smaller than zero for a large sample size (Yen 1984). A large deviation from zero is a sign of LD. Display 10.4 demonstrates the Stata codes for this, by retrieving the saved result "e(beta)" from raschtest. In the SCT dataset, the pairwise correlation coefficients between the items ranged from 0.031 to −0.162, showing little evidence of LD.

10.1.2.4.4 Unidimensionality

A usual rule in measurement is to measure one characteristic at a time. The development of a multi-item test usually groups items into different domains during the test development stage. If a set of items that is supposed to measure one domain does indeed measure one domain, it is said to be *unidimensional*.

Whether unidimensionality is a prerequisite in using the Rasch model is a controversial topic. If the items involve more than one aspect of ability that are highly correlated, or if there is a dominant dimension and some secondary dimensions that are minor in determining the test results, the Rasch model may provide a useful estimate about the overall level of ability despite the violation of unidimensionality (e.g., Embretson and Reise 2000; PROMIS 2005).

The presence of multidimensionality can be assessed using residuals, that is, the square root of the quantity in Equation (10.17) (Linacre 1998; Wright

```
** assess local dependence

* estimate Rasch model and save ability parameter as variable "latent"
raschtest m1-m12,meandif genlt(latent)

* retrieve the saved difficulty parameters from the matrix e(beta)
matrix beta=e(beta)

* create a loop over the 12 items,
* generate item difficulty parameters as variables beta1 to beta12
* generate expected values as exp1 to exp12
* generate residuals as res1 to res12
forvalues i=1(1)12{
  gen beta`i'=beta[1,`i']
  gen exp`i'=exp(latent-beta`i')/(1+exp(latent-beta`i'))
  gen res`i'=m`i'-exp`i'
  }

* correlation coefficients
corr res1 - res12
```

DISPLAY 10.4
Stata codes for assessing local dependence by correlation between residuals in the example of 12 items.

1996). Similar to the examination of LD, if the unidimensionality assumption holds, the pairwise correlation coefficients between residuals should be close to zero. The residuals can be further interrogated using *factor analysis* (see Section 10.2). The failure to find a correlation pattern or factor structure suggests unidimensionality. Simulation studies have shown that for a test of 30 items and a sample size of 500 persons, if the Rasch model assumption holds, the *eigenvalues* (Section 10.2) in a factor analysis are likely to be smaller than 1.5 (Raiche 2005). However, some simulations of other sample sizes and number of items have shown that the largest eigenvalue can center around 2.0 (Raiche 2005). So the indiscriminate use of the eigenvalue 1.5 cutoff may give a high rate of false alarm. Raiche (2005) suggested using simulations based on correct models for the specific sample size and number of items concerned to define the cutoff for eigenvalue in factor analysis.

Example 10.1 (Continued)

The 12 items in the SCT dataset was simulated based on a single dimension. A factor analysis of the standardized residuals from the 12 items and 1500 persons shows that the largest eigenvalue was 1.19. That the eigenvalue was close to one suggested unidimensionality.

10.1.2.5 Two-Parameter Logistic Model

The Rasch model is a one-parameter logistic (1PL) model in the item-response theory family of models because each item is characterized by only a difficulty parameter, β. On the log odds scale, the ICCs are parallel to each other. The equal slopes of the ICCs on the log odds scale imply equal discriminative power of the items. If items are expected to have different slopes, a discrimination parameter, α, can be introduced to form a two-parameter logistic (2PL) model

$$P(Y_{ij} = 1 | \theta_i, \beta_j, \alpha_j) = \frac{\exp\left[\alpha_j(\theta_i - \beta_j)\right]}{1 + \exp\left[\alpha_j(\theta_i - \beta_j)\right]} \tag{10.20}$$

Implicitly, the Rasch model sets all α_j to 1.

Figure 10.8 introduces a discrimination parameter to the three ICCs shown in Figure 10.4. The α's are 1, 2, and 1 for the curves with β's –1, 0, and 1, respectively. So, only the ICC in the middle has changed. A higher value for the discrimination parameter means this item is more strongly associated with the underlying ability, as reflected by a steeper rise.

From the viewpoint of developing a test with multiple items, the data pattern described by the 2PL model can pose a conceptual problem. A test designer may want to know which item is more difficult. Data described by the Rasch model offers unambiguous answers. However, as illustrated in

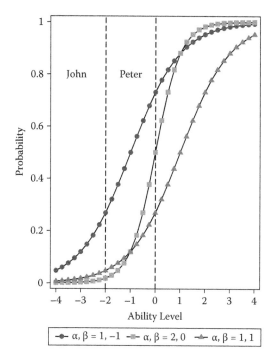

FIGURE 10.8
Log odds of passing three hypothetical items with combinations of discrimination and difficulty parameters (1,–1), (2,0), and (1,1).

Figure 10.8, in a 2PL model, the ICCs may cross over. For Peter, the item with difficulty α and β parameters (1,1) is most difficult. He has a 27% chance of passing this item. But for John, it is the item with parameter (2,0) that is most difficult. His chance of passing this item is about 2%.

Items with very low discrimination may suggest that it is not reflecting the latent trait and represents random noises. On the other hand, items with very high discrimination may sometimes suggest problems in the measurement process. For example, if a set of genuinely equal discriminative items is ordered in a test according to their difficulty level, but by mistake one easy item is put near the end of the test. People with higher ability who complete the test in time tend to pass this item. Those with lower ability tend to fail this item because they may not have time to attempt it at all. This would artificially accentuate the item's discriminative ability (Embretson and Reise 2000). So, sometimes the test designer may want to consider other treatments of the problem instead of using a 2PL model.

In contrast, from the viewpoint of estimating the ability of the individuals, there is no compelling reason to discard a genuinely more discriminative item from the test. The use of the 2PL model can provide better accuracy and precision in this estimation.

10.1.2.6 Three-Parameter Logistic Model

Some items can be attempted by guessing, such as the multiple-choice items in Raven's progressive matrices. If there are four options to each question and respondents answer by pure guessing, even the subjects with extremely low ability levels would have about a 25% chance of passing each item. So the ICC never approaches zero. A third parameter γ, which represents the lower asymptote at a low value of ability, may be added to form a three-parameter logistic (3PL) model

$$P(Y_{ij} = 1 | \theta_i, \beta_j, \alpha_j, \gamma_j) = \gamma_j + (1 - \gamma_j) \frac{\exp[\alpha_j(\theta_i - \beta_j)]}{1 + \exp[\alpha_j(\theta_i - \beta_j)]} \qquad (10.21)$$

If item j is a question with four options to choose from, $\gamma_j = 0.25$. In practice, the estimation of the guessing parameter is complicated by people who do not attempt to guess, so that γ_j may be smaller than 0.25. An important issue in the data collection process is to record clearly what a nonresponse means and therefore whether the nonresponse should be considered a failure or a genuine missing value. For instance, if a child who was not tested on some difficult items near the end of a test because the caregiver has to hurry to leave, the nonresponses should be missing values and excluded from the analysis.

10.1.2.7 Closing Remarks on Item-Response Theory

Embretson and Reise (2000) is an introductory reading suitable for nonstatisticians who are comfortable with equations. Currently general statistical analysis software packages may not cover IRT. However, specialist software packages are available. See Embretson and Reise (2000) for a review. Furthermore, quite a few well-documented user-written macros are available. Some examples include the Stata macro raschtest (Hardouin 2007) and the SAS® (SAS Institute Inc., Cary, North Carolina) %AnaQol macro (Hardouin and Mesbah 2007).

10.2 Summary Index Based on Quantitative Variables

10.2.1 Inferential Norming

When there is a quantitative score, such as the number of seconds a subject needs to perform a task or an intelligence quotient (IQ) test score derived from a multi-item inventory, the test developers and users may want to develop age-related norms, similar to that for height- and weight-for-age. Conventionally, psychologists tend to divide their reference sample into many subsamples according to age. Then for each age interval the mean and SD are calculated. Standardized scores and percentiles can then be calculated from the mean and SD separately for each age interval. This practice needs a huge sample size because each age interval requires a sufficient sample size in order for the estimates of mean and SD to be stable.

More recently, psychologists are beginning to use *inferential norming*, or *continuous norming* (Zhu and Chen 2011). For each age interval, the mean and SD are calculated. But instead of calculating standardized scores and percentiles for each age interval separately, the inferential norming fits a polynomial regression model to the means and to the SDs in relation to age. Using the smooth, predicted mean and SD functions, standardized scores and percentiles are calculated. This is actually the old way of producing growth charts. For example, the former version of the Swedish birth weight-for-gestational age chart was developed by calculating the mean and SD of birth weight for each gestation week and then fitting a third-degree natural polynomial to the means and to the SDs in relation to gestation week (Niklasson et al. 1991). If one is willing to use the inferential norming approach, there is no reason why the methods described in Chapter 9 should not be used instead. Admittedly, developmental tests usually give discrete, not continuous scores. If a development test is based only on a small number of binary items and the score represents the number of passes, the distribution of the scores may be far from continuous. On the other hand, if the test includes a large number of items, the distribution can be approximated by a continuous distribution. The use of inferential norming or the methods in the previous chapter are justified when the distribution of the scores is approximately continuous.

10.2.2 Factor Analysis

10.2.2.1 Model and Estimation

Factor analysis is a method to explain the correlation structure between quantitative measures in terms of a parsimonious number of underlying factors. If a dominant factor is found to explain most of the variations in the measures, it can provide an overall summary as the primary endpoint of a study. Another usage of factor analysis is to identify the structure of a relationship between variables. In this section we focus on the former usage of factor analysis.

Example 10.2

Jukes et al. (2006) examined the impact of early childhood malaria prophylaxis (versus placebo) on cognitive outcomes. They administered six tests to the subjects: three of them measured memory and attention, two measured general intelligence, and one measured verbal ability. A factor analysis was performed to create a cognitive function variable that summarized the six measures. This was the primary endpoint in the study.

The six measures individually were secondary endpoints. By using a single primary endpoint, the study avoided the problem of an inflated type I error rate due to multiple significance testing.

Let Y_j be a standardized variable with mean 0 and SD 1. Factor analysis postulates that each of p standardized quantitative variables is a linear function of m common factors (F_1 to F_m), $m \leq p$, and a specific factor (e_1 to e_p), as described in Equation (10.22)

$$Y_1 = L_{11} \times F_1 + L_{12} \times F_2 + \ldots + L_{1m} \times F_m + e_1$$

$$Y_2 = L_{21} \times F_1 + L_{22} \times F_2 + \ldots + L_{2m} \times F_m + e_2$$

$$\vdots$$

$$Y_p = L_{p1} \times F_1 + L_{p2} \times F_2 + \ldots + L_{pm} \times F_m + e_p$$

(10.22)

where L_{jk}, $j = 1, 2, \ldots, p$ and $k = 1, 2, \ldots, m$ are coefficients known as *factor loadings*. The common factors F's are latent variables. The observed variables Y's are indicators of the latent variables. In matrix notation, Equation (10.22) can be written as

$$\begin{array}{ccccc} \mathbf{Y} & = & \mathbf{L\,F} & + & \mathbf{e} \\ (p \times 1) & & (p \times m)(m \times 1) & & (p \times 1) \end{array}$$

(10.23)

Each factor's ability to explain the observed variables is quantified by a measure called *eigenvalue*. The eigenvalue divided by the number of variables, p, is the proportion of variability in the data explained by the factor. An eigenvalue of one means the factor explains only as much variability as a single observed variable does. Such a factor is not more useful than a single variable and is not summarizing a lot of information from across variables. Conventionally, any factor that does not have eigenvalue ≥1 is ignored. However, as noted in the previous section, this conventional criterion tends to overestimate the number of common factors.

If the variables are results of tests that are measuring the same aspect or strongly correlated aspects of development, there should be a factor F_1 that is strongly related to the Y's, meaning its eigenvalue is clearly larger than one, whereas the other factors' loadings are close to zero, such that Equation (10.22) can be simplified to

$$Y_1 = L_{11} \times F_1 + e_1$$

$$Y_2 = L_{21} \times F_1 + e_2$$

$$\vdots$$

$$Y_p = L_{p1} \times F_1 + e_p$$

Based on the following set of assumptions,

- $Cov(F_k, e_j) = 0$ ($k = 1, 2, \ldots, m; j = 1, \ldots, p$), meaning the common and specific factors are uncorrelated
- $E(F_k) = E(e_j) = 0$, meaning expected values of the common and specific factors are zero
- $Var(F_k) = 1$ and $Cov(F_k, F_t) = 0$ for $k \neq t$, meaning each common factor has variance 1 and the common factors are uncorrelated
- $Var(e_j) = \psi_j$ and $Cov(e_j, e_h) = 0$ for $j \neq h$, meaning the variance of e_j is ψ_j and the specific factors are uncorrelated

Equation (10.22) implies a variance structure that

$$\underbrace{\sigma_j^2}_{\text{variance}} = \underbrace{L_{j1}^2 + L_{j2}^2 + \cdots + L_{jm}^2}_{\text{commonality}} + \underbrace{\psi_j}_{\text{specific variance}} \tag{10.24}$$

and a covariance structure that

$$Cov(Y_j, Y_h) = L_{j1}L_{h1} + \ldots + L_{jm}L_{hm} \tag{10.25}$$

$j \neq h$. The variance–covariance matrix is

$$\underset{(p \times p)}{\Sigma} = \underset{(p \times m)}{L} \underset{(m \times p)}{L'} + \underset{(p \times p)}{\Psi} \tag{10.26}$$

where Ψ is a $p \times p$ diagonal matrix with the jth diagonal element being Ψ_j.

If F and e are jointly normally distributed, a likelihood function based on the multivariate normal distribution can be specified (Johnson and Wichern 2002). The likelihood includes the variance–covariance structure, Σ, as a component and so is dependent on L and Ψ via Equation (10.26). Maximization of the likelihood with respect to L and Ψ is implemented by many statistical packages, giving estimates $\hat{L}$ and $\hat{\Psi}$. If the variables are not normal, transformation of variables may be needed before performing the standardization and factor analysis. There are alternative estimation methods, such as the principle-component factor method (Johnson and Wichern 2002). The different estimation methods usually give similar results. Sensitivity analysis using more than one method to confirm the factor structure is advisable.

10.2.2.2 Factor Scores

Having estimated the factor structure and factor loadings, the next step is to estimate the value of the latent factor values for each subject, denoted by $\hat{F}_i = (F_{i1}, F_{i2}, \ldots, F_{im})'$, where the subscript i indices the subject. This is comparable to the estimation of the ability parameter in the Rasch analysis. Bartlett (1937) proposed a weighted least-squares (WLS) method to estimate the latent factor values. The method assumes that the estimated factor loadings and specific variances are the true values. Note that Equation (10.22) resembles least-squares regression equations with no intercept and with regression coefficients L_{jk}. Treating the estimates of L_{jk} as if they are true values, Equation (10.22) now resembles a least-squares regression equation with $\hat{F}_i$ being the regression coefficients to be estimated and L_{jk} being the independent variables. The Bartlett's method uses the inverse of the specific variances as the weights. Given the estimates $\hat{L}$ and $\hat{\Psi}$ and the observed variables $Y_i = (Y_{i1}, \ldots, Y_{ip})'$, the method finds the estimates for $\hat{F}_i$ that minimizes the sum of the squared errors inversely weighted by their unique variances. The solution is, in matrix notation,

$$\hat{F}_i = (\hat{L}'\hat{\Psi}^{-1}\hat{L})^{-1}\hat{L}'\hat{\Psi}^{-1}Y_i \tag{10.27}$$

where Y_i and $\hat{F}_i$ are $(p \times 1)$ vector of the p (standardized) scores and $(m \times 1)$ vector of the estimates of the m latent factor scores of subject i, respectively. The merit of the Bartlett's method is that the estimates $\hat{F}_i$ are unbiased. The estimates $\hat{F}_i$ in Equation (10.27) is the product of the *scoring coefficients*

$$S = (\hat{L}'\hat{\Psi}^{-1}\hat{L})^{-1}\hat{L}'\hat{\Psi}^{-1}$$

and the data Y_i. The scoring coefficients estimated from a large sample may serve to score data in future studies. The sum of the product

$$\text{Factor score } (i,k) = S_{k1}y_{i1} + S_{k2}y_{i2} + \ldots + S_{kp}y_{ip} \tag{10.28}$$

is the estimate of the *i*th subject's *k*th factor score, $k = 1, 2, \ldots, m$, where S_{kj} is the element in the *k*th row and *j*th column in the scoring coefficient matrix and y_{ij} is the *i*th subject's score on the *j*th standardized variable.

Example 10.2 (Continued)

In the study of 579 Gambian children, Jukes et al. (2006) used the maximum likelihood method to estimate the factor structure. The six variables were examined for normality. One of them (visual search test) was not normally distributed and was therefore transformed to obtain normality. The six variables were standardized before factor analysis. There was only one factor that had an eigenvalue larger than one; the eigenvalue was 2.32. It explained $2.32/6 = 0.39$, or 39%, of the variance. The factor loadings ranged from 0.45 to 0.79, suggesting that the six measures were roughly equally representative of the latent factor. The Bartlett method was used to obtain the factor scores.

Display 10.5 shows the Stata codes for, first, simulating a dataset according to the published findings of Jukes et al. (2006), and then factor analyzing the simulated data.

The maximum likelihood analysis showed a dominant factor with eigenvalue 2.32. No other factor had an eigenvalue larger than 1. Dividing the eigenvalue by 6 (because there were six variables) indicated that 39% of the variance was explained by this factor.

Since the second factor had an eigenvalue smaller than one, we refitted a model that retained only one factor, using the "factor(1)" option. Based on this simplified and final model, a factor score was obtained by the Bartlett method. This could be used as the primary endpoint in an analysis.

```
clear
version 12.1
set seed 123

*** simulate data according to the parameters published by
*** Jukes et al. (2006)

* make sample size very large to minimize the influence of randomness
set obs 100000

* generate a true underlying score
gen cog=rnormal()

* generate the six standardized measures
gen digit_span =.51*cog+sqrt(1-.51^2)*rnormal()
gen verbal_fluency =.45*cog+sqrt(1-.45^2)*rnormal()
gen visual_search =.71*cog+sqrt(1-.71^2)*rnormal()
gen raven_matrices =.60*cog+sqrt(1-.60^2)*rnormal()
gen vocabulary =.79*cog+sqrt(1-.79^2)*rnormal()
gen proverbs =.60*cog+sqrt(1-.60^2)*rnormal()

*** perform factor analysis following Jukes et al.

* MLE estimation of the model
factor digit-proverbs,ml

* MLE estimation restricting to have 1 factor
factor digit-proverbs,ml factor(1)

* obtain latent factor scores for each participant
predict score1,bartlett

* model diagnostic

estat residuals, obs fit
estat factors
```

DISPLAY 10.5
Stata codes to simulate the data of Jukes et al. (2006) and implement factor analysis.

The vector of scoring coefficients was given by the "predict *varname*, bartlett" postestimation command:

$$(0.1572\ 0.1287\ 0.3265\ 0.2121\ 0.4730\ 0.2153)$$

We use subject 1 in the simulated dataset to illustrate the scoring. The standardized variable values for subject 1 were

$$(0.4586\ 1.2085\ 0.6335\ 1.7623\ 1.5105\ 1.4970)$$

So, for the first subject, the factor score calculated by Stata's "predict *varname*, bartlett" command was

$$(0.1572 \times 0.4586) + (0.1287 \times 1.2085) + (0.3265 \times 0.6335) + (0.2121 \times 1.7623)$$
$$+ (0.4730 \times 1.5105) + (0.2153 \times 1.4970) = 1.845$$

as in Equation (10.28). The scoring coefficients could be used to score data in other studies if they do not have a large sample size to develop their own scoring coefficients.

10.2.2.3 Diagnostics

A factor analysis model implies a matrix of correlation between the observed items, Σ, as reflected by Equation (10.26). A comparison of the observed correlation matrix, C, and the fitted correlation matrix, $\hat{\Sigma}$, shows whether the model fitted the data sufficiently. A residual correlation matrix, $C - \hat{\Sigma}$, can also be calculated. Values close to zero suggest the factor model sufficiently captures the correlation between items, or vice versa.

For factor models estimated by maximum likelihood, the Akaike information criterion (AIC) can also be used to examine the number of underlying factors. In particular, if the AIC for a single-factor model is the smallest, it supports the presence of a single dominant factor.

If substantial residual correlations are present, or a model with a larger number of factors has a smaller AIC, it suggests the possibility of some more substantial common factor(s). In that case one may want to reconsider whether it is appropriate to find a single summary index, or whether it is better to partition the measures into different domains and find one summary index per domain. However, strong residual correlations can also arise from correlated measurement errors, such as in the situation of one interviewer performed some measurements but another interviewer performed some other measurements. Without a specific study design for evaluating measurement errors, it is difficult to tell which is which. So we need to evaluate the residual correlation with care and utilize knowledge about the subject matter and study designs.

11

Defining Growth and Development: Longitudinal Measurements

Growth and development may be measured at multiple time points (or ages). Although researchers in different fields have a tendency to use different study designs and statistical methods, many of the techniques and ideas for repeated measurements are generic and applicable to both the studies of growth and development. Therefore, this chapter covers both areas. This chapter uses the word *scores* to represent both growth and developmental variables.

11.1 Expected Change and Unexplained Residuals

11.1.1 More on Change Score

The use of change scores is often considered a way to adjust for baseline status. In parallel group comparisons, if the exposed and unexposed groups have different distributions of the variables of interest at baseline, there is a concern about the validity of the comparison at the end of follow-up. It is hoped that the use of Δy instead of y_t as the outcome variable would make the comparison fair. There are several issues in this line of argument.

First, the scores at baseline are usually negatively correlated with changes. If at baseline the scores are unbalanced so that one group is expected to have higher scores at the end of the follow-up period, the use of change scores does not remove the bias. It only reverses the direction of the bias. The group that has a higher mean score at baseline is expected to have a smaller mean change.

Second, the use of a change score may improve or reduce the precision of the analysis. The variance of the change score is smaller than the variance of the score at time t if

$$r > \frac{\text{SD}(y_0)}{2\text{SD}(y_t)} \tag{11.1}$$

where r is the Pearson's correlation coefficient of the baseline y_0 and outcome assessment y_t (Cameron et al. 2005). If y is a z-score instead of the raw score, the SDs at the two time points should be both approximately equal to 1.

Then, there is a gain in precision only if the correlation coefficient is larger than 0.5. As the age gap between baseline and outcome assessment widens, the correlation weakens and the use of change score can reduce precision.

Example 11.1

A longitudinal study measured the height of 495 South African children at age 2 and 5 years (Cameron et al. 2005). The correlation between height in centimeters at the two ages was 0.72, larger than half of the ratio of SDs at 2 years (SD = 3.6) and 5 years (SD = 4.3):

$$r = 0.72 > \frac{SD(y_2)}{2SD(y_5)} = \frac{3.6}{2 \times 4.3} = 0.4$$

Similarly, the correlation between the z-scores at two ages was also larger than 0.5 and larger than half of the ratio of the SDs. In this case the use of a change score helps to improve the precision of the analysis.

Implicitly, the calculation of a change score assumes that for every one unit increase in baseline score, there will be a one unit increase in the score at the end of follow-up. In other words, it is implicitly assumed that the regression coefficient between the outcome and baseline is one. This is usually not the case. A better way to assess the effect of an exposure on an outcome with adjustment for the baseline score is to fit the regression equation of the form

$$\hat{y}_t = a + b_1 E + b_2 y_0$$

where E stands for the exposure status. This model will provide better fit of the data because the use of the change score implies that the coefficient b_2 is constrained to 1. In contrast, a regression will find the coefficient that fits the data best. The adjustment is also fairer than using the change score as a means to adjust for baseline score, for the reason explained earlier (Frison and Pocock 1992).

11.1.2 Expected Change

Regression to the mean implies that some of the change over time is expected. Or one may consider some of the change the consequence of the baseline score. As shown in Cameron et al. (2005), the expected change in y is

$$E(\Delta y) = y_0 \left[r \frac{SD(y_t)}{SD(y_0)} - 1 \right] + c \tag{11.2}$$

where r is the Pearson's correlation coefficient between y_0 and y_t and c is a constant. If z-scores are used, the expected change simplifies to

$$E(\Delta z) = z_0(r-1) \tag{11.3}$$

Usually, r is between 0 and 1. Equation (11.3) indicates that those who were below average at baseline (negative z-score) are expected to have a positive change, or vice versa. The interpretation of a change over time should take this expectation into account. That there is a positive gain may mean nothing more than a deficit at baseline.

11.1.3 Unexplained Residuals

Note that the regression coefficients b_0 and b_1 in the simple least-squares regression equation

$$y_{it} = b_0 + b_1 y_{i0} + e_i \tag{11.4}$$

where e_i is the residual in the prediction of the *i*th subject's score, can be expressed in terms of the elements of Equation (11.2) as $b_0 = c$ and

$$b_1 = r\frac{SD(y_t)}{SD(y_0)}$$

Hence, a change in the *i*th subject beyond prediction by baseline score is

$$\Delta y_{unexpected,i} = \Delta y_i - E(\Delta y_i) = (y_{it} - y_{i0}) - [y_{i0}(b_1 - 1) + b_0] = e_i \tag{11.5}$$

This is simply the regression residuals in Equation (11.4). In the pediatric literature it is variably called *unexplained residuals*, *conditional gain*, or *conditional growth* (Gandhi et al. 2011; Keijzer-Veen et al. 2005; Martorell et al. 2010). The advantage of using the unexplained residuals as a variable in statistical analysis is that it is uncorrelated with baseline value, y_0. Analysis of the baseline scores and the unexplained residuals are not confounded by each other. Similarly, a change in z-score beyond expectation in the *i*th subject can be defined as

$$\Delta z_{unexpected,i} = \Delta z_i - E(\Delta z_i) = (z_{it} - z_{i0}) - z_{i0}(r-1) = z_{it} - rz_{i0} \tag{11.6}$$

The unexplained residuals can be generalized to the case of three or more time points. Let i indices the subjects and j = 1, 2, ..., J be the *j*th measurement ordered in time sequence. The following least-squares regression equation can be fitted to predict the *i*th subject's value at the *j*th measurement:

$$\hat{x}_{ij} = b_0 + b_1 x_{i1} + ... + b_{(j-1)} x_{i(j-1)} \tag{11.7}$$

where x's are measures at fixed time points, the b's are regression coefficients to be estimated, and $\hat{x}_{ij}$ is the regression predicted value for the jth measurement. The unexplained residuals using all previous point measures are

$$e_{ij} = x_{ij} - \hat{x}_{ij} = x_{ij} - \left[b_0 + b_1 x_{i1} + \ldots + b_{(j-1)} x_{i(j-1)} \right]$$

They are unexplained by any of the previous point measures. Since they are residuals, their mean is zero.

Example 11.2

Martorell et al. (2010) analyzed longitudinal weight data of children measured at birth, 24 months, and 48 months. They derived the unexplained residuals (which they called conditional weight gain) at age 24 months by regressing weight at 24 months upon birth weight, and then subtracting the regression predicted weight from the observed weight at 24 months.

$$e_{i,24mo} = x_{i,24mo} - \left[b_0 + b_{birth} x_{i,birth} \right]$$

Then, the unexplained weight at age 48 months was derived by regressing weight at 48 months upon birth weight and weight at 24 months, and then subtracting the regression predicted weight from the observed weight at 48 months.

$$e_{i,48mo} = x_{i,48mo} - \left[b_0 + b_{birth} x_{i,birth} + b_{24mo} x_{i,24mo} \right]$$

Their multivariable regression analysis of schooling outcomes included birth weight and the two unexplained residuals as independent variables. The three independent variables were uncorrelated with each other. The regression coefficients represented the outcome degrees of association with birth weight, the weight gain up to age 24 months not explained by birth weight, and the weight gain up to age 48 months not explained by birth weight and weight at 24 months.

11.2 Reference Intervals for Longitudinal Monitoring

Most references for growth and development are based on cross-sectional data, to which each subject only contributed one measurement. Longitudinal data presents both challenges and opportunities. As discussed in Chapter 8, the correlation between repeated measurements invalidates some statistical methods. In longitudinal studies, usually there is some variation in the number of measurements per subject. If methods for constructing cross-sectional references are used to develop references based on longitudinal data, the reference intervals will gravitate toward subjects who contributed more measurements. Furthermore, the number of measurements each subject contributed may depend on the variables of interest. An elderly person whose cognitive decline is serious enough to make him drop out early from a longitudinal study will be underrepresented in the longitudinal data. The reference development is at risk of biasing toward an overestimation of the cognitive scores unless proper statistical methods are used.

Example 11.3

Landis et al. (2009) measured fetal size at 4-week intervals by ultrasound. A total of 144 pregnancy and 755 ultrasound scans were obtained. The number of scans per pregnancy varied from two to eight. Application of methods for analysis of cross-sectional data to this study would gravitate toward those pregnancies that were monitored more frequently.

The past offers opportunities to better define the present. Cross-sectionally, a 24-month-old girl whose supine length is 83 cm appears normal (at about 15th percentile in WHO 2006 reference). However, knowing that the girl was 74 cm at 12 months of age (50th percentile) should intuitively lead to some concern about her health in recent times. Also knowing that she was 72 cm at 9 months (about 75th percentile) should raise an even bigger concern. *Conditional references* can take past data into account. The mixed models and quantile regression are useful for constructing confidence references.

11.2.1 Fitting Mixed Models

11.2.1.1 Random Coefficient Model

The random coefficient model in Chapter 8, Equation (8.7) forms the basis of the development of longitudinal references. Here we have

$$y_{ij} = (b_0 + \beta_{0i}) + (b_1 + \beta_{1i})t_{ij} + e_{ij} \tag{11.8}$$

where y_{ij} is the jth score of the ith subject, t_{ij} is the time (or age) at which y_{ij} is measured, and β_{0i} and β_{1i} are random intercept and random slope with variance $\text{var}(\beta_0)$ and $\text{var}(\beta_1)$ and covariance $\text{cov}(\beta_0, \beta_1)$. If the relationship between y_{ij} and t_{ij} is linear and the residuals are normally distributed, the estimated regression coefficients and the variances and covariances of the random effects and residuals from the mixed model in Equation (11.8) are readily usable for the construction of reference intervals. However, nonnormal distribution and nonlinear relation need to be taken into account. Royston (1995) proposed a model in which the (possibly transformed) outcome for each subject is a linear function of the (possibly transformed) time variable. The model can be written as

$$z_{ij} = (b_0 + \beta_{0i}) + (b_1 + \beta_{1i})g(t_{ij}) + e_{ij} \tag{11.9}$$

where

$$z_{ij} = y_{ij}^{(\lambda)} = \begin{cases} y_{ij}^{\lambda} & \text{if } \lambda \neq 0 \\ \ln(y_{ij}) & \text{if } \lambda = 0 \end{cases}$$

$g(t_{ij})$ is a transformation of t_{ij} based on the fractional polynomial, and the variance–covariance matrix for the random effects is unstructured. $\lambda > 1$ (<1) suggests negative (positive) skewness in the dependent variable.

11.2.1.2 Transformation of Y

The transformation aims to reduce nonnormality and unequal variances in the residuals. If the variance of the random slope, $var(\beta_1)$, is close to zero, one may constrain it to zero for simplicity. Then the transformation also serves to improve parallelism of the response curves between subjects in the constrained model. Log transformation and square root transformation are likely options for positively skewed data like weight and skinfold thickness. Nontransformation ($\lambda = 1$) is likely for height and IQ data. Other values for λ can be obtained by experimentation and graphical examination, or by maximum likelihood estimation. Since an apparent skewness in y may be due to skewness in t, the choice of transformation parameter should take t into account. For example, birth weight may appear to be somewhat negatively skewed because of the negative skewness in gestational duration. But within each gestational week, birth weight is fairly normally distributed. So, graphical exploration should be conducted by intervals of t. Maximum likelihood estimation can include t as a third-degree natural polynomial as the explanatory variables (Royston 1995).

Example 11.4

The Simulated Longitudinal Study (SLS) dataset include height in centimeters measured at 2-month intervals between birth and 24 months. Figure 11.1a shows the bivariate distribution of height and age in the girls. Display 11.1 shows the transformation by maximum likelihood using the "boxcox" command of Stata. The Stata output names the λ coefficient as theta though. Without taking age in month into account, the coefficient that maximized the normal probability model likelihood was 2.66 (or round to integer 3 since the 95% confidence interval (CI) included 3). This apparent negative skewness was an artifact. Height gain in the first 2 months of life was rapid. So the values near

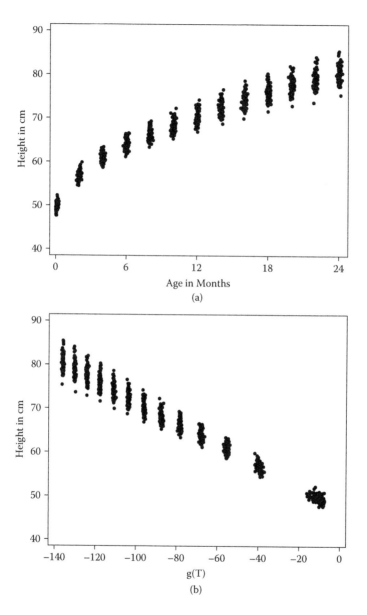

FIGURE 11.1
Height plotted against age (a) and FP function of age (b).

birth were much lower than the values after the age of 2 months. This created an impression of negative skewness. At each age, height was approximately normally distributed. When a third degree natural polynomial of age in months was included, λ was found to be 0.77, and the 95% CI covered 1.0. No transformation was necessary.

```
. ** transformation
. boxcox heightcm

(output omitted)

------------------------------------------------------------------
 heightcm |     Coef.    Std. Err.      z     P>|z|   [95% Conf. Interval]
----------+-------------------------------------------------------
   /theta |   2.66307    .2240446    11.89   0.000    2.22395    3.102189
------------------------------------------------------------------

(output omitted)

. gen visitage2=visitage^2

. gen visitage3=visitage^3

. boxcox heightcm visitage visitage2 visitage3

(output omitted)

------------------------------------------------------------------
 heightcm |     Coef.    Std. Err.      z     P>|z|   [95% Conf. Interval]
----------+-------------------------------------------------------
   /theta | .7683328    .1216525     6.32   0.000    .5298984   1.006767
------------------------------------------------------------------

(output omitted)
```

DISPLAY 11.1
Stata codes and outputs: Transformation of height (cm) in girls in the SLS dataset, without and with adjustment for age (month).

11.2.1.3 *Fractional Polynomial for T*

After the transformation for the dependent variable is decided, the linearizing function g(T) in Equation (11.9) is obtained from fitting the dependent variable to a first- or second-degree fractional polynomial of the independent variable T. For the first-degree fractional polynomial (FP),

$$g(T) = T^{(p_1)} \tag{11.10}$$

For the second-degree FP,

$$g(T) = \begin{cases} T^{(p_1)} + (b_2/b_1)T^{(p_2)} & \text{if } p_1 \neq p_2 \\ T^{(p_1)}\left[1 + (b_2/b_1)\right]\ln(T) & \text{otherwise} \end{cases} \tag{11.11}$$

where (p_1), (p_2), b_1, and b_2 are as described in Section 9.1.1. Recall that if $p_1 = 0$, $T^{(p1)} = \ln(T)$.

Example 11.4 (Continued)

Having decided that the dependent variable needed no transformation, a FP2 model was fitted to the untransformed data, see Display 11.2. Using Equation (11.11) and $p_1 = 0$ and $p_2 = 0.5$, $g(T)$ was given by

$$g(T) = \ln(age) + \left[6.944219/(-0.2442821)\right]\sqrt{age}$$

Note that in Stata the variable $g(T)$ can be generated using the temporary variables and coefficients saved by the "fracpoly" procedure (see the last line of Display 11.2). Figure 11.1b plots the height values against $g(T)$. The two were related in a linear manner. Display 11.3 shows the mixed model in Equation (11.9). The variance of the random effects intercept, $var(\beta_0)$, and slope, $var(\beta_1)$, were 0.7997189 and 0.0001699, respectively. The covariance of the intercept and slope, $cov(\beta_0, \beta_1)$, was 0.0017775, and the variance of the residual, $var(e_{ij})$, was 0.1028244.

The finding of the nonlinear relation $g(T)$ can be distorted by early dropouts or late entry. In this data the number of subjects with incomplete follow-up was small. Should the number be substantial, one may want to limit the FP fitting to subjects with complete or near complete data before applying the $g(T)$ from the FP fitting to the mixed model analysis of all subjects.

```
. ** g(T)
. fracpoly regress heightcm visitage,degree(2) adjust(no) noscaling
compare
........
-> gen double Ivisi__1 = ln(visitage) if e(sample)
-> gen double Ivisi__2 = visitage^.5 if e(sample)
```

Source	SS	df	MS			
Model	90128.5937	2	45064.2969			
Residual	2379.46228	1096	2.17104223			
Total	92508.056	1098	84.2514171			

	Number of obs =	1099
	F(2, 1096) =	20756.99
	Prob > F =	0.0000
	R-squared =	0.9743
	Adj R-squared =	0.9742
	Root MSE =	1.4734

heightcm	Coef.	Std. Err.	t	P>|t|	[95% Conf. Interval]	
Ivisi__1	-.2442821	.0614718	-3.97	0.000	-.3648979	-.1236663
Ivisi__2	6.944219	.0755486	91.92	0.000	6.795983	7.092455
_cons	47.08703	.1477492	318.70	0.000	46.79712	47.37693

Deviance: 3967.78. Best powers of visitage among 44 models fit: 0.5.

(output omitted)

```
. gen gt=Ivisi__1+(_b[Ivisi__2]/_b[Ivisi__1])*Ivisi__2
```

DISPLAY 11.2
Stata codes and outputs: Finding and generating $g(T)$ for girls in the SLS dataset.

```
. ** estimate mixed model
. xtmixed heightcm gt || id: gt,cov(unstructured) var

(output omitted)
```

```
-------------------------------------------------------------------------
heightcm |      Coef.   Std. Err.      z    P>|z|   [95% Conf. Interval]
---------+---------------------------------------------------------------
     gt  |  -.2443604   .0014012  -174.39   0.000   -.2471067   -.2416141
   _cons |   47.07029   .0936272   502.74   0.000    46.88678    47.25379
-------------------------------------------------------------------------
```

```
-------------------------------------------------------------------------
Random-effects Parameters |  Estimate  Std. Err. [95% Conf. Interval]
--------------------------+----------------------------------------------
id: Unstructured          |
              var(gt)     |  .0001699  .0000262   .0001256    .0002298
              var(_cons)  |  .7997189  .1228765   .5917668    1.080747
           cov(gt,_cons)  |  .0017775  .0013067  -.0007836    .0043386
--------------------------+----------------------------------------------
           var(Residual)  |  .1028244  .0048155   .0938065    .1127094
-------------------------------------------------------------------------
```

```
(output omitted)

. * save estimates
. local cov_b0b1 =///
> tanh([atr1_1_1_2]_b[_cons])*exp([lns1_1_1]_b[_cons])* ///
> exp([lns1_1_2]_b[_cons])

. local var_b1   =exp([lns1_1_1]_b[_cons])^2

. local var_b0   =exp([lns1_1_2]_b[_cons])^2

. local var_e    =exp([lnsig_e]_b[_cons])^2


. * unconditional reference
. gen Ezx=_b[_cons]+_b[gt]*gt

. gen Varzx=`var_b0'+(gt^2)*`var_b1'+2*gt*`cov_b0b1'+`var_e'

. gen p95=Ezx+invnorm(0.95)*sqrt(Varzx)

. gen p05=Ezx-invnorm(0.95)*sqrt(Varzx)
```

DISPLAY 11.3
Stata codes and outputs: Mixed model and generation of unconditional reference for girls in the SLS dataset.

11.2.2 Unconditional References

Given the results of the mixed model, the mean and variance of $z = y^{(\lambda)}$ at time t are

$$E(z_{g(t)}) = b_0 + b_1 g(t) \tag{11.12}$$

$$\text{var}(z_{g(t)}) = \text{var}(\beta_0) + g(t)^2 \text{var}(\beta_1) + 2g(t)\text{cov}(\beta_0, \beta_1) + \text{var}(e) \qquad (11.13)$$

As shown in Equation (11.13), the variance of the dependent variable is dependent on the time variable, a common phenomenon in age-related reference intervals.

The z-score is calculated by

$$\text{z-score} = \left[y^{(\lambda)} - E(z_{g(t)}) \right] / \sqrt{\text{var}(z_{g(t)})} \qquad (11.14)$$

The pair of percentiles with $100(1 - \alpha)\%$ coverage at $g(t)$ is

$$
\begin{aligned}
&\left[E(z_{g(t)}) \pm \Phi^{-1}(1 - \alpha/2)\sqrt{\text{var}(z_{g(t)})} \right]^{1/\lambda} \quad \text{if } \lambda \neq 0 \\
&\exp\left[E(z_{g(t)}) \pm \Phi^{-1}(1 - \alpha/2)\sqrt{\text{var}(z_{g(t)})} \right] \quad \text{if } \lambda = 0
\end{aligned}
\qquad (11.15)
$$

where $\Phi^{-1}(\cdot)$ is the inverse cumulative standard normal distribution function. For obtaining, for example, the 5th and 95th percentiles, which jointly define 90% coverage, $\alpha = 0.1$ and $\Phi^{-1}(0.95) = 1.645$.

Example 11.4 (Continued)

To estimate the 5th and 95th percentiles of height at age 12 months using the FP2 models shown in Display 11.2,

$$g(t) = \ln(12) + \left[6.944219 / (-0.2442821) \right]\sqrt{12} = -95.989$$

Plugging the mixed model coefficients (Display 11.3) into using Equations (11.12) and (11.13),

$$E(z_{g(t)}) = 47.0703 - 0.24436 \times (-95.9893) = 70.526$$

$$\text{var}(z_{g(t)}) = 0.7997 + (-95.9893)^2 \times 0.00017 + 2(-95.9893)(0.001778) \\ + 0.1028 = 2.127$$

Note that Display 11.3 saved the variance–covariance estimates from the mixed model to "local macros" named "cov_b0b1", "var_b1", etc., for subsequent generation of the mean and variance variables. The mixed model estimates the log-transformation of the standard deviations so as to make sure that they are larger than zero and the hyperbolic tangent function of the correlation to make sure that the correlation is between –1 to 1. Hence, the local macros had to back-transform them to the variances and covariance.

Using Equation (11.15), the 5th and 95th percentiles for $\lambda = 1$ were $70.526 \pm 1.645 \times \sqrt{2.127} = (68.127, 72.925)$. Had a transformation been used ($\lambda \neq 1$), the above values would need a back-transformation as indicated in Equation (11.15). Conversely, if a 12-month-old girl was 68.127 cm tall, the z-score is $(68.127 - 70.526)/\sqrt{2.127} = -1.645$. The percentiles at other ages were similarly calculated. Figure 11.2 plots the 5th, 50th, and 95th percentiles. The raggedness in the first few months was due to gaps in the data between measurement time points and the percentiles were connected linearly. Had the data collection been for the purpose of developing growth reference, the study should be designed to have more measurements in the first few months of life when the rate of change was changing rapidly.

Example 11.3 (Continued)

Landis et al. (2009) used the parametric modeling approach to develop ultrasound-derived fetal weight reference intervals. In their study, fetal weight in grams was log-transformed ($\lambda = 0$). A second-degree FP model was used to estimate g(T). A mixed model for log(weight) in relation to g(T) was estimated. The results included, for example, the 10th and 90th percentiles for fetal weight at 36 weeks of gestation equaled

$$\exp\left[5.7534 \pm 1.282\sqrt{0.00506}\right] = (288, 345)$$

Since their analysis used log-transformation, the generation of percentiles required a back-transformation step.

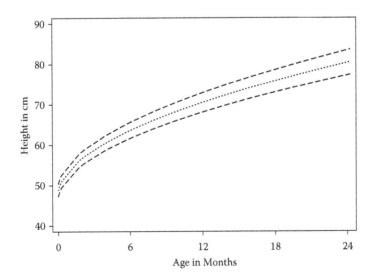

FIGURE 11.2
Unconditional height reference estimated using longitudinal data: 5th, 50th, and 95th percentiles.

11.2.3 Conditional References

The mixed model can provide not only references for scores at different time points but also the rate of change. Consider two time points t_1 and t_2, $t_1 < t_2$, and corresponding transformed times $g(t_1)$ and $g(t_2)$, and transformed scores z_1 and z_2. Building on Equation (11.9),

$$E(\Delta) = b_1 \left[g(t_2) - g(t_1) \right] \tag{11.16}$$

$$var(\Delta) = \left[g(t_2) - g(t_1) \right]^2 var(b_1) + 2var(e) \tag{11.17}$$

Similar to Equations (11.14) and (11.15),

$$z\text{-score} = \left[(z_2 - z_1) - E(\Delta) \right] / \sqrt{var(\Delta)}$$

and the pair of percentiles with $100(1 - \alpha)\%$ coverage is

$$E(\Delta) \pm \Phi^{-1}(1 - \alpha/2)\sqrt{var(\Delta)}$$

If there was no transformation of y, the percentiles for change scores concern differences in the original metrics and is easy to understand. If the log-transformation was used (i.e., $\lambda = 0$), $\Delta = \ln(y_2/y_1)$. Taking exponentiation of the percentile values gives the reference for the proportional change, y_2/y_1. If other transformation is used, unless further assumption about the distribution of the proportional change is made, there is no intuitive interpretation. (See Royston, 1995, for the case of assuming log-normal distribution.)

Conditional reference takes a subject's past record into account. Given z_1, z_2 follows a normal distribution with mean and variance

$$E(z_2|z_1) = E(z_2) + \frac{\left[z_1 - E(z_1) \right] cov(z_2, z_1)}{var(z_1)} \tag{11.18}$$

$$var(z_2|z_1) = var(z_2) - \frac{cov(z_2, z_1)^2}{var(z_1)} \tag{11.19}$$

where $E(z_1)$, $E(z_2)$, and $var(z_1)$ are defined in Equations (11.12) and (11.13), and

$$cov(z_2, z_1) = var(\beta_0) + \left[g(t_1) + g(t_2) \right] cov(\beta_0, \beta_1) + g(t_1)g(t_2)var(\beta_1) \tag{11.20}$$

The conditional distribution of the change score $(z_2 - z_1)$ has mean $E(z_2|z_1) - z_1$ and the same variance $var(z_2|z_1)$ of the conditional score. Then the references for the conditional score and conditional change scores can be established following Equation (11.15).

Example 11.4 (Continued)

Consider height at 12 months conditional on height at 6 months. Based on Display 11.2, $t_1 = 6$ and $t_2 = 12$ correspond to

$$g(t_1) = \ln(6) + \left[6.944219/(-0.2442821)\right]\sqrt{6} = -67.8400$$

and, similarly, $g(t_2) = -95.9893$. Based on the results in Display 11.3, for these two time points

$$E(z_1) = 47.0703 + (-0.24436)(-67.8400) = 63.648$$

$$E(z_2) = 47.0703 + (-0.24436)(-95.9893) = 70.526$$

$$\text{var}(z_1) = 0.7997 + (-67.8400)^2 \times 0.00017 + 2(-67.8400)(0.001778) + 0.1028$$
$$= 1.443$$

$$\text{var}(z_2) = 0.7997 + (-95.9893)^2 \times 0.00017 + 2(-95.9893)(0.001778) + 0.1028$$
$$= 2.127$$

$$\text{cov}(z_2, z_1) = 0.7997 + (-67.8400 - 95.9893)(0.001778)$$
$$+ (-67.8400)(-95.9893)(0.00017)$$
$$= 1.6154$$

Using Equations (11.18) and (11.19), the conditional mean and variance for height at age 12 months for a girl whose height was given 65 cm at age 6 months was

$$E(z_2 \mid z_1 = 65) = 70.526 + (65.000 - 63.648) \times 1.615/1.443 = 72.039$$

$$\text{var}(z_2 \mid z_1) = 2.127 - 1.615^2/1.443 = 0.319$$

Since there was no transformation for the height data ($\lambda = 1$ and $z = y$), the 5th and 95th percentiles were $72.039 \pm (1.645 \times \sqrt{0.319}) = (71.110, 72.968)$. Had a transformation been used ($\lambda \neq 1$), the aforementioned values would need a back-transformation as indicated in Equation (11.15).

It was earlier shown that the unconditional 5th percentile of height at 12 months was 68.127 cm. However, for a girl who was taller than average at age 6 months (65.0 cm versus mean 63.648 cm), the conditional distribution had a higher mean and smaller variance than the unconditional distribution. If her height at 12 months was below 71.110 cm, it would cause a concern despite the unconditional 5th percentile value. By taking individual history into account, the conditional distribution is more powerful in picking up faltering.

Given that the subject was 65.0 cm at 6 months, the conditional distribution of the change score had mean $72.039 - 65.0 = 7.039$, and its variance is the same as the conditional distribution of the scores, that is, 0.319. The 5th and 95th percentile of the conditional change score can be obtained by those of the conditional score minus the preceding score: (71.110 – 65.0, 72.968 – 65.0) or (6.110, 7.968). Equivalently, they could be obtained by (72.039 – 65.0) ± (1.645 × $\sqrt{0.319}$).

11.3 Conditional Scores by Quantile Regression

The mixed model approach is powerful for fully characterizing longitudinal data and producing references accordingly. The quantile regression is a less informative but more convenient approach to estimate percentiles conditional on past data. It is not for the modeling of longitudinally measured dependent variables, but it can include longitudinally measured predictors. The quantile regression equation takes the form

$$\hat{y}_{ij} = a + b_1 y_{i1} + b_2 y_{i2} + \ldots + b_{j-1} y_{i(j-1)} + c_1 g_1(t_{ij}) + \ldots + c_k g_k(t_{ij}), \qquad (11.21)$$

where a, b's, and c's are regression coefficients to be estimated; j is the total number of scores measured up to and including time t_j; g(t)'s are functions of the time at which the measurements are taken. If all subjects are measured for y_{ij} exactly at the same time, g(t) is constant and drops out of the equation. The coefficients b's capture the effect of the previous scores and the coefficients c's capture the effect of the time of the *j*th measurement. In Example 9.1, each subject had body weight measured at one point below the age of 5 years. The quantile regression modeling was based on a cubic spline with 3 knots. Totally, six terms were used on the right-hand-side of the quantile regression equation to capture the nonlinear relation. That is a special case of Equation (11.21), with j = 1 and k = 6. Suppose additionally that birth weight was taken and one would like to find the child weight percentiles conditional on birth weight, then j = 2, y_2 (child weight) was measured at variable time t in the range 0 to 5 years. This is the same as the quantile regression in Example 9.1, but additionally conditioning on birth weight.

$$\text{child weight} = a + b_1 (\text{birth weight}) + c_1 g_1(t) + \ldots + c_k g_k(t)$$

Example 11.4 (Continued)

For a small sample size, it is not advisable to use quantile regression to estimate the percentiles near the margins of the distribution. For a sample size smaller than 100, less than 5 data points are expected below (above) the 5th (95th) percentile. The method is unreliable in such a situation. The convenience of quantile regression comes with its limitations.

```
. keep id visit heightcm

. reshape wide heightcm,i(id) j(visit)

(output omitted)

. qreg heightcm12 heightcm6,q(.5)

(output omitted)
```

```
-------------------------------------------------------------------------
 heightcm12 |    Coef.   Std. Err.     t    P>|t|   [95% Conf. Interval]
------------+------------------------------------------------------------
   heightcm6 |  1.164506   .058296   19.98  0.000    1.048516   1.280497
       _cons | -4.111895  3.722918   -1.10  0.273  -11.51933   3.295544
-------------------------------------------------------------------------
```

DISPLAY 11.4
Stata codes and output: Conditional distribution by quantile regression for girls in the SLS
dataset.

For illustration purposes, we continue with the investigation about the
distribution of height at 12 months conditional on height at 6 months,
that is, j = 2, despite small sample size.

Since the height values were taken at narrow time windows centering
at the target age, there was no need to include g(t) as a predictor. Display
11.4 shows the analysis using Stata's "qreg" command. To begin with,
the data needs to be reshaped to the wide format, such that the 6-month
measure could be used as a predictor of the 12-month measure. The estimated quantile regression equation was

50th percentile at 12 months = −4.112 + (1.165 × Height at 6 months)

As such, for the girl who was 65 cm tall at 6 months, the 50th percentile
at 12 months was −4.112 + (1.165 × 65 cm) = 71.613 cm.

11.4 Trajectory Characteristics

Longitudinal measurements show how things change over time. Particular
aspects of change and the age at the change are characteristics of interest. For
example, Figure 1.4 shows the trajectory of cognitive abilities during adulthood in the Seattle Longitudinal Study. One interest is the peak level of cognitive ability and the age at which the peak is reached and the decline begins.
An intervention would aim at raising the peak level or delaying the age at
onset of cognitive decline. It is more general to describe the characteristics of
trajectories in terms of first derivative (rate of change) and second derivative

TABLE 11.1

Examples of Trajectory Characteristics

Age Range	Characteristics	First Derivative	Second Derivative
Childhood	BMI trough and rebound	0	>0
Puberty	Peak height velocity	Maximum	0
Adulthood	Cognition peak and onset of decline	0	<0

(change in the rate of change). The peak corresponds to the point where the first derivative is zero and second derivative is negative. Conversely, a trough corresponds to the point where the first derivative is zero but second derivative is positive. Table 11.1 gives some examples of trajectory characteristics that are of interest.

Despite a relatively smooth population average trajectory, observed individual values over time can show much more variability. It is therefore important to smooth the individual trajectories before estimating the derivatives. The fractional polynomials and cubic splines can be embedded in a mixed model to perform the smoothing. For polynomials, the derivatives can be analytically obtained easily (Goldstein et al. 1993). Let $f'(x)$ be the first derivative of the function f(x) with respect to x, and $f(x) = a(x) + b(x)$,

$$f'(x) = a'(x) + b'(x) \tag{11.22}$$

Let $f(x) = ax^p$,

$$f'(x) = pax^{(p-1)} \tag{11.23}$$

For fractional polynomials with repeated powers, the *product rule* and the result that the derivative of ln(x) is 1/x are used to find the derivatives (Goldstein et al. 1993). Let $f(x) = ax^p \ln(x)$,

$$f'(x) = pax^{(p-1)} \ln(x) + ax^p / x \tag{11.24}$$

Alternatively, the numerical derivatives can be used to estimate the true derivatives (Gould 1997). The following examples will illustrate the two methods to obtain derivatives.

11.4.1 Fractional Polynomials

Section 8.4 has introduced the mixed models and the best linear unbiased prediction (BLUP). Section 11.2 has introduced a strategy of first fitting an FP model to identify the overall pattern, followed by the mixed model estimation given the FP. These methods can be combined for the present purpose:

- Fit an FP model to the whole sample
- Fit a mixed model with an unstructured covariance matrix between the random coefficients, using the transformed age variables from the FP model
- Obtain the BLUPs for the random coefficients
- Calculate the individual trajectories using the mixed model results and the BLUP
- Obtain the derivatives analytically or numerically
- Calculate the age and level of the trajectory at the point of interest

Example 11.5

Two hundred individuals were simulated to resemble the pattern in Figure 1.4 using the following data structure:

$$\text{Cognition Score}_{ij} = (30 + \beta_{0i}) + (1 + \beta_{1i}) \times \text{age}_i - (0.01 + \beta_{2i}) \times \text{age}_i^2 + e_{ij}$$

where β_0, β_1, β_2, and e were random effects and residuals that followed independent normal distributions with standard deviations 5, 0.05, 0.002, and 1, respectively. Furthermore, there was a weak correlation (0.05) between the three random effects. See Display 11.5 for the codes for simulating this dataset. The underlying trend peaks at 50 years, reaching a score of 55. The data was in the long format, with a measure every 7 years from 25 to 88. The solid line in Figure 11.3 shows the observed trajectory of subject number 1 for example.

The codes for the analysis by FP model were also shown in Display 11.5. Fixed effects FP3 and FP2 models gave similar fit to the data. The difference in deviance was less than 1. In contrast, an FP1 model gave a significantly worse fit (difference in deviance 115.98; $P < 0.0001$). Therefore, the FP2 model, which had power terms 1 and 2, was chosen. (The codes for fitting the FP3 and FP1 are not shown in the display.) For illustration purpose, the "fracgen" command was used to generate the transformed age variables, but we may simply use, for example, "gen age2 = age^2" instead.

A mixed model in the form

$$y_{ij} = b_0 + (b_1 + \beta_{1i})\text{age}_i + (b_2 + \beta_{2i})\text{age}_i^2 + e_{ij}$$

was fitted, with an unstructured covariance matrix to allow the random effects to be correlated. The BLUPs for the random effects were then estimated, using the "predict *prefix**,reffects" command. The new variable "beta3" was renamed as "beta0" so that it agreed with the notation used here.

Based on the above regression equation and Equations (11.22) and (11.23), the first derivative with respect to age is

```
** simulate data similar to the Seattle Longitudinal Study

version 12.1
clear
set more off
set seed 123

set obs 200
gen id=_n

matrix m= (0,0,0)
matrix sd= (5,.05,.002)
matrix c= (1,.05,.05\.05,1,.05\.05,.05,1)
drawnorm B0 B1 B2, means (m) sds (sd) corr (c) seed (123)

expand 10
gen age=.
sort id
qui by id: replace age=25+(_n-1)*7

gen y=min(,max(((30+B0)+(1+B1)*age-(.01+B2)*age^2+rnormal(0,1),0),100)

** FP model

fracpoly regress y age,degree(2) adjust(no) noscaling
fracgen age 1 2,noscaling
xtmixed y age_1 age_2 || id: age_1 age_2,covariance(unstr)
gen b0=_b[_cons]
gen b1=_b[age_1]
gen b2=_b[age_2]
predict beta*,reffects
rename beta3 beta0
gen peakage=-(b1+beta1)/(2*(b2+beta2))
gen peaky=(b0+beta0)+(b1+beta1)*peakage+(b2+beta2)*peakage^2
gen d2=2*(b2+beta2)

sum peak* d2 if age==25
```

DISPLAY 11.5
Stata codes: Estimating age and score at peak by a fractional polynomial.

$$(b_1 + \beta_{1i}) + 2(b_2 + \beta_{2i})\text{age}_i$$

Setting the first derivative to zero, the peak (or trough) occurred at

$$\text{peak age}_i = \left[\frac{-(b_1 + \beta_{1i})}{2(b_2 + \beta_{2i})} \right]$$

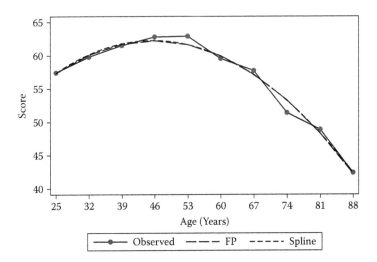

FIGURE 11.3
Observed values and fitted trajectories for subject 1.

By plugging the fixed and random effects estimate into the above equations, the age and score at the peak were obtained. The second derivative at the point of the estimated peak age was

$$2(b_2 + \beta_{2i})$$

We can verify that the turning point found was indeed the peak, not the trough, by checking that the second derivatives at the turning point was negative for all subjects. The mean age and score at the peak were 50.5 and 55.8, respectively. The peak found for subject 1 was score 62.2 at age 45.8 years. The smoothed trajectory was included in Figure 11.3.

11.4.2 Penalized Cubic Spline Model

If the expected changes over time are less regular than described by FP, the cubic splines introduced in Chapter 9 can provide more flexibility for the curve fitting. Here we rewrite Equation (9.6) (see Chapter 9) in slightly different notations that are more convenient to further develop in the present context:

$$y_{ij} = b_0 + b_1 x + b_2 x^2 + b_3 x^3 + \sum_{k=1}^{K} \mu_k I(x > K_k)(x - K_k)^3 + e_{ij} \qquad (11.25)$$

where K is the total number of knots. If K is large, it is desirable to put constraints on the regression coefficients μ_k to improve stability. The *penalized cubic spline* constrains the regression coefficients μ_k to follow a normal distribution, that is, $\mu_k \sim N(0,\sigma_\mu^2)$, where the variance σ_μ^2 is to be estimated. Note that μ_k does not have a subscript i. All subjects share this set of constrained regression coefficients. This can be seen as a mixed model with the whole sample defined as one "super subject" and μ_k as random effects with an identity variance–covariance matrix. Equation (11.25) can be further extended to become a full mixed model for smoothing longitudinal data.

$$y_{ij} = (b_0 + \beta_{0i}) + (b_1 + \beta_{1i})x + (b_2 + \beta_{2i})x^2 + (b_3 + \beta_{3i})x^3$$
$$+ \sum_{k=1}^{K}(\mu_k + v_{ki})I(x > K_k)(x - K_k)^3 + e_{ij}$$

where $(\beta_{0i},\ldots,\beta_{3i}) \sim N(0,\Sigma)$, $\mu_k \sim N(0,\sigma_\mu^2)$, $v_{ki} \sim N(0,\sigma_v^2)$, and Σ is an unstructured covariance matrix. This is called a *penalized spline model* (Durbán et al. 2005; Silverwood et al. 2009). Having fitted a mixed model and obtained the BLUPs, the individual smoothed curve can be fitted by plugging the estimates for the fixed and random effects into Equation (11.26). The derivatives can be obtained by numerical approximation.

Example 11.6

Silverwood et al. (2009) studied the age at peak body mass index (BMI) in infancy. Subjects with fewer than three observations of BMI between birth and age 3 years were excluded. The median number of BMI observations was 13. They used a penalized spline model with 12 knots, placed at the $(1/13)$th, $\ldots$, $(12/13)$th quantiles of the ages. The estimated age and BMI at BMI peak were used as independent variables in regression analysis of later BMI z-scores. It was found that both age and BMI at BMI peak during infancy were positively associated with later BMI z-scores.

Example 11.5 (Continued)

We continue to illustrate with the cognition peak example, using cubic splines and numerical derivatives this time. See Display 11.6 for the Stata codes. The observations were made once every 7 years. To facilitate the later use of numerical derivatives, we expand the age variable to run from 25 to 88 in single years. The precision of results using numerical derivatives depends on the precision of the x-axis values in the data file. If a higher level of precision is needed, we can change the age in years to age in months before the expansion. Nevertheless, only the original cognition scores were kept so that the regression models have the correct data.

For illustration, three knots were used for the splines. In reality, cubic splines probably include more knots. If only a small number of knots is needed, one may consider using the FP model instead. Note the use of a variable "one" to define the whole sample as one "super subject"

```
capture drop peak* d2

* generate cublic splines
expand 7 if age<88,gen(expand)
replace y=. if expand==1
sort id age expand
bys id (age): replace age = 1+age[_n-1] if _n>1
gen age2=age^2
gen age3=age^3
foreach i in 40 60 80{
gen age3k`i'=cond(age>`i',(age-`i')^3,0)
}

* generate a single group
gen one=1

* penalized cubic spline model
xtmixed y age age2 age3 || one: age3k40 age3k60 age3k80, nocons cov(id)///
                        || id: age age2 age3, cov(unstr) ///
                        || id: age3k40 age3k60 age3k80, nocons cov(id)
predict fit,fitted

gen peakage=.
gen peaky=.

qui forvalues i=1(1)200{
dydx fit age,gen(d1), if id==`i'
dydx d1 age,gen(d2), if id==`i'
egen mind1 = min(abs(d1)) if d2<0 & id==`i'
sum age if mind1==abs(d1) & id==`i'
replace peakage=r(mean) if id==`i'
sum fit if mind1==abs(d1) & id==`i'
replace peaky=r(mean) if id==`i'
drop d1 d2 mind1
}

sum peak* if age==25
```

DISPLAY 11.6
Stata codes: Estimating age and score at peak by the penalized spline model.

for the purpose of fitting the μ_k. Also note that the random effects for $(\beta_{0i}, \ldots, \beta_{3i})$ were allowed to have an unstructured (unstr) covariance matrix, whereas the other random effects terms were constrained to have equal variance and be uncorrelated as defined by the "identity" (id) covariance matrix.

Having estimated the model, the fitted values were obtained for each subject at each age value by the "predict" command. The predict command can provide the fitted values for all the x-axis values, including, for example, 26, 27, 28, ..., even though they were not included in the estimation model because the y-axis values were only observed for age 25, 32, and so on. Furthermore, the "fitted" option of the predict command is based on both the fixed and random effects, as opposed to the "xb" option that is based only on the fixed effects. It saves us the trouble of obtaining the BLUPs before plugging them into the mixed model regression equation to obtain the fitted values.

A "forvalues" loop was created to calculate the derivatives and targets of interest for each subject. The numerical derivatives for the fitted values with respect to the x-axis were estimated by the "dydx" command. Then the year with the minimum absolute first derivative where the second derivative was negative was identified. The age and score at that year was saved into the variables "peakage" and "peaky".

The estimated mean age and score at the peak were again 50.5 and 55.8, respectively. Subject 1's age and score at the peak were 46.0 and 62.3, respectively.

11.4.3 Graphical Assessment

Some trajectory characteristics are marked by abrupt changes that are difficult to be clearly and uniquely identified by quantitative methods. Furthermore, there is no guarantee that the FP or penalized cubic spline models will be able to find the characteristics concerned for every subject. Sometimes the researchers have to plot the individual trajectories one by one and visually inspect them and locate the characteristics. In that case, there should be at least two raters who perform the reading independently and the interrater reliability should be examined (see Chapter 12).

Example 11.7

According to the Infancy–Childhood–Puberty (ICP) growth model, the onset of the childhood phase of height gain is marked by an abrupt change in growth velocity. A delay in the onset is a risk factor for childhood growth stunting. In their studies of Pakistani and Swedish children, Liu et al. (1998, 2000) plotted the individual height gains by time intervals to read the age at onset one by one. Two of the authors independently read the graphs. But details of the interrater reliability were not reported.

12

Validity and Reliability

In the context of measurement, *validity* refers to the extent an instrument actually measures what it is supposed to measure. The word *instrument* is used in its broad sense here. It includes not just the physical measurement tool but also how the tool is used. An imaging technology may be supposed to measure the body surface area of a person. If the technology fails in distinguishing the person from the background, it ends up measuring the surface of a mixture of both the person and the background objects. School examinations may be supposed to measure knowledge and competency. But a concern is that such examinations may end up measuring short-term memory. The measurement of human growth tends to be more straightforward and involves less concern about validity than the measurement of development, which is relatively subjective and difficult to define to begin with. *Reliability* refers to the degree of consistency between multiple measurements of the same target. In the presence of systematic or random measurement errors, they disagree. Technically speaking, reliability is the ratio of the variance of the true values to the variance of the observed values. The observed values consist of the true values plus errors.

The term *validation* usually refers to the processes of establishing the validity and reliability of an instrument. There is no absolutely valid and reliable instrument. In practice we are talking about the *degree* of validity and reliability. It is an oversimplification to regard a validated instrument as having the best possible validity and reliability. It is more appropriate to regard a validated instrument as having at least some degree of validity and reliability. It may still need significant improvement. In the planning stage of a study, the researchers need to review and interpret the prior evidence about the validity and reliability of the instruments. Before the data collection starts, usually there is a need to provide training to standardize the application of the instruments. The data from the training activities are examined for validity and reliability. Continuous training and monitoring of data quality may be required. If a new instrument is being developed, it needs to be examined thoroughly for the measurement properties. The final data from a study also need to be examined, to the extent possible, for unexpected deviation from valid and reliable measurement.

There is a lack of standardized terminology. Other terms such as accuracy and reproducibility are sometimes used to refer to validity and reliability. This chapter mainly follows the terminology of Fayers and Machin (2007). Although validity and reliability can be subdivided into many aspects, some

TABLE 12.1

Statistical Methods Widely Used in the Assessment of Validity and Reliability

Measurement Properties	Correlation and Regression	ANOVA and Related Methods			Bland-Altman	Kappa
		Parallel Groups	ICC	TEM		
Validity						
Concurrent	✓		✓		✓	✓
Convergent/divergent	✓					
Known group		✓				
Criterion	✓					✓
Reliability						
Test–retest			✓	✓	✓	✓
Intraobserver			✓	✓	✓	✓
Interobserver			✓	✓	✓	✓

of them can be assessed by the same statistical methods (Table 12.1). Many of these statistical methods are not new to the readers who have attended an introductory statistics course and read the previous chapters. Although statistics plays an important role in the evaluation of instruments, qualitative methods also play a role. They will be described briefly next.

12.1 Concepts

Many concepts are difficult to define and measure. Since there may not be a gold standard, validity is usually examined from multiple angles. Consistency of findings from different angles reinforces the conclusion. Reliability is a matter of variance due to errors. In different contexts there are concerns about different sources of errors. If errors associated with the passage of time are of concern, test–retest reliability may be examined. If errors associated with the use of different observers are of concern, intraobserver reliability is examined. Different estimates of validity and reliability address somewhat different concerns.

12.1.1 Aspects of Validity

12.1.1.1 Face Validity

Face validity refers to whether an instrument looks valid. It answers the question: On the face of it, does the instrument appear to measure the target it is intended to measure? If an instrument was originally developed in a different culture, elements that are culturally irrelevant may be removed at this stage. Simple though it may be, face validity is an important starting point of the development of a new instrument and an important step to understand the properties of an existing instrument. The evaluation typically requires a qualitative review of the instrument by subject-matter experts or the members of the target population.

12.1.1.2 Content Validity

A concept may contain many domains and elements. Broadly speaking, for example, child motor development contains two major domains, namely, gross motor and fine motor development. Within each domain, there is an infinite number of observable elements. It is not feasible to measure all the elements. Content validity is about the adequacy of the instrument in capturing important domains of the target concept and representative samples of the elements within the domains. The evaluation typically requires judgment by subject-matter experts and *focus group discussions*. During the evaluation of content validity, irrelevant or unimportant domains or elements may be removed while new ones may be added.

Example 12.1

The Developmental Milestones Checklist (DMC) was developed for the monitoring of development of children aged 24 months or below in a rural African setting (Abubakar et al. 2010). The final physical tool was a checklist that consisted of 66 items covering three broad domains, namely, motor, language, and personal-social development. Each item was scored on a three-point scale (0 = not observed, 1 = emergent, and 2 = established). The evaluation was by a caregiver report, as opposed to the more commonly used approach of direct observation/assessment by a trained researcher. Clearly the use of a caregiver report has the advantage of simplicity and a potential to perform timely monitoring for more children. But its validity and reliability needed evaluation.

The development of the DMC began with a review of the literature. It included items from the Griffith Mental Development Scale and the Vineland Adaptive Behavior Scale. To build on the previously established items was one way to strengthen content validity. The study then conducted four focus groups of mothers randomly selected from the community, one focus group of teachers, and one focus group of pediatric nurses for the purpose of assessing the face and content validity of the set of identified items.

12.1.1.3 Concurrent Validity

Concurrent validity means agreement with the true value as measured by a "gold standard" (Fayers and Machin 2007). A gold standard is not always available. In practice, sometimes it means agreement with the value indicated by an established instrument that is considered most valid. Sometimes a gold standard exists, but researchers prefer less resource-intensive and more portable instruments. Hence the need to assess the concurrent validity of the alternative instruments.

The assumption that the established instrument is sufficiently valid needs a critical appraisal. The development of a new instrument is sometimes motivated by the limited validity of the existing instrument. The existing instrument might be the best among all currently available instruments. But if the validity is not sufficiently strong, it will not be sensible to use it as a target for examining concurrent validity. In that case, support of validity will need to be sought, either instead or in addition, from the criteria described in the subsequent sections.

Agreement implies correlation, but correlation does not necessarily imply agreement. Say a new instrument persistently overestimates the true value by one unit, while the gold standard accurately measures the true value. The two instruments do not agree, but they have a perfect correlation. If the instruments use the same metrics, concurrent validity can be and should be based on agreement. For quantitative data, the statistical methods include the intraclass correlation coefficient and the Bland-Altman method. For categorical data, the kappa statistic is the major tool. Sometimes the instruments concerned may use different metrics. Association may be examined instead, such as by the Pearson's and Spearman's correlation coefficients.

Example 12.1 (Continued)

The Kilifi Developmental Inventory (KDI) was an established measure of psychomotor functioning administered by trained researchers. It consisted of a locomotor skill domain and an eye–hand coordination domain. The two scores can be added to give an overall psychomotor skill score. The motor score of the DMC via caregiver report was compared against the KDI for concurrent validity. The two scales have different metrics. The Pearson's correlation coefficient was used to assess the concurrent validity. A strong correlation (0.83; $P < 0.01$) between the DMC motor score and KDI overall score was found and supported the validity of the DMC motor score.

12.1.1.4 Convergent and Divergent Validity

Based on subject-matter knowledge, one can theorize some pattern of association or lack of association between variables. Comparing the observed pattern versus the theorized pattern provides information about the validity of the instruments. *Convergent validity* is demonstrated if the variable concerned is correlated with other variables as it is theoretically expected. Conversely, *divergent validity* is demonstrated if the variable concerned has no or weak association with other variables as it is theoretically expected. However, if two variables are supposed to represent two different phenomena but they are very strongly correlated, it casts doubt on whether there are indeed two distinct phenomena. A matrix of correlation coefficients is usually shown as evidence for convergent and divergent validity.

12.1.1.5 Known-Group Validity

Certain groups of people are known to be different in the variables of interest. *Known-group validity* is demonstrated if the instrument can detect the differences between groups. In order to establish known-group validity, one needs strong prior knowledge about differences between groups. For instance, in the studies of child development, there is very strong evidence that stunted children performed more poorly than nonstunted children. As such, it is quite common in this field to use stunting status to define known groups (e.g., Abubakar et al. 2010).

Statistical methods for a comparison of central tendency between groups, such as the two-sample t-test, form part of the assessment of known-group validity. Since even a tiny difference can be statistically significant given a large sample size, it is important to show that an instrument can demonstrate both practically and statistically significant differences between groups. The magnitude of difference between group means and the effect size, defined as the difference between two group means divided by a pooled estimate of within-group standard deviation, should be presented.

12.1.1.6 Criterion Validity

Known-group validity concerns a small number of groups. If people who are different on a quantitative/continuous variable are also known to have a gradient in the variables of interest, we can extend the concept of known-group validity to "known-criterion" validity. For example, stunting is known to be a predictor of child development. Instead of dichotomizing children into stunted and nonstunted groups, height-for-age z-score as a quantitative variable can be used as a criterion to assess the validity of measurements of child development (e.g., Cheung et al. 2008). Age is another criterion commonly used in the validation of developmental assessment instruments. But seemingly nobody has ever used the term known-criteria. This chapter will call it *criterion validity*. Correlation and regression analysis are the means to asses criterion validity.

12.1.2 Aspects of Reliability

12.1.2.1 Test–Retest Reliability

Reproducibility is an important criterion of scientific research. Two measurements taken at different time points by a single instrument on the same subject should give the same test result provided that the subject and the conditions under which measurements were taken have not changed. In reality, most if not all instruments have some degree of measurement error, so the test results will not be exactly the same. The extent that the two measurements agree is referred to as *test–retest reliability*.

If the subject or the conditions have changed between the test and retest, the reliability estimate is biased. It is important to carefully define and select the subjects and conditions for test–retest reliability assessment as well as the duration between the test and retest. For example, some developmental tests induce a learning effect or a short-term memory. For the retest to be valid, the researchers need to conduct the retest after a sufficiently long time has lapsed so that the effect of the first test is gone. On the other hand, the time interval cannot be too long, otherwise the ability of the subject may have changed over time. Although in principle there is no limitation to the number of retests in the assessment of test–retest reliability, usually there is only one retest because having multiple retests over an extended duration of time may suffer bias due to the changes in subjects or conditions.

Test–retest reliability and *intraobserver* and *interobserver reliability* are similar and overlapping concepts. They differ in the focus of what the source of errors is.

12.1.2.2 Intraobserver Reliability

Intraobserver reliability concerns the performance of the observers using a given measurement tool. The observers measure each subject multiple times. Intraobserver reliability is the degree of agreement between the multiple measurements by the same observer on the same subject. Since the aim is to evaluate the performance of the observers, the measurements are usually taken at almost the same time to rule out disagreement due to changes in subjects or conditions. For example, height tends to change a bit in relation to the time of the day due to the compression of the spine. Intraobserver reliability in the measurement of height should be based on measurements taken at the same encounter or the same time of the day.

12.1.2.3 Interobserver Reliability

Interobserver reliability refers to the degree of agreement between measurements taken by multiple observers on the same subject under the same conditions. It is possible that one of the observers is the trainer of observers, or gold standard. The purpose of the evaluation of interobserver reliability is to standardize measurement practice and prevent noises in the data due to the deployment of multiple observers.

12.2 Statistical Methods

12.2.1 Correlation and Regression

12.2.1.1 Pearson's and Spearman's Correlation Coefficients

Convergent, divergent, criterion, and, to some extent, concurrent validity are usually established by correlation analysis (Table 12.1). Spearman's correlation coefficient is an application of Pearson's correlation coefficient to the ranks of data points. This limits the influence of outliers. In the absence of outliers, the two methods tend to give similar results.

Example 12.1 (Continued)

The KDI is a previously established measure of motor development. If all three domains of the DMC were valid, we would expect a strong correlation between the KDI scores and the DMC motor score but only a weak correlation between the KDI scores and the DMC language and personal-social scores. Examination of convergent–divergent validity requires details of the concepts. So it is common to focus on the domain scores instead of the total (summary) score. Abubakar et al. (2011) looked at the correlation pattern for the purpose of examining the validity of the DMC, although they did not make explicit theorization about the pattern. The Pearson's correlation coefficients between the DMC motor score and the KDI locomotor and eye–hand coordination domain scores were 0.84 and 0.73, respectively. The data supported the convergent validity of the motor score. The Pearson's correlation coefficients between the DMC language and personal-social scores and the KDI locomotor and eye–hand coordination domain scores ranged from 0.50 to 0.63. They supported the divergent validity in that the language score and personal-social score had visibly weaker correlation with the KDI domain scores than the motor score did.

This example also illustrates some of the typical ambiguity in validity assessment. First, there is no definite differentiation between concurrent validity and convergent validity. On the face of it, it makes sense to consider the correlation coefficient between the DMC motor score and the KDI locomotor score (0.84) evidence of concurrent validity. But it is not clear whether the correlation coefficient with eye–hand coordination (0.73) should be seen as evidence of concurrent validity or convergent validity. Second, many phenomena share similar underlying factors, such as nutrition and parental education. Even if the language and personal-social domains were phenomena truly distinct from motor development and the DMC provided a valid measurement of them, there would still be some correlation with the KDI scores. As such, divergent validity is often considered in terms of stronger versus weaker correlation, instead of presence versus absence of correlation. Third, there is no clear definition of what a strong or weak correlation is.

How strong a correlation is strong depends on context. Fayers and Machin (2007) suggested that during the initial development phase of an instrument, a correlation of at least 0.3 is a sufficiently convincing level of convergent validity. However, at the final stage of validation, convergent validity should be demonstrated by a correlation of 0.4 or above. A correlation of 0.7 is considered high as evidence of concurrent validity.

12.2.1.2 Regression Analysis

Many instruments are based on a number of items scored as pass or fail. A quantitative test score is then obtained by adding the total number of items passed. In addition to evaluating the quantitative test score, it is also useful to evaluate the criterion validity of each individual item, so that one can identify problematic items for exclusion or modification. Such evaluation can be based on the logistic regression. For example, in the development of child development assessment tools, age usually can be used as a criterion in the evaluation. The probability (P) of a child passing an item is expected to increase with age. Statistical evidence of the association gives support to the item's validity. Goodness-of-fit as described in Chapter 6 should also be examined. This helps to understand whether the pattern of association is as formulated in the regression equation, for example, whether the probability increases monotonically with age. Covariates and interaction can be included in the logistic regression models. To continue the child development example, items usually are expected to be gender neutral. That is, they are developed to measure child development, not boy's development only or girl's development only. Let the gender codes 0 and 1 represent boys and girls, respectively, in the following logistic regression model:

$$\ln\left(\frac{P}{1-P}\right) = a + b_1 \text{Age} + b_2 \text{Gender} + b_3 (\text{Age} \times \text{Gender})$$

Validity would be supported if $b_1 > 0$ and $b_2 = b_3 = 0$. To use the terminology in Chapter 10, $b_3 \neq 0$ indicates differential item functioning. If the item is about whether a child can "turn a door knob" but many of the local households do not use door knobs, the item is not valid in the local population and b_1 would be close to zero. Suppose the item is about whether a child can "comb a doll," which tends to be more likely to occur in girls in many societies. The gender bias would be reflected in b_2 not equal to 0 and the gender-specific validity would be reflected in b_3 not equal to 0. Other covariates can be included in the model to examine criterion validity.

Example 12.2

The Malawi Developmental Assessment Tool is a recently developed instrument based on a number of binary items scored as pass or fail.

Gladstone et al. (2010) reported, among other analyses, the use of logistic regression to assess the individual items in relation to age and gender and other covariates.

12.2.1.3 Correlation As Reliability Coefficient

Reliability is technically speaking a ratio of variances. The numerator is the variance of the true values (Z) and the denominator is the variance of the observed values (Y), which includes measurement errors (e).

Let Y_1 and Y_2 be two independent measurements of Z. $Y_1 = Z + e_1$ and $Y_2 = Z + e_2$, where e_1 and e_2 are independent measurement errors (i.e., $Cov(e_1, e_2) = 0$), with no systematic bias (i.e., $E(e_1) = E(e_2) = 0$) and equal variances (i.e., $SD(e_1) = SD(e_2)$). The Pearson's correlation coefficient between Y_1 and Y_2 is

$$r = \frac{Cov(Z+e_1, Z+e_2)}{SD(Y_1)SD(Y_2)} = \frac{Cov(Z,Z)}{SD(Y)SD(Y)} = \frac{Var(Z)}{Var(Y)} \tag{12.1}$$

Under the aforementioned assumptions, r estimates the reliability, the ratio of the variance of the true values (Z) to the variance of the observed values (Y). Even though correlation analysis is typically used in validity assessment and frowned at as a measure of agreement and reliability, under some assumptions it is one approach to estimate reliability.

12.2.2 ANOVA and Related Methods

12.2.2.1 ANOVA for Parallel Group Comparison

A quick review of the *analysis of variance* (ANOVA) is useful here. Let y_{ij} be the data value of subject j ($j = 1, 2, \ldots, n_i$) of group i ($i = 1, 2, \ldots, k$), n_i be the sample size of group i, n be the total sample size, s_i^2 be the variance within group i, and $\bar{y}$ be the overall mean of y_{ij}. The *between-group mean square* (MS_b) and *within-group mean square* (MS_w) are

$$MS_b = \frac{\sum n_i (\bar{y}_i - \bar{y})^2}{k-1}$$

$$MS_w = \sum \frac{(n_i - 1)s_i^2}{n-k} \tag{12.2}$$

$$F = \frac{MS_b}{MS_w}$$

The F-value is compared to the F-distribution with $(k - 1)$ numerator and $(n - k)$ denominator degrees of freedom. An F-value that is more extreme than the critical value leads to the rejection of the null hypothesis of no difference between groups. When $k = 2$, the F-test and the two-sample t-test are equivalent to each other. Indeed, in this situation the F-value is the square of the t-value.

The ANOVA can be seen as a signal-to-noise ratio, the MS_b being the signal and the MS_w the noise. It is straightforward to see that ANOVA can contribute to the evaluation of known-group validity. In addition, some measure of the magnitude of difference is also required in the analysis of known groups.

Example 12.2 (Continued)

Gladstone et al. (2010) compared children known to have neurodisabilities versus normal children, and also children with malnutrition versus normal children. Table 12.2 shows the means and standard deviations (SDs) from the normal children and children with neurodisabilities. The methods section of Gladstone et al. (2010, p. 7) stated: "We used paired t-tests to compare the numeric scores." It is unclear to what the paired t-test was applied. For the comparison of two groups of subjects, it should need a two-sample t-test, which is equivalent to the ANOVA for $k = 2$. Using the "immediate" command "ttesti" of Stata, we can calculate the P-values of the ANOVA based on the means and SDs. The P-values in the recalculation are the same as those reported in Gladstone et al. (2010). The difference between the two groups was clearly statistically significant (each $P < 0.001$). The children with neurodisabilities, on average, passed $25.4 - 9.2 = 16.2$ gross motor items fewer than the normal children. The effect size is calculated as the difference in means divided by the mean of the two SDs, for example, effect size for gross motor development $(25.4 - 9.2)/[(6.1 + 6.9)/2] = 2.5$.

The effect sizes ranged from 2.2 to 2.8 SD for the domain and total scores. The two groups clearly could be differentiated by the Malawi Developmental Assessment Tool (MDAT) scores, demonstrating

TABLE 12.2

Comparison of MDAT Scores between Normal Children and Children with Neurodisabilities

Domain	Neurodisabilities Mean (SD)		Normal Mean (SD)		P-Value	Difference	Effect Size
Gross motor	9.2	(6.9)	25.4	(6.1)	<0.001	16.2	2.5
Fine motor	7.9	(7.9)	25.2	(6.3)	<0.001	17.3	2.4
Language	7.4	(5.9)	22.7	(8.1)	<0.001	15.3	2.2
Social	10.4	(7.5)	26.4	(7.1)	<0.001	16.0	2.2
Overall	35.0	(20.4)	99.0	(25.9)	<0.001	64.0	2.8

Source: Gladstone, M., Lancaster, G. A., Umar, E., et al., 2010, The Malawi Developmental Assessment Tool (MDAT): The creation, validation, and reliability of a tool to assess the child development in rural African settings, *PLoS Medicine* 7:e1000273, Table 5.

known-group validity. For equal sample size (n = 80 in each group in this example), a simple average of the two SDs forms the denominator in the effect size. For unequal sample size, one may want to use a weighted average to obtain the pooled SD, see Chapter 3, Equation (3.1).

It is quite extreme to compare normal children versus children with neurodisabilities. Gladstone et al. (2010) also included a set of comparisons using a less extreme criterion: malnutrition defined as weight-for-height <80% of expectation. A valid and precise instrument should be able to differentiate groups that are less extreme but nevertheless known to be different. It is desirable to use multiple criteria to examine known-group validity.

12.2.2.2 Intraclass Correlation Coefficient

The agreement between two sets of quantitative values can be examined by statistical and graphical means. Possibly one of the two is from an established instrument and the other from a new instrument. The major statistical method in this context is the *intraclass correlation coefficient* (ICC). There are various estimators of the ICC (Donner 1986). This section discusses the *ANOVA estimator*, which is simply an application of the ANOVA method with the "groups" being the "subjects." The ICC is estimated by partitioning the total variance into two components, namely, between-subject and within-subject variances:

$$ICC = \frac{\sigma^2_{between}}{\sigma^2_{between} + \sigma^2_{within}} \tag{12.3}$$

This is comparable to the right-hand side of Equation (12.1). The variance of the observed values is known. The variances of the true values and errors are unknown and the ratio has to be estimated. The two equations represent different approaches to the estimation. The ICC must be within the range 0 to 1. If most of the variances are due to differences between subjects, the ICC is close to 1. If most of the variances are due to disagreement between measurements within subjects, the ICC is close to 0. A larger ICC indicates stronger concurrent validity or reliability, depending on the context of application. The variance components in Equation (12.3) are estimated by using subjects as a factor in an ANOVA model, using Equation (12.2). Now, k is the number of subjects and n_i is the number of replicates (measurements) per subject. Σn_i is the total number of measurements. In the context of concurrent validity, usually $n_i = r = 2$ for all i and $\Sigma n_i = kr$. That is, each subject is measured on an alternative instrument and a gold standard. In the context of inter- and intraobserver reliability analysis, it is quite common to have more than two assessors or more than two measurements made by the same assessor. So, r may be larger than two. The expected value of the between-group mean square (MS_b) and within-group mean square (MS_w) in Equation (12.2) are (Gleason 1997)

$$E(MS_b) = g\sigma^2_{between} + \sigma^2_{within} \qquad (12.4)$$

and

$$E(MS_w) = \sigma^2_{within} \qquad (12.5)$$

where

$$g = \frac{\sum_i n_i - \sum_i n_i^2 / \sum_i n_i}{k-1}$$

It is easy to verify that g = r if the number of replicates is constant across subjects. By estimating the ANOVA with subjects as a factor, we can obtain the ICC from the mean squares using Equations (12.2) to (12.5). Table 12.3 illustrates the common case of equal number of replicates within subjects.

Example 12.3

Bland and Altman (1986) presented data on peak expiratory flow rate measured with a Wright flow rate meter and a mini-Wright flow rate meter, in random order. Two measurements were made with each meter, but they only analyzed the first measurement. Seventeen subjects were measured. That is, k = 17 and $n_i = r = 2$. The data are shown in Table 12.4.

The mini-Wright meter was a convenient alternative. Concurrent validity needed to be established, using the Wright meter as the gold standard. Display 12.1 shows the procedures and outputs from Stata. First, the wide format data (two measurements per row) were reshaped to the long format, which was required by the ANOVA command. Then the "large one-way ANOVA" command "loneway" was used to produce not only the ANOVA results but also calculate the ICC. The ICC was 0.946. This was considered a very strong agreement.

For illustration purposes, note that the "one-way ANOVA" command "oneway" gave the same mean squares that could be used to calculate

TABLE 12.3

ANOVA Table to Estimate the Intraclass Correlation Coefficient

Component	Sum of Squares	Degrees of Freedom	Mean Squares	Variances
Between subjects	SS_b	k − 1	$MS_b = \dfrac{SS_b}{k-1}$	$\sigma^2_{between} = \dfrac{MS_b - MS_w}{r}$
Within subjects	SS_w	k × r − k − r + 1	$MS_w = \dfrac{SS_w}{k \times r - k - r + 1}$	$\sigma^2_{within} = MS_w$

Notes: k = number of subjects and r = number of replicates (measurements) per subject.

TABLE 12.4

Peak Expiratory Flow Rate (l/min)

ID	Wright (Observer 1)	Mini-Wright (Observer 2)	D^{2*}	Wright + $e^†$
1	494	512	324	514
2	395	430	1225	395
3	516	520	16	557
4	434	428	36	447
5	476	500	576	450
6	557	600	1849	529
7	413	364	2401	417
8	442	380	3844	417
9	650	658	64	681
10	433	445	144	452
11	417	432	225	364
12	656	626	900	733
13	267	260	49	236
14	478	477	1	538
15	178	259	6561	186
16	423	350	5329	392
17	427	451	576	447
		Sum of D^2 =	24120	

Source: Wright and mini-Wright readings from Bland, J. M., and Altman, D. G., 1986, Statistical methods for assessing agreement between two methods of clinical measurement, *Lancet* 1(8476):307–310.

* D = difference between Wright and mini-Wright readings.

† e is a random error drawn from a normal distribution with mean 0 and SD(50).

the ICC. The difference between the two Stata commands is that loneway can fit ANOVA model with over 376 levels (subjects) and provides the ICC estimate after providing the mean squares (Gleason 1997). From the outputs of either command, the variance between subjects was (25572.265 − 709.412)/2 = 12431.427. The variance within subjects is 709.413. ICC = 12431.427/(12431.427 + 709.413) = 0.946.

The preceding example illustrates the use of ICC to address concurrent validity. The same procedures can be used to address intraobserver reliability if the k subjects were assessed by an observer r times each, or to address interobserver reliability if the k subjects were assessed by r observers once each.

Fayers and Machin (2007) suggested that an ICC of 0.70 or above is needed for applications that concern the averages in groups. For applications that concern decision making for individual persons, they quoted Nunnally and Bernstein as saying that an ICC of 0.90 or above is required.

```
. rename wright m1

. rename mini m2

. reshape long m, i(id) j(meter)

(output omitted)

. loneway m id
```

<pre>
                   One-way Analysis of Variance for m:

                                              Number of obs =   34
                                                 R-squared =   0.9714

       Source   SS   df   MS   F   Prob > F

    Between id   409156.24   16   25572.265   36.05   0.0000
     Within id   12060   17   709.41176

    Total   421216.24   33   12764.128

         Intraclass   Asy.
         correlation   S.E.   [95% Conf. Interval]

            0.94601   0.02587   0.89530   0.99673

         Estimated SD of id effect   111.4963
         Estimated SD within id   26.63478
         Est. reliability of a id mean   0.97226
                   (evaluated at n=2.00)
</pre>

```
. oneway m id
```

<pre>
                      Analysis of Variance
       Source   SS   df   MS   F   Prob > F

    Between groups   409156.235   16   25572.2647   36.05   0.0000
     Within groups   12060   17   709.411765

       Total   421216.235   33   12764.1283
</pre>

DISPLAY 12.1
Intraclass correlation coefficient of Wright and mini-Wright meter readings.

12.2.2.3 Technical Error of Measurement

The *technical error of measurement* (TEM) is a commonly used indicator of intraobserver and interobserver reliability in the studies of anthropometric measures (Mueller and Martorell 1988). Despite the specialist name, this quantity is just the square root of the σ^2_{within} in Equation (12.3) and Equation (12.5). In Display 12.1, it is labeled as "Estimated SD within". TEM is nothing but an application of ANOVA to the estimation of the within-subject variance.

Several alternative formulas exist for the estimation of TEM for two or more than two replicates per subject (Mueller and Martorell 1988; Ulijaszek and Kerr 1999). In particular, there is a simple form for the case of two replicates per subject:

$$\text{TEM} = \sqrt{\frac{\sum_i D_i^2}{2k}} \tag{12.6}$$

where D_i is the difference between the two replicates within subject i and k is the number of subjects assessed. If the two replicates are from two observers, the TEM estimates the interobserver reliability. If the two replicates are from the same observer who has measured each subject twice, the TEM estimates the intraobserver reliability. It can be shown that the square of the TEM in Equation (12.6) is equivalent to the MS_w in Equation (12.2).

Example 12.3 (Continued)

For illustration purposes, suppose the two columns in Table 12.4 were the measurements taken on 17 subjects by two observers using the same instrument. The interest was to estimate the interobserver TEM. The fourth column of Table 12.4 shows the calculation of

$$\sum_i D_i^2$$

using Equation (12.6):

$$\text{TEM}_{inter} = \sqrt{24120/(2 \times 17)} = 26.63$$

We had earlier obtained this result by using the ANOVA (Display 12.1).

The TEM is the within-subject standard deviation. Assuming approximate normal distribution, its interpretation is that about two-thirds of the time a measurement should come within ±1 TEM and about 95% of the time a measurement should come within ±2 TEM of the true value.

The *coefficient of reliability* (R) estimates the total variance that is not due to measurement error (WHO MGRS Group 2006a). It is estimated as

$$R = 1 - \frac{\text{TEM}^2}{\text{SD}^2} \tag{12.7}$$

where the SD value is taken from a population reference. For example, the WHO Multicentre Growth Reference Study Group (2006a) reported that the interobserver TEM for birth length (cm) measurement in India was 0.35. The SD of length at week 0 was about 1.88 in the WHO Child Growth Standards (boys 1.893; girls 1.863). The R was $[1 - (0.35^2/1.88^2)] = 0.97$. Ulijaszek and Kerr (1999) recommended anthropometric studies to target a reliability level of 0.95 where possible. Since SD varies across age, the TEM required to achieve $R \geq 0.95$ also varies.

12.2.3 Bland-Altman Method

The intraclass and the Pearson's correlation coefficients have the disadvantage of being dependent on the level of heterogeneity in the sample in addition to the degree of agreement. Other factors being the same, they tend to be closer to 1 if the subjects are more heterogeneous.

Another way to look at agreement is to use a scatterplot with a 45-degree line. If the two measures agree well, the data points should scatter closely around the 45-degree line. If the data points mostly lie below (above) the 45-degree line, the measure shown on the y-axis tends to under- (over) estimate as compared to that shown on the x-axis. The scatterplot with a 45-degree line is a useful and intuitive way to visually identify systematic bias. However, it is not easy to identify subtle differences or individual random errors from this plot.

Bland and Altman (1986) proposed to plot the difference (y-axis) versus the mean of two measures (x-axis). They maintained that the method is for comparison with an established measurement rather than with the true quantity. In addition, they proposed to add three horizontal lines on the scatterplot: the mean of the differences ($\bar{d}$) and the "limits of agreement" defined as

$$\bar{d} \pm 2SD_d \tag{12.8}$$

where SD_d is the standard deviation of the differences. This is called the *Bland-Altman plot*. Assuming that the differences are normally distributed, most (95%) data points will fall between the limits of agreement.

Example 12.3 (Continued)

The Wright and mini-Wright readings are shown in Figure 12.1. A 45-degree line is superimposed. The data points scatter around the 45-degree line. The plot is useful in showing that there was no systematic pattern of over- or underestimation. However, at the lower-left quadrant, there was a data point that deviated quite a bit from the 45-degree line. The (x,y) coordinate was (178, 259). The disagreement was 81. It was not easy to visually pick up this difference because both the x- and y-axis ranged from 100 to 700. A deviation of 81 can be a clinically important error but it was numerically minor on the scale from 100 to 700.

Figure 12.2 shows the Bland-Altman plot for the same data. The data point (178, 259) was now the (average, difference) = (218.5, −81). The y-axis had a much smaller range than the 45-degree line plot and it magnified the disagreement. The mean and standard deviation of the

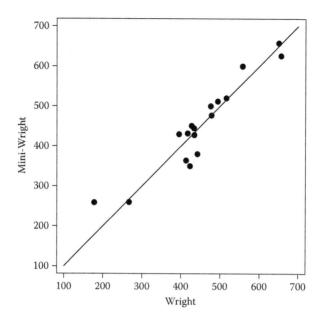

FIGURE 12.1
Scatterplot of mini-Wright versus Wright readings of peak expiratory flow rate in 17 people.
(Data from Bland, J. M., and Altman, D. G., 1986, Statistical methods for assessing agreement
between two methods of clinical measurement, *Lancet* 1(8476):307–310.)

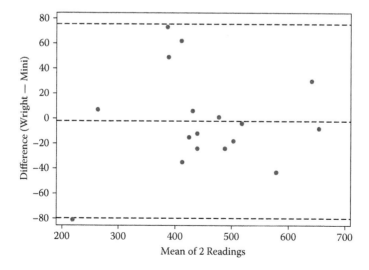

FIGURE 12.2
Bland-Altman plot of mini-Wright and Wright readings of peak expiratory flow rate in 17 people. (Data from Bland, J. M., and Altman, D. G., 1986, Statistical methods for assessing agreement between two methods of clinical measurement, *Lancet* 1(8476):307–310.)

differences were –2.12 and 38.77, respectively. The limits of agreement
were –79.65 and 75.41.

There seems to be a popular misunderstanding that if most data points fall
between the limits of agreement, the degree of disagreement is minor. This
is not true. The limits of agreement are defined such that most data points
do fall between them, no matter if the disagreement is minor or material.
The usage of the limits is for comparison against one's definition of what
an acceptable level of disagreement is. If the limits are within an acceptable level of disagreement defined according to subject-matter knowledge,
it means the two measures are practically exchangeable because most of the
times the differences are acceptable.

The Bland-Altman plot also has a nice property for revealing the relative
variability of two measures. Other factors being the same, a smaller variance
(due to measurement error) is preferred because it gives smaller confidence
intervals and requires a smaller sample size to achieve a target power. If one
measure is more variable than the other, the data points would show a nonzero slope. As such, a test of the correlation between the differences and the
averages is also a test of difference in variance. This is known as the Pitman's
test (Seed 2001). To illustrate, we consider a "fake" measurement generated
by adding a random error that is normally distributed with mean 0 and SD
50 to the Wright PEFR data of Bland and Altman (1986). The fake readings
are included in Table 12.4. Since the mean of the random error is zero, there
is no systematic bias in the fake measure. Figure 12.3 shows the differences
defined as (Wright – fake) versus the averages of the two readings. Consider

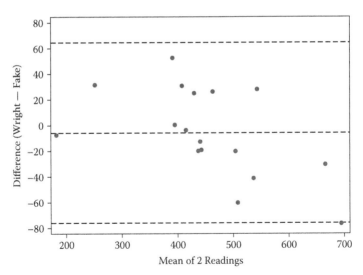

FIGURE 12.3
Bland-Altman plot of "fake" and Wright readings.

the data point near the coordinate (700, –80) as an example. The true value (Wright) was 656 and the random error was 77. The positive random error pulled up the average. Also consider the data point near (400, 60) as another example. The true value was 417 and the random error was –53. The negative random error pulled down the average. A negative correlation between (Wright – fake) versus the average was therefore induced.

Example 12.3 (Continued)

The standard deviations of the Wright and mini-Wright measures were 116 and 113, respectively. On the Bland-Altman plot there was no obvious gradient seen. The Pearson's correlation coefficient between the difference and average of the Wright and mini-Wright readings was 0.084 (P = 0.750). There was no evidence of unequal variances.

12.2.4 Kappa

12.2.4.1 Unweighted Kappa

The kappa is for estimating agreement between categorical variables. Consider the case of two measurements on each subject. Every subject can be classified as one of k categories, $k \geq 2$. Let p_{ij} be the proportion of subjects classified as category i by the first measure and j by the second measure, and

$$p_{i\bullet} = \sum_{j} p_{ij}$$

and

$$p_{\bullet j} = \sum_{i} p_{ij}$$

be the proportions classified by the first and second measures as category i and j, respectively, $i = 1, \ldots, k$ and $j = 1, \ldots, k$. The observed proportion of agreement between two measures is

$$p_{obs} = \sum_{1 \leq i \leq k} \sum_{j=i} p_{ij} \tag{12.9}$$

However, some of the agreement is likely the result of chance. Even if the two measures are purely random, some observations are expected to agree. The expected proportion of agreement is

$$p_{ex} = \sum_{1 \leq i \leq k} \sum_{j=i} p_{i\bullet} p_{\bullet j} \tag{12.10}$$

TABLE 12.5

Hypothetical Example of Kappa Statistics Calculation: Frequency in Cells and Proportions in Margins

Measure 2 \ Measure 1	Not Observed	Emergent	Established	$p_{\cdot j}$
Not observed	20	5	5	$p_{\cdot 1} = 0.30$
Emergent	10	20	10	$p_{\cdot 2} = 0.40$
Established	5	5	20	$p_{\cdot 3} = 0.30$
$p_{i \cdot}$	$p_{1 \cdot} = 0.35$	$p_{2 \cdot} = 0.30$	$p_{3 \cdot} = 0.35$	$N = 100$

The kappa statistics is the unexpected proportion of agreement divided by the complement of the expected proportion

$$\text{kappa} = \frac{p_{obs} - p_{ex}}{1 - p_{ex}} \tag{12.11}$$

Table 12.5 shows a hypothetical example of k = 3. Since the total number of subjects is 100, the cell frequencies are also the percentage distribution in each cell. Similar to Abubakar et al. (2010), the three categories are labeled as "not observed," "emergent," and "established." The observed proportion of agreement is

$$p_{obs} = 0.2 + 0.2 + 0.2 = 0.60$$

The expected proportion of agreement is

$$p_{ex} = 0.35 \times 0.30 + 0.30 \times 0.40 + 0.35 \times 0.30 = 0.33$$

Therefore,

$$\text{kappa} = \frac{0.60 - 0.33}{1 - 0.33} = 0.403$$

A much cited reference of Landis and Koch (1977) suggested that kappa 0 to 0.2, 0.21 to 0.40, 0.41 to 0.60, 0.61 to 0.80, and 0.81 to 1.00 could be interpreted as poor, slight, moderate, good, and almost perfect level of agreement, respectively.

12.2.4.2 Weighted Kappa

The kappa only counts exact agreement. For an ordinal level of measurement, it is possible to estimate a weighted version of kappa such that mild disagreement will also contribute somewhat toward the index. Fleiss and Cohen (1973) recommended the weights

$$w_{ij} = 1 - \left[(i-j)/(k-1)\right]^2 \qquad (12.12)$$

which is implemented in Stata as the option "wgt(2)". Equation (12.9) and Equation (12.10) generalize to

$$p_{obs} = \sum_{1 \le i \le k} \sum_{j=i} w_{ij} p_{ij} \qquad (12.13)$$

and

$$p_{ex} = \sum_{1 \le i \le k} \sum_{j=i} w_{ij} p_{i\bullet} p_{\bullet j} \qquad (12.14)$$

For exact agreement, $i = j$ and the weight is 1. For a $(k - 1)$ step disagreement out of k categories, the weight is zero. In the example in Table 12.5, for a case of one step disagreement out of three categories, the weight is $1 - (1/2)^2 = 0.75$. A one step difference is counted as 75% agreement. Applying the weights to Equation (12.13) and Equation (12.14), $p_{obs} = 0.825$ and $p_{ex} = 0.675$ and the weighted kappa is 0.462. This is slightly larger than the unweighted kappa 0.403.

12.3 Further Topics

12.3.1 Cronbach's Alpha

Cronbach's alpha, α, is an estimate of *internal consistency* (Bland and Altman 1997; Fayers and Machin 2007). That is, the extent the items in a multi-item scales measure the same concept. It is given by

$$\alpha = \frac{m}{m-1}\left(1 - \frac{\sum_{i=1}^{m} V_i}{V_S}\right) \qquad (12.15)$$

where m is the total number of items in a scale whose score is a sum of the item scores, V_i is the variance of the item score i, $i = 1, 2, \ldots, m$, and V_S is the variance of the scale score. The variance of the scale score consists of the variances and covariances of the items. If the items are not correlated at all, $\sum V_i / V_s = 1$ and $\alpha = 0$. The more correlated the items are, the larger their covariances, the smaller the $\sum V_i / V_s$. Cronbach's alpha ranges from 0 to 1. A value close to one means a high level of inter-item correlation. That indicates internal consistency, that is, the items measure the same underlying concept. Cronbach's alpha can be applied to binary, ordinal, or quantitative variables. Fayers and Machin (2007) suggested that a Cronbach's alpha value of 0.7, 0.8, and 0.9 represent acceptable, good, and excellent levels of internal consistency for group analysis, but 0.9 or above is needed for using the instrument to make decisions about individuals.

There are other ways to interpret Cronbach's alpha. One of them is *split-half reliability*. A scale consisting of multiple items may be split into two sets of items. If the items are measuring the same concept, the scores from the split halves should correlate. There are many ways to split the scale into halves. Cronbach's alpha can be regarded as a mathematical approximation to the correlation coefficients averaged across all possible pairs of split halves.

Cronbach's alpha is sometimes referred to as an estimate of reliability (e.g., Fayers and Machin 2007). But there seems to be no consensus about this. For example, Bland and Altman (1997) did not use the term reliability at all. They only called this a coefficient of internal consistency. Recall from Equation (12.1) that under certain assumptions a correlation coefficient between two measures is equivalent to a ratio of true variance to observed variance. Given that we accept the assumptions and given the interpretation of Cronbach's alpha as the average correlation of all split halves that are supposed to measure the same target, it can be interpreted as an estimate of reliability. Here, the measurement errors and inflation of the observed variance are due to the use of different items in the scale. There is a true underlying phenomenon to be measured. Each item makes some measurement errors. Cronbach's alpha estimates the ratio of the true variance of the underlying phenomenon to the variance of the scale score as a sum of the item scores.

12.3.2 Acceptable Levels

In the assessment of validity and reliability, questions are often asked about how large a coefficient needs to be for the instrument to be considered acceptable. It is paramount to appreciate that all such rules are arbitrary and depends on context. Although it is unwise to use instruments that have little validity and reliability, it is not practicable to always attempt an extremely high target. For example, a reliability of 0.95 is sometimes cited as the ideal for anthropometric measures (Ulijaszek and Kerr 1999). However, the measurement of skinfold thickness is more difficult than some other anthropometric measurements. Practically, different expectations have to be set for different measures (Perini et al. 2005; WHO MGRS Group 2006a). Conversely, some instruments that are said to have demonstrated acceptable levels of validity and reliability may not satisfy your requirements.

One way to look at the importance of reliability is from the angle of the impact on sample size or power. For example, consider the situation of a clinical trial testing for a 0.5 cm difference in body length and the SD of

TABLE 12.6

Illustration of Sample Size and Power under Different Levels of Reliability: Detecting a Difference of 0.5 cm between Group Means, Assuming True SD without Measurement Error Is 2.0 cm

Reliability	SD of Observed Values	Sample Size per Group for 80% Power	Power Based on N = 252
1.0	2.000	252	80%
0.9	2.108	280	76%
0.8	2.236	314	71%
0.7	2.390	359	65%

body length is 2 cm. A sample size of 252 per group is needed to achieve 80% power in detecting this difference at 5% 2-sided type I error rate (calculated using Stata's "sampsi" command). If the reliability of the measurement is 0.8, it means the variance is inflated by $1/0.8 = 1.25$, or equivalently the SD is inflated by $\sqrt{1.25} = 1.118$. Other factors being the same, a sample size of 314 is required to maintain 80% power. Conversely, given the same sample size of 252 per group, the power is only 71% if the ICC is 0.8. Table 12.6 shows the sample size for 80% power and the power given a sample size of 252 under various levels of reliability. Such a perspective gives an intuitive feel about the impact of reliability.

13

Missing Values and Imputation

13.1 Introduction

13.1.1 Patterns of Missing Data

Missing data is almost always present in research studies. The pattern of missing data may be monotone or nonmonotone. Table 13.1 illustrates the patterns. A monotone pattern is usually due to dropout from the study, such that some subjects are measured for the exposure variables and short-term outcomes but not the long-term outcomes. Apart from reducing the effective sample size, missing values may introduce a selection bias. This chapter will discuss some analytic strategies for handling missing values. No matter if there is a selection bias or not, sometimes researchers want to replace the missing values by reasonable guesstimates for the purpose of exploration or visualization of data. We will also discuss some methods for this purpose.

13.1.2 Mechanisms of Missing Data

In statistics, *missing completely at random* (MCAR) means that the mechanisms that lead to the data being missing are independent of both the observed and the unobserved (missing) data. *Missing at random* (MAR) means that the missingness depends only on the observed data. For example, a subject may

TABLE 13.1

Patterns of Missing Values

Monotone					Nonmonotone				
A	B	C	D	...	A	B	C	D	...
✓	✓	✓	✓		✓	✓	X	✓	
✓	✓	✓	X		X	✓	✓	X	
✓	✓	X	X				⋮		
✓	X	X	X		✓	X	X	✓	

Notes: A, B, C, and so on, are variables. ✓ is observed; X is missing.

drop out as cognitive function declines linearly to a state that it is no longer feasible for him or her to continue with the study. The linearly declining cognitive function was observed until the dropout and the missingness is completely explained by the observed decline. The terminology MAR may be confusing because in lay terms if something is not completely random then it is not random. So, in communications to general readers, it may be easier to spell out "missingness depends only on the observed data." *Missing not at random* (MNAR) means that the missingness depends on the unobserved data and possibly the observed data as well. For example, the observed cognitive function of a subject was stable over time, but then the cognitive function suddenly dropped sharply. The decline was not observed by the study, and the subject dropped out of the study due to poor cognitive function. The observed data could not predict this dropout. It was the unobserved decline that led to the dropout. No method can handle MNAR without making assumptions unverifiable by the data.

It is important to examine missing data mechanisms before proceeding to analysis. One way to do this is to create a subject-level binary response variable R, where R = 1 for a subject with missing values in at least one of the variables the intended analysis requires, and R = 0 otherwise. We can then compare the distribution of the variables concerned between the two groups defined by R, by either tabulating the univariate summary statistics for each variable one by one or performing a multivariable logistic regression analysis.

Example 13.1

Gandhi et al. (2013) tested the hypothesis that developmental assessment scores at the age of 5 years was predictive of mathematics ability at age 12 years and highest school grade completed and number of school grades repeated by age 12 years. About one-third of the eligible subjects had missing values. The reason for missingness was mainly due to loss of follow-up, so the pattern was near monotone. It was near but not exact monotone because there were missing values in the baseline variables for a small number of subjects who did participate in the age 12 follow-up.

Before the main analysis was presented, a comparison of observed data values between subjects with and without missing values was presented. In particular, the key concern was whether missingness depends on the main exposure variables (developmental scores at age 5) and outcome variables (mathematics and schooling at age 12). Comparing the cohort members with complete and incomplete data, it was found that the mean language domain score at 5 years was lower in the incomplete cases by 0.22 SD. Due to the near monotone pattern, there were very few outcome data for those with missing values (except for a small number of subjects who were measured at age 12 but had missing values in the first 5 years of the study). Nevertheless, height-for-age z-score in childhood was known to be a predictor of cognitive functions in adolescence and early adulthood. This was used as a proxy of the outcome in this exploration. Height-for-age z-score at 5 years was 0.34 z-score lower among those with incomplete data. Furthermore, proportion of preterm birth was 6% higher and weight-for-age near birth was 0.27 z-score lower in the incomplete cases. It was clear that the data were not MCAR. Such a pattern raised a concern about selection bias.

13.2 When Is Missing Data (Not) a Problem?

Missing data is always a problem. At the very least it reduces power and precision. So it is important to plan the study to prevent missing data. It is important to realize that the amount of missing data in a study is a function of the resources invested and that resources are not just financial resources (Bradburn 1992). The more resources you put in, the fewer missing values you will get. It is controllable, not fate.

In randomized trials, it is useful to compare the proportion of subjects with missing values between the intervention and control group. If the proportion is about the same between groups, it does not mean MCAR. The missingness may still be dependent on some other observed or unobserved data. However, if missingness is to confound the relationship between the intervention and outcome, it has to relate to both of them. That missingness is not related to the intervention gives some reassurance that the *relationship* is not confounded. Having said that, an estimate of the *level* of the outcome may still be biased. Randomized trials usually mainly concern the estimation of the relationship, but the level may sometimes be a concern.

Regression analysis of subjects with no missing values, or complete-case analysis, is valid if the missingness depends only on the variables that are included on the right-hand-side of the regression equation (Fairclough 2010; van Buuren et al. 1999). This provides some but not much comfort. First, it is difficult to be sure that all the variables concerned are known, measured, and included in the regression. Second, even if we are confident about that, we cannot obtain unbiased estimates for regression equations that exclude these variables. The problem of including these variables is that it may give the right answer to the wrong question. For example, regression models with or without including an intermediate variable (a variable that lies on the causal pathway) as a covariate answers different research questions. This will be discussed in Chapter 15. The point here is that if we know the missingness depends on the intermediate variable, we are still unwilling to include it as an explanatory variable in the analysis because this changes the question that we want to answer.

13.3 Interpolation and Extrapolation

Strictly speaking, analysis of data with missing values replaced by interpolated or extrapolated values are not valid. However, if these simple methods are used judiciously, they can make data exploration convenient and provide practically valid results. Actually, in Chapter 7 we saw that interpolation can provide a simple alternative to more principled analysis for interval-censored data, which is partially missing data.

13.3.1 Interpolation

If a missing value is flanked by valid observations, *linear interpolation* may be used:

$$\hat{y}_{ij} = y_{i(j-1)} + \frac{(t_{ij} - t_{i(j-1)})(y_{i(j+1)} - y_{i(j-1)})}{(t_{i(j+1)} - t_{i(j-1)})} \tag{13.1}$$

where the value y is missing for subject i at age t_{ij}, $y_{i(j-1)}$ and $y_{i(j+1)}$ are the observed values before and after the missing observation, and $t_{i(j-1)}$ and $t_{i(j+1)}$ are the ages at the preceding and following measurements. Linear interpolation assumes that change in y over the age range is linear. Growth and development may be nonlinear depending on the age range. Hence, the preceding interpolation using raw values may not be accurate unless the time interval is quite narrow. Z-scores are supposed to be horizontal over age. Linear interpolation based on z-scores therefore tends to be more appropriate.

Example 13.2

A clinical trial enrolled infants at the age of about 6 months and then assessed them at age 10, 14, and 18 months. If a child has a body length record of 73.28 cm, missing value, and 80.91 cm for the three visits, respectively, the missing value can be imputed by

$$\text{Length}_{14 \text{ months}} = 73.28 + \frac{(14-10)(80.91-73.28)}{18-10} = 77.10$$

According to the World Health Organization (WHO) growth standards (2006), the child's length z-scores at the 10 and 18 months are 0.0 and −0.50, respectively. Liner interpolation based on the z-scores gives

$$\text{Length z-score}_{14 \text{ months}} = 0.0 + \frac{(14-10)(-0.50-0.00)}{18-10} = -0.25$$

Back-transforming to the original value as per WHO standards (2006) would give 77.44. The difference between 77.44 and 77.10 is the result of assuming linearity where the trend in that age range is nonlinear. At the age of 14 months, the SD is 2.48 cm as per the WHO standards. So, the error is about $(77.44 - 77.10)/2.48 = 0.13$ SD.

Events are sometimes classified based on continuous variables. For instance, height-for-age z-score <-2 is considered stunting. The z-scores may be measured at age t_1, t_2, and so on. The interest is to estimate the age when the event occurs. Ashorn (personal communication, May 2012) suggested a simple yet reasonable modification to the linear interpolation in Equation (13.1):

$$\hat{t}_{event} = t_{i(j-1)} + \frac{(z_{cutoff} - z_{i(j-1)})(t_{ij} - t_{i(j-1)})}{(z_{ij} - z_{i(j-1)})} \tag{13.2}$$

where the z-score crossed over the cutoff point for defining the event between the assessments at ages t_{j-1} and t_j.

Example 13.3

If height-for-age z-score $= -1.5$ at age 6.0 months and z-score $= -3.0$ at age 10.0 months, age at stunting, defined as z-score below -2, occurred between 6.0 and 10.0 months. Since at age 6 months the z-score was already quite close to the cutoff value and at age 10 months the z-score was much lower than the cutoff, intuitively the age at stunting is closer to 6 than 10 months. Using Equation (13.2)

$$\text{age at event} = 6.0 + \frac{\left[-2.0 - (-1.5)\right](10.0 - 6.0)}{-3.0 - (-1.5)} = 7.3$$

13.3.2 Extrapolation

If a participant drops out from follow-up, all data values after the dropout are missing. Interpolation will not be possible because it requires flanking by observed data values. *Last observation carried forward* (LOCF) is a form of *extrapolation* and is one option to be considered. Following the earlier example, consider if a child has a record of 0.00, -0.25, and missing length z-score at age 10, 14, and 18 months, respectively. The LOCF approach would carry the -0.25 at age 14 months to impute the missing value at 18 months of age as -0.25 z-score. This approach assumes that the trend is stable, which needs careful considerations on a case-by-case basis.

LOCF uses only one data value to extrapolate. For repeated measures over time, a linear regression line can be obtained to relate the data values for each subject in relation to time, along the line of analysis of subject-level summary statistics we have discussed in Chapter 8, Section 8.3. The regression is then used to extrapolate a data value at the time measurement is intended.

Interpolation and extrapolation are convenient tools for simple situations. Their accuracy partly depends on the distance between flanking measurements and the distance to extrapolate beyond the last observation.

13.4 Mixed Models

An important thing about imputation for missing values is that it is not always necessary. The mixed model is valid for both MCAR and MAR. The mixed models can be used to prevent the bias resultant of MAR. Table 13.2 uses a simple example to illustrate. All participants are measured for a quantitative variable at baseline (time 1). A proportion (p) of the participants are also measured at end of study (time 2). Denote the means of the outcome at time 1 and time 2 of the completers by y_1^C and y_2^C, respectively. Denote the mean of the outcome at time 1 of the dropouts by y_1^D. This is MAR because missingness is determined only by the observed y at baseline. Fit a mixed model with a random intercept and a binary exposure variable (time). The mixed model estimate for the baseline mean (b_0) is a weighted average of the means of the completers and dropouts, with p and $(1 - p)$ for the weights for the two groups, respectively. This is equivalent to the simple mean at baseline. The mixed model estimate for the mean at time 2, $(b_0 + b_1)$, consists of the mean of the completers at time 2 plus an adjustment factor. The adjustment factor is a product of three elements. The first element is $(1 - p)$. The smaller the proportion completing the study is, the larger the adjustment factor. The second element is an estimate of the correlation between y at the two time points among the completers ($\hat{\rho}$). The larger the correlation, the larger the adjustment factor. The third element is the difference between means at baseline between the dropouts and completers. If the dropouts have a

TABLE 13.2

Observed Means and Mixed Model Estimates of Means in a Simple Case with Two Time Points, Missing Values Only at Time 2, and Equal Variances

Time	Completers	Dropouts	Overall
1	$\bar{y}_1^C$	$\bar{y}_1^D$	$p \times \bar{y}_1^C + (1-p) \times \bar{y}_1^D$
2	$\bar{y}_2^C$		$\bar{y}_2^C + (1-p) \times \hat{\rho} \times \left(\bar{y}_1^D - \bar{y}_1^C\right)$

Source: Fairclough, D., 2010, *Design and Analysis of Quality of Life Studies in Clinical Trials,* 2nd ed., Boca Raton, FL: Chapman & Hall/CRC.

lower mean at baseline, the adjustment is downward, or vice versa. If the two means at baseline are the same, this is MCAR and the adjustment is 0. The mixed model adjusts for the effect of MAR without imputing for missing values by using the correlation between the repeated measurements.

13.5 Multiple Imputation

13.5.1 Univariate Imputation

Linear interpolation and extrapolation do not reflect the uncertainty in guessing what the missing values may be. Some researchers may call them "substitution" instead of imputation. Ignoring the uncertainty leads to underestimation of standard errors in the analysis.

Let Y be a variable with missing values, $X = (X_1, X_2, \ldots, X_p)$, the predictors without missing values, and $\theta = (\beta, \nu)$, where β is the $(p + 1)$ vector of regression coefficients including the intercept and ν is the auxiliary parameter(s), such as the log variance in the case of ordinary least-squares regression for an outcome variable assumed to be normally distributed. Furthermore, Y is partitioned to (Y_{obs}, Y_{mis}) where the two elements are the observed and missing values, respectively. X can be partitioned to (X_{obs}, X_{mis}) where the two elements are the predictor values from the subjects with complete and missing data, respectively. Assuming MAR, the posterior predictive density of the missing values Y_{mis} is given by

$$p(Y_{mis}|X) = \int p(Y_{mis}|X, \theta) p(\theta|X) d\theta \qquad (13.3)$$

A multiple imputation strategy for a single variable with missing values follows the general process described next. The details differ depending on the type of Y variables and imputation algorithms.

1. Fit a regression model of Y in relation to X using the complete cases to obtain the estimates $\hat{\theta}$ and their variance–covariance matrix, $\hat{\Sigma}$.

2. Randomly draw a new set of parameters θ^* from the posterior distribution $p(\theta|X)$, usually approximated by the multivariate normal distribution $MVN(\hat{\theta}, \hat{\Sigma})$.

3. Obtain one set of imputed values by randomly drawing from the posterior predictive distribution $p(Y_{mis}|X, \hat{\theta})$ characterized by $X_{mis}\beta^*$ and ν^*.

4. Repeat steps 2 and 3 to obtain the total m sets of imputed values.

Step 4 makes it a *multiple imputation* (MI). Each of the m datasets with missing values filled in is different. The differences between them reflect the uncertainty in the imputation.

For comparison, LOCF (an intercept model) and linear extrapolation (an intercept and slope model) may be seen as imputation methods that use only information from $\hat{\beta}$ to obtain a single imputation $X_{mis}\hat{\beta}$.

13.5.2 Combinations of Results

Statistical estimation is performed on each of the m datasets containing observed and imputed values. Let $\hat{\beta}_j$ and $\text{Var}(\hat{\beta}_j)$ be the point estimate and variance to a regression coefficient from analysis of the jth dataset. The *Rubin's rule* (Rubin 1987) has it that the MI point estimate and variance are, respectively,

$$\hat{\beta}_{MI} = \frac{1}{m}\sum_{j=1}^{m}\hat{\beta}_j$$

$$\text{Var}(\hat{\beta}_{MI}) = W + B + \frac{B}{m} \tag{13.4}$$

where W and B are estimates of the within- and between-imputation variances

$$W = \frac{1}{m}\sum_{j=1}^{m}\text{Var}(\hat{\beta}_j)$$

$$B = \frac{1}{m-1}\sum_{j=1}^{m}(\hat{\beta}_j - \hat{\beta}_{MI})^2$$

Confidence interval is calculated by

$$\hat{\beta}_{MI} \pm t_\tau\sqrt{\text{Var}(\hat{\beta}_{MI})}$$

on

$$\tau = (m-1)\left[1 + \frac{W}{B + \frac{B}{m}}\right]^2 \tag{13.5}$$

degrees of freedom. Note that the degree of freedom does not have to be an integer. The hypothesis test is similarly based on the t-statistics. The likelihood ratio test is not usable.

13.5.3 Choice of m

As shown in Equation (13.4), the precision of the MI estimate depends on m. As m increases, the precision converges to W + B. There are different perspectives on how to choose m. According to the efficiency consideration, the variance of the MI estimate usually levels off as m increases beyond 20 (Rubin 1987). Based on this argument, usually m ≤ 20 is sufficient. Another perspective is reproducibility. You want to be quite sure that essentially the same results will be produced if you change the random number seed. Based on this consideration, White et al. (2011) suggested a rule of thumb that m should be at least equal to the percentage of incomplete cases. With high-speed computers widely available nowadays, there is little reason to limit to m ≤ 20.

13.5.4 Multivariate Imputation

The previous section only considers the case where there is a single variable y with missing values. The variables on the right-hand-side of the regression equation are always predictors and they have no missing values. If there are several variables with missing values and the pattern is monotone, we can perform univariate imputation in a sequential manner. For a nonmonotone pattern, a more general approach is needed. Van Buuren et al. (1999) developed a multivariate imputation approach based on *switching regression*, also known as *multiple imputation by chained equation*, MICE (Royston 2004, 2007). Here, more than one variable has missing values. The pattern may be monotone or nonmonotone. The variables take turns to predict and be predicted. Let the set of variables with missing or complete data be x_j, j = 1, 2, ..., p. The variables can be a mixture of quantitative, binary, or other variables. The procedure works iteratively as follows:

1. Impute the missing values in each x_j by randomly drawing from the observed x_j values. This initializes the iterative procedure. The variable $x_j^{(0)}$ is a mixture of observed and initially imputed data.
2. For each $x_j^{(0)}$, in turn, perform univariate imputation for the missing values using the other (p − 1) predictors from step 1.
3. Repeat the above step K cycles. In each cycle, update the previously imputed values in $x_j^{(k-1)}$ by the latest imputed values $x_j^{(k)}$, k = 1, 2, ..., K. Van Buuren et al. (1999) recommended K = 20 to allow the updating to stablize. This provides a single imputation dataset.
4. Repeat the above procedure m times to obtain multiple imputation datasets.

The variables need not be all of the same type. For quantitative variables, the prediction uses least-squares regression. For binary variables, the predication uses logistic regression. The same principle applies to other types of variables. Identical analysis is then performed on each of the m datasets and the Rubin's rule is used to combine the results.

Example 13.1 (Continued)

Gandhi et al. (2013) studied whether developmental assessment scores at the age of 5 years was related to mathematics ability at age 12 years and highest schooling outcomes. Depending on the outcome measure, 30% to 34% of the cases had missing data. The investigators used MICE with nine variables measured by age 5 and three variables measured at age 12. According to the rule of thumb based on reproducibility concerns, m should be at least 34. The investigators used m = 50.

In an analysis of the complete cases only, the least-squares regression coefficient for predicting mathematics ability at age 12 by developmental score at age 5 was 1.77 (SE = 0.93; P = 0.058), without covariate adjustment. Using the MI approach, the coefficient was 1.89 (SE = 0.87; P = 0.031). In this particular case the MI approach result only slightly differed from the complete case analysis. The possible bias appeared to be small.

13.5.5 Imputation and Analytic Models

The imputation and analytic models should be consistent. First, a variable that is to be included in the analysis model should be included in the imputation model, even if there is no missing values in this variable. Otherwise, the association pattern between this and the other variables are not reflected in the imputation. This inconsistency between the two models leads to a bias toward the null value in the analysis. The converse is not a problem. An imputation model can include more variables than the analysis model without compromising the validity of the analytic results (Royston 2007; Rubin 1996). Furthermore, to improve precision, variables that can predict the outcome variable should be included in the imputation model as well. Study design variables, such as stratification factors, should also be included.

The functional form of the prediction equations also need to be consistent with the analytic model. If the analysis intends to study the interaction between x_1 and x_2, the interaction term should be included in the imputation model. This requires *passive imputation* (Royston 2004, 2007). That is, in each cycle of the estimation of the chained equations, when x_1 or x_2 are to be predicted, the interaction term is not included as a predictor. When the other variables are being predicted, however, the interaction term is calculated using the most updated x_1 and x_2, and the three terms are all included in the prediction equation, among others. Similarly, if the analysis model will use polynomial terms, they should also be handled by passive imputation.

13.5.6 Further Remarks

Multiple imputation is not the only principled approach to handling missing values. The use of mixed models is one of the alternatives. Multiple imputation by chained equation is an attractive option because of its flexibility. The trade-off is that it does not have a complete theoretical basis, although case studies and simulation studies suggested that the method works (van Buuren et al. 2006; van Buuren 2007). In particular, there is no guarantee that step 3 described in Section 13.5.4 must converge. Researchers should check the convergence by comparison of imputation results at different number of cycles.

There are various adaptations of imputation algorithms. For example, one may constrain the imputed values to be within the range of the observed values. For details and practical issues in MI, Royston (2004, 2007) and White et al. (2011) are excellent references.

13.6 Imputing Censored Data

Chapter 7 discussed methods for the regression analysis of censored data. However, for presentation purposes, it is sometimes useful to replace the censored values by imputed value.

Age at achievement of a developmental milestone is usually interval-censored between the ages at two assessments. The WHO Multicentre Growth Reference Study (MGRS) provided percentiles of the age at achieving motor milestones by randomly imputing specific ages based on the assumption of uniform distribution within each interval (WHO MGRS Group 2006b). This simple imputation is reasonable only for narrow measurement intervals; monthly assessment in the MGRS case.

Royston et al. (2008) discussed visualization of censored data to highlight variability. Regression coefficients are some kind of average values. They underplay the variability. A regression coefficient and its confidence interval may clearly suggest that the averages between two groups separate, but the individual data points may still hugely overlap. Although the regression can rightly demonstrate an intervention effect on average, the plotting of the individual data points will highlight that an individual who received the intervention may still have an outcome worse than those who did not.

Royston (2007) proposed to embed a normal interval censoring regression equation in the MICE model. This assumes that the interval-censored variable follows a normal distribution, possibly after transformation. The log-transformation is a likely choice for some outcomes. For example, as discussed in Chapter 7, ages at achievement of motor milestones tend to follow log-normal distributions. Assessing normality in interval-censored data is

difficult. An approximate way to assess normality of the original or transformed data is to assign a group mean for censored observations at each
interval. For right-censored data, one can make an informed guess about
the mean. For example, if age at independent walking is right-censored
at 14 months, based on the literature we may guess that the mean among
the right-censored cases is probably about 15 to 16 months. So we assign a
value of 15.5. Then apply the graphical and statistical diagnostic methods
described in the previous chapters. The choice of transformation will need
to be guided by external knowledge and prior information.

For right-censored data, a finite upper boundary should be chosen carefully
in the imputation. Without an upper boundary, a subject may be imputed to
have an unrealistic value, for example, starting to walk independently at the
age of 60 years. It is particularly easy to encounter such odd predictions when
the imputation is for log-transformed data, as the back-transformed data are
highly skewed. Imputation for censored data is mainly useful for outcomes
that would eventually occur in most if not all subjects. Adults starting to experience cognitive decline and children in a resource-poor setting experiencing
mild stunting (<-1 z-score) are cases in point. It is possible to specify realistic
upper limits for the age at these events. Some events will not occur in a siz-
able proportion of the population. Time to moderate or severe stunting is an
example. I do not recommend imputation for such right-censored variables.

The basic imputation process is the same as that described in Section 13.5.1.
However, step 3 is modified to draw values from a truncated normal distribution for each incomplete observation. This ensures that the imputed data
will respect the observed lower and upper bounds of the censored value
(Royston 2007).

Example 13.4

The Simulated Longitudinal Study (SLS) data is used for illustration. The
data resemble the pattern of growth faltering in some developing countries, where at birth height-for-age (HAZ) is normal but then it declines
roughly linearly until stabilizing at about -2 in the second year of life.
As such, most children are expected to have experienced at least mild
stunting at some point during childhood.

An interval-censored variable on age at mild stunting (HAZ<-1) was
generated in a way similar to the generation of the time-to-stunting variable in Appendix B. Using midpoint for left- and interval-censored data
and assigning 30 months for right-censored data, initial assessment suggested that the assigned values were somewhat positively skewed, but a
log transformation would overcorrect. A Box-Cox transformation suggested a square-root transformation, as the point estimate (95% CI) of
the power term was 0.435 (0.296 to 0.574). A worm plot showed that the
square-root transformed data were approximately normally distributed
(Figure 13.1). The unusual pattern of data points aligning like straight
lines was due to near-tied values in the variable. They were "near" but not

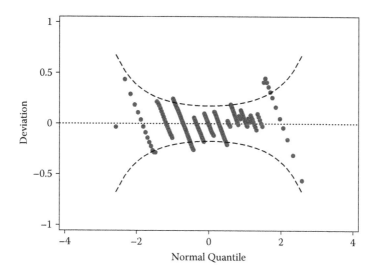

FIGURE 13.1
Worm plot of initially assigned age at first mild stunting.

exactly tied because the age at assessment was slightly different between individuals and therefore the midpoint was also slightly different.

Having accepted the assumption that square-root transformed age at first mild stunting was approximately normal, multiple imputation as described in Section 13.5 was used to impute the censored data. Display 13.1 shows the Stata codes and outputs. A realistic upper bound at 30 months was included for right-censored data. Variables with square-root transformed lower bound and upper bound were created. A variable *ageimp* was specified in the "mi" command to hold the imputed values. Socioeconomic status (SES) and gender were used as the predictors. Twenty new datasets were created with the original data, and the variable *_mj_m* indicated the original (0) and the imputed datasets (1 – 20).

The imputed age was back-transformed to the original age scale. The "mi estimate" command applied Rubin's rule to a least-squares regression analysis comparing age to mild stunting between the low and high SES groups. The estimated mean age at first mild stunting was 7.296 months in the low SES group. The high SES group had a mean age of 5.502 months later (P < 0.001). For comparison with accelerated failure time model, the time ratio was (7.296 + 5.502)/7.296 = 1.754. For comparison purposes, a generalized gamma model for the original interval censored data gave a regression coefficient of 0.685 (P < 0.001), or a time ratio of exp(0.685) = 1.984. The two analyses gave a roughly comparable result. The regression coefficients and their P-values and confidence interval (CI) conveyed a sense of clear difference between groups. But that was misleading in the sense that they were only about a clear difference in the averages. Figure 13.2 compares the imputed age by means of a box-whisker plot. Acknowledging this is only to give a feel about the variability, only one imputed dataset (*_mi_m* = 1), was used in the graphing. The

```
. gen lb = sqrt(mild_lbage)

. gen ub = sqrt(mild_ubage)
(42 missing values generated)

. replace ub=sqrt(30) if mild_ubage==.
(42 real changes made)

.
. mi set flong

. mi impute intreg ageimp = SES male, ll(lb) ul(ub) add(20) rseed(123)

(output omitted)

. gen mild_age=ageimp^2
(200 missing values generated)

. mi estimate: regress mild_age SES

(output omitted)
```

mild_age	Coef.	Std. Err.	t	P>|t|	[95% Conf. Interval]
SES	5.502322	.9474284	5.81	0.000	3.629652	7.374993
_cons	7.296311	.5703931	12.79	0.000	6.168706	8.423916

DISPLAY 13.1
Stat codes and outputs: Multiple imputation of interval-censored time-to-mild stunting.

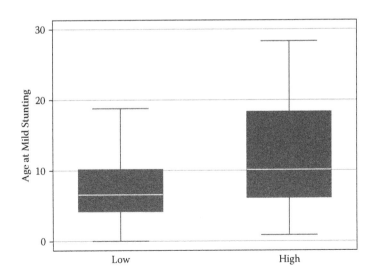

FIGURE 13.2
Box-whisker plot comparing distribution of age at first mild stunting between groups.

distribution of age at mild stunting substantially overlapped between the two groups. It was by no means deterministic that an individual in the low SES group must have an earlier onset of mild stunting than an individual in the high SES group.

14

Multiple Comparisons

14.1 The Problem

Seldom does a study answer only one single question and perform only one statistical test. The problem of *multiple comparisons*, or *multiplicity*, may arise because of the involvement of multiple endpoints, pairwise comparisons among three or more groups, multiple subgroup analyses, interim analyses, and so on. This chapter mainly uses multiple endpoints and pairwise comparisons to illustrate the methods discussed, but the methods are applicable to other situations.

If $P < 0.05$ is used as the level of statistical significance for rejecting a null hypothesis, it means that there is a 5% chance of falsely rejecting a null hypothesis. In other words, the chance of making one mistake is $1 - 0.95 = 0.05$. If k statistical tests are performed and all the null hypotheses are truly correct, the chance of falsely rejecting at least one null hypothesis is not 5%. Instead, it is $1 - 0.95^k$. The solid line in Figure 14.1 shows the probability of rejecting at least one null hypothesis by chance in relation to the number of statistical tests performed. This illustration is based on the *universal null hypothesis* that all null hypotheses are true. If k = 14 tests are performed, it is more likely than not (51% chance) that at least one null hypothesis will be falsely rejected. If 100 tests were performed, it is almost certain that there will be some statistically significant findings. The universal null hypothesis is not always plausible. It is sometimes reasonable to expect that some null hypotheses are true but some are false. In either case, the issue of multiple comparisons raises concerns about how to interpret statistical test results and how to control the risk of errors in decision making based on statistical significance tests. Whether it is essential to make any statistical adjustments for multiplicity has been a heated debate over decades. Some examples of diverse views are shown in Table 14.1.

A *familywise error rate* (FWER) is the probability of making at least one type I error in the testing of a set (family) of hypotheses, when at least some of the null hypotheses in the family are true. Some statistical procedures provide *weak control* of error rate, or *control in the weak sense*. They concern only the global null hypothesis. For example, the analysis of variance (ANOVA)

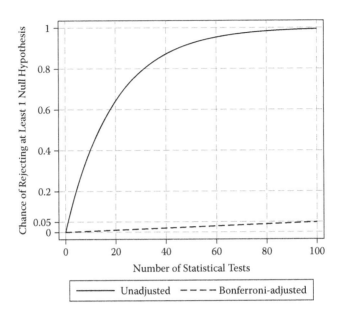

FIGURE 14.1
Probability of falsely rejecting at least one true null hypothesis if all null hypotheses are true.

TABLE 14.1

Quotations on Need for Statistical Adjustment for Multiplicity

For
- "If nonadjustment for multiple comparisons became acceptable … It would be a license to publish coincidences with a pseudoscientific gloss" (Thompson 1998, p. 804).
- "Type I error accumulates with each executed hypothesis test and must be controlled by the investigators" (Moyé 1998, p. 354).

Against
- "No adjustments for multiple comparisons are needed for multiple comparisons" (Rothman 1990, p. 43).
- "Simply describing what tests of significance have been performed, and why, is generally the best way of dealing with multiple comparisons" (Perneger 1998, p. 1236).

controls the probability of falsely rejecting the global null hypothesis of $\mu_1 = \mu_2 = \ldots = \mu_K$ at the significance level α. However, it does not answer which group has a mean that is different from the others. In contrast, *strong control*, or *control in the strong sense*, refers to the control of the error rate for each member (hypothesis) of the family. In the ANOVA example, strong control of the FWER concerns $\mu_1 = \mu_2, \ldots, \mu_1 = \mu_K, \ldots$, and $\mu_{K-1} = \mu_K$. Rejecting a global null hypothesis is usually not the end of a study. It prompts questions about which groups or which endpoints are different. Therefore, strong control of FWER is often wanted.

14.2 When Not to Do Multiplicity Adjustment

14.2.1 Different Families of Hypothesis

The control for FWER begins with considering what constitutes a family. If each hypothesis to be tested belongs to a different family, there is no need to consider any statistical adjustment despite multiple comparisons. To be considered a family, they should be related in terms of their contents and intended usage (Hochberg and Tamhane 1987). In practice, there is no easy definition about what a family of hypotheses is. For instance, suppose a randomized control trial of nutritional supplementation versus placebo would conclude that the supplementation is efficacious as long as either the participants in the intervention group gain more weight or they gain more height during the study period. The null hypotheses (1) no more weight gain and (2) no more height gain in the intervention group than the placebo group clearly belong to the same family, because they are used to answer the same question about the efficacy of the intervention and decide on whether the product is to be recommended. Statistical adjustment is needed to control the type I error rate in this case. In contrast, if there is one primary efficacy endpoint and one primary side effect endpoint, the two endpoints shed light on the understanding of different biological aspects of the intervention. They may be regarded as belonging to different families, each with a single hypothesis. Then no statistical adjustment is needed.

> **Example 14.1**
>
> The Lungwena Antenatal Intervention Study (LAIS) compared three groups: standard care (two doses of sulfadoxine-pyrimethamine [SP] during pregnancy), monthly SP, and azithromycin plus monthly SP (AZI-SP) (Luntamo et al. 2010). There could be three null hypotheses, each concerning one pairwise comparison. It was fairly clear that the comparisons of (1) standard care versus monthly SP and (2) standard care versus AZI-SP belong to the same family, because the rejection of each of the null hypothesis would lead to the recommendation of switching from the standard care to an alternative. Control of FWER was preferred. But it was less clear whether the comparison of monthly SP versus AZI-SP should be seen as belonging to the same or a different family. It was possible to see this as a study of the effect of AZI, a scientific question unrelated to the standard care and unrelated to the effect of SP. From this angle, this comparison belonged to a different family.

14.2.2 Reject All Null Hypotheses

The *intersection-union test* maintains that if each null hypothesis within a family needs to be rejected at significance level α in order to make a conclusion, then the probability of wrongly concluding is α at most (Berger and Hsu 1996). Suppose a randomized trial of a nutritional supplementation group versus a control group must show that the participants in the intervention group gain more weight and more height to claim efficacy. The hypothesis about weight and the hypothesis about height can each be tested at the significance level 0.05. The overall significance level is not more than 0.05 as long as a conclusion of efficacy is made only if both null hypotheses are rejected. There is no additional statistical procedure required.

14.2.3 To Err on the Side of False Positive

The risk of type I error (false positive) and type II error (false negative) are inversely related to each other. Statistical procedures that aim to prevent inflation of type I errors may increase the risk of type II errors. Whether it is better to err on one side than the other depends on context. The ICH Guidelines E9 (1998) does recognize that P-values may be used informally as a "flagging device" to highlight safety and tolerability signals that may be worth further investigation. From this angle, one may prefer to err on the side of making type I errors.

Example 14.2

Nauta (2010) discussed the case of monitoring the side effects of vaccines. There can be many safety variables to analyze, giving rise to a multiplicity problem. In this context, regulatory authorities may consider false negatives more undesirable than false positives. If the intention is to err on the safe side in the sense of not missing a true safety signal, it is better not to do any multiplicity adjustment. The risk of false-positive findings is understood by regulatory authorities; each safety signal needs to be further reviewed carefully. Regulatory bodies tend to discourage multiplicity adjustment in the analysis of safety data. This is in contrast to their requirements for the analysis of efficacy data.

14.3 Strategies of Analysis to Prevent Multiplicity

14.3.1 Statistical Analysis Plan

It is important that statistical analysis is planned in advance. Justifications about what analyses to perform, and why, should be given according to specific research aims and scientific principles. The reasons of not attempting to adjust for multiplicity such as those discussed in Section 14.2 should be justified, if applicable. If the researchers try to provide a justification after seeing the analysis results, it will be difficult for them to claim objectivity.

The statistical analysis plan should specify which variable(s) is the primary endpoint(s) in the confirmatory analysis upon which the main conclusion is to be made. If there are multiple primary endpoints, what roles do they serve and what conclusion is to be made? For example, is the rejection of one null hypothesis sufficient to conclude efficacy or does it require the rejection of all null hypotheses? Exploratory analyses are for hypothesis generating. There is usually no need to consider multiplicity adjustment for exploratory analyses, but the hypothesis-generating nature should be clearly stated.

14.3.2 Defining a Primary Analysis

The use of a single primary endpoint, if possible, is the simplest way to avoid the multiplicity problem arising from multiple endpoints. The use of factor analysis to generate a single summary measure that represents the multitude of variables is a convenient strategy. See Chapter 10, Example 10.2 for a case study that applied factor analysis to create a single variable to summarize six developmental outcomes. As only one variable was analyzed as the primary endpoint, there was no issue about multiple testing of variables. The disadvantage of this strategy is that it obscures differences in opposite directions, if any, among the endpoints.

In time-to-event analysis, there can be distinct events such as achievement of different developmental milestones. Assuming that the impact of an exposure is the same on each type of event in terms of hazard ratio, one can perform a stratified Cox regression analysis of the multiple events (Wei and Glidden 1997). In this analysis, each type of event is treated as one stratum. The number of records per person equals the number of types of event. The stratified analysis constrained the hazard ratio to be the same for each stratum. The hazard ratio may be seen as a form of average of the hazard ratios across types of events (strata). Since there are multiple records per person, the robust standard error is used for statistical inference. By using this summary hazard ratio, the issue of multiplicity is circumvented. It has the disadvantage that differences in opposite directions, if any, are obscured.

In the context of multiple groups defined according to dosages or exposure, the researchers may consider defining the estimation and testing of a dose-response relation across groups as the primary analysis (Schulz and Grimes 2005).

14.4 P-Value Adjustments

14.4.1 Bonferroni Adjustment

The Bonferroni adjustment is probably the most widely known but also most criticized method of statistical adjustment for multiplicity (Perneger 1998;

Rothman 1990). It is based on the universal null hypothesis. If all K null hypothesis are true, the chance of making at least one type I error can be controlled at α by using the rule of rejecting an individual hypothesis only if the P-value is smaller than α/K. Equivalently, the *adjusted P-value* of an individual hypothesis is the minimum of the raw P-value multiplied by K and 1:

$$\text{Bonferroni adjusted } P = P^B = \min(P \times K, 1) \tag{14.1}$$

where K is the total number of hypotheses within the family. If P × K is larger than 1, which is not sensible, the adjusted P-value is capped at 1. The broken line in Figure 14.1 shows the probability of falsely rejecting at least one null hypothesis if all the 100 hypotheses were true using $P = 0.05/100 = 0.0005$ as the rejection rule. With 100 hypotheses tested, the probability was controlled at 5%.

Example 14.3

Table 14.2 shows an example of a comparison of five endpoints between two experimental groups. The five endpoints represent different aspects of child development. The raw P-values calculated by a two-sample t-test ranged from 0.0033 to 0.0622. Four out of five null hypotheses were rejected according to the raw P-value if there was no intention to control for FWER.

In this example, K = 5 and the Bonferroni adjusted P-values were the raw P-values multiplied by 5. They ranged from 0.0165 to 0.3110. Only two out of five null hypotheses were rejected according to the Bonferroni adjustment.

The Bonferroni adjustment tends to be conservative in the sense that it rejects fewer null hypotheses than most other adjustment methods. Critics of the Bonferroni adjustment point out that the universal null hypothesis on which the method is based is unrealistic.

TABLE 14.2

Comparison of Multiplicity Adjustment Methods

Endpoint	$P_{(k)}$	Raw P-Value	Bonferroni	Holm	FDR
Locomotor	$P_{(1)}$	0.0033	0.0165	0.0165	0.0165
Personal-social	$P_{(2)}$	0.0095	0.0475	0.0380	0.0192
Hearing-language	$P_{(3)}$	0.0115	0.0575	0.0380	0.0192
Coordination	$P_{(4)}$	0.0301	0.1505	0.0602	0.0376
Performance	$P_{(5)}$	0.0622	0.3110	0.0622	0.0622

14.4.2 Holm Adjustment

The Holm adjustment method is more powerful than the Bonferroni method (Aickin and Gensler 1996). It never rejects fewer null hypotheses than the Bonferroni method. Yet it is relatively simple to implement. With the Holm adjustment, the P-values $P_{(k)}$ are ordered such that $P_{(1)} \leq P_{(2)} \leq \ldots \leq P_{(K)}$, $k = 1, 2, \ldots, K$. Correspondingly, the null hypotheses are numbered as $H_{(1)}$, $H_{(2)}, \ldots, H_{(K)}$. The procedure begins with testing $H_{(1)}$ at α/K. If $H_{(1)}$ is not rejected, this and all subsequent hypotheses are accepted. If $H_{(1)}$ is rejected, $H_{(2)}$ is tested at $\alpha/(K - k + 1)$, where $k \geq 2$. The process stops until $H_{(k)}$ is not rejected; all remaining hypotheses are not rejected. As such, the Holm adjustment is called a stepwise procedure. Equivalently, the Holm adjusted P-values are

$$\text{Holm adjusted } P_{(k)} = P_{(k)}^H = \begin{cases} \min(P_{(1)} \times K, 1) & \text{if } k = 1 \\ \min\left\{\max\left[P_{(k-1)}^H, P_{(k)} \times (K - k + 1)\right], 1\right\} & \text{if } k \geq 2 \end{cases} \quad (14.2)$$

The max(.) operation in Equation (14.2) ensures that if the jth hypothesis with a smaller raw P-value is not rejected, any subsequent hypothesis with a larger raw P-value will not be rejected either. The min(.) operation capped the adjusted P-values at 1.

Example 14.3 (Continued)

In Table 14.2, the P-values were sorted in ascending order. The Holm adjusted $P_{(1)}$ was $0.0033 \times 5 = 0.0165$. This was the same as the Bonferroni adjusted P-value. The adjusted $P_{(2)}$ was $0.0095 \times 4 = 0.0380$. This was smaller than the Bonferroni adjusted P-value. In considering $H_{(3)}$, however, we note that $0.0115 \times 3 = 0.0345$. This was smaller than the adjusted P-value for the preceding hypothesis. The correct Holm adjusted $P_{(k)}$ here that maintained the stepwise nature of the procedure was the maximum of the preceding adjusted P-value (0.0380) and 0.0345, as described in Equation (14.2). As such, the Holm adjusted $P_{(3)}$ equals 0.0380. The adjusted $P_{(4)}$ and $P_{(5)}$ are .0602 and 0.0622, respectively. The Holm procedure rejects three hypotheses at the 0.05 level here, whereas the Bonferroni only rejected two.

14.4.3 False Discovery Rate

Although the control of type I error is important in the context of confirmatory analysis, its value in exploratory analysis is not so clear. Another angle to look at the issue of multiple comparisons is to shift the perspective from type I error rate to the *false discovery rate* (FDR) (Benjamini and Hochberg 1995).

TABLE 14.3

Truth and Decision in Testing Multiple Hypotheses

	Decision		
Truth	Reject Null Hypothesis	Accept Null Hypothesis	Total
Null hypothesis is true	A	B	(A + B)
Null hypothesis is false	C	D	(C + D)
Total	(A + C)	(B + D)	

Source: Motulsky, H., 2010, *Intuitive Biostatistics*, Oxford: Oxford University Press, Table 22.1.

Motulsky (2010) provided an intuitive summary of the different approaches. He tabulated the number of null hypotheses according to the truth and the decision to reject a null hypothesis, as in Table 14.3. Among all hypotheses tested, (A + C) are rejected and (B + D) are accepted. The (unadjusted) type I error rate concerns keeping A/(A + B) at the significance level α. The FWER concerns keeping Prob(A>0) at α. In contrast, the FDR concerns A/(A + C). Rejecting a null hypothesis that is in fact true is a false discovery. The FDR approach aims to control the proportion of false discovery among all discoveries. This is especially useful in exploratory analysis that involves a very large number of exploratory statistical tests.

The adjusted "P-value" in the FDR approach is called the *Q-value*. In contrast to the calculation of the Holm adjusted P-values, which starts at $P_{(1)}$, the calculation of the Q-values starts from the largest P-values, $P_{(K)}$, and works backward:

$$Q_{(K)} = P_{(K)}$$

$$Q_{(k)} = \min\left[Q_{(k+1)}, P_{(k)}(K/k)\right], \text{ for } k < K \tag{14.3}$$

The min(.) operation in Equation (14.3) ensures that a null hypothesis with a smaller raw P-value will not be accepted if a null hypothesis with a larger raw P-value has been rejected.

Example 14.3 (Continued)

The Q-value for the fifth hypothesis was 0.0622 according to Equation (14.3). The Q-value for the fourth hypothesis, $Q_{(4)}$, was 0.0301×(5/4) = 0.0376. $Q_{(3)} = 0.0192$. In the calculation of $Q_{(2)}$, note that 0.0095×(5/2) = 0.0238, which was larger than $Q_{(3)}$. Therefore, $Q_{(2)}$ was the minimum of $Q_{(3)}$ (0.0192) and 0.0238 as per equation (14.3). Hence, $Q_{(2)} = 0.0192$. Without this min(.) operation, it would be possible to give a non-sensible result of rejecting a null hypothesis with a larger raw P-value but accepting a null hypothesis with a smaller raw P-value. Finally, $Q_{(1)} = 0.0033×5 = 0.0165$. The FDR approach rejected four of the five null hypotheses at the 0.05 level.

14.5 Close Testing Procedure

14.5.1 Multiple Groups

The *close testing procedure* introduced by Marcus et al. (1976) may look difficult at the beginning. But its applications, especially in the comparisons among three groups, are quite simple. Suppose there are K elementary null hypotheses, $H_{(k)}$, k = 1, 2, ..., K, to be tested. The closed testing procedure rejects any one of the $H_{(k)}$ if all null hypotheses containing $H_{(k)}$ can be rejected at the significance level α by a valid test. It provides strong control of the FWER at α.

In a simple case of a three-arm experiment, there can be three elementary null hypotheses, $H_{(1)}$, $H_{(2)}$, and $H_{(3)}$, concerning pairwise comparisons of the group means, $\mu_A = \mu_B$, $\mu_A = \mu_C$, and $\mu_B = \mu_C$, as shown in the upper panel of Table 14.4. There are three intersection hypotheses that contain two of the three elementary hypotheses. Farther up there is a global null hypothesis. Note that in the case of a three-arm experiment any intersection of two elementary null hypotheses leads to the global null hypotheses and vice versa (Bauer 1991). So, the middle level is redundant in this case. It follows that if one can reject the global null hypothesis at significance level α before rejecting an elementary null hypothesis also at significance level α, the FWER is controlled at α. The global null hypothesis of all groups being the same may be tested by ANOVA, the chi-squares test, the (c-by-r extended version of) Fisher's exact test, the log-rank test, and so on, depending on the types of variables and appropriateness of model assumptions. These tests need to be valid themselves for the close testing procedure to be valid. But also note that in the case of four groups instead of three, the intersection hypotheses are not redundant and they need to be tested also at significance level α.

TABLE 14.4

Three Levels of Hypotheses in Close Testing Involving Three Elementary Null Hypotheses

Compare One Endpoint between Three Groups (Subscript A, B, C)	
Global null hypothesis	$(\mu_A = \mu_B = \mu_C)$
Intersection null hypotheses	$(\mu_A = \mu_B \cap \mu_A = \mu_C)$ $(\mu_A = \mu_B \cap \mu_B = \mu_C)$ $(\mu_A = \mu_C \cap \mu_B = \mu_C)$
Elementary null hypotheses	$\mu_A = \mu_B$ $\mu_A = \mu_C$ $\mu_B = \mu_C$

Compare Three Endpoints (Subscript 1, 2, 3) between Two Groups (Subscript A, B)	
Global null hypothesis	$(\mu_{A1} = \mu_{B1} \cap \mu_{A2} = \mu_{B2} \cap \mu_{A3} = \mu_{B3})$
Intersection null hypotheses	$(\mu_{A1} = \mu_{B1} \cap \mu_{A2} = \mu_{B2})$ $(\mu_{A1} = \mu_{B1} \cap \mu_{A3} = \mu_{B3})$ $(\mu_{A2} = \mu_{B2} \cap \mu_{A3} = \mu_{B3})$
Elementary null hypotheses	$\mu_{A1} = \mu_{B1}$ $\mu_{A2} = \mu_{B2}$ $\mu_{A3} = \mu_{B3}$

Example 14.1 (Continued)

The LAIS (Luntamo et al. 2010) reported that "to prevent inflated type I errors caused by multiple comparisons, we began hypothesis testing with global null hypotheses of all three groups being identical before doing pairwise comparisons. We tested the global null hypotheses either with Fisher's exact test (for binary endpoints) or analysis of variance (quantitative endpoints)." The primary endpoint was the proportion of preterm deliveries. The global null hypothesis of equal proportion across all three groups was rejected at the 0.05 level (P = 0.038). The hypothesis of equal proportion between standard care and AZI-SP was also rejected at the 0.05 level (P = 0.013). This was a valid conclusion with strong control of FWER at the 0.05 level.

14.5.2 Multiple Endpoints

The same principle applies to the case of multiple endpoints in a comparison between two groups. The lower panel of Table 14.4 shows an example of three elementary null hypotheses concerning three endpoints. Each of them is contained in two intersection null hypotheses and a global null hypothesis. A generic method to test the intersection null hypothesis and the global null hypothesis is to use the Bonferroni adjustment. That is, each of the K elements in the global null hypothesis is to be tested at the significance level α/K. In Table 14.4, the global null hypothesis includes K = 3 elements, or equivalently the raw P-values are multiplied by K (3 in the present example) and compared with α. If at least one of the three elements is rejected after the Bonferroni adjustment, the global null hypothesis is rejected. In the intersection of two hypotheses, each of the two elements is to be tested at the significance level α/K, K = 2. Or equivalently the raw P-values are multiplied by 2 and compared with α. If at least one of the two elements is rejected after the Bonferroni adjustment, the intersection hypothesis is rejected. Finally, each individual elementary null hypothesis is tested at the significance level α. If all the global, intersection, and elementary null hypotheses concerning an endpoint are rejected, the FWER is controlled at α. Equivalently, the smallest P-values among all hypotheses concerning an elementary null hypothesis is the adjusted P-value of that elementary null hypothesis.

Example 14.1 (Continued)

The LAIS has no multiple endpoints issue because it defined only one primary endpoint (preterm delivery). The study demonstrated that AZI-SP was better than standard SP in terms of preterm delivery (<37weeks), very preterm delivery (<35 weeks), and low birth weight (<2500 grams). The P-values were 0.013, 0.019, and 0.021, respectively. Suppose the study compared AZI-SP and standard SP, and used all three variables as coprimary endpoints. Rejection of any one of the three null hypotheses would lead to recommendation of AZI-SP. Then there would be a multiple end-points issue.

Table 14.5 shows the elementary, intersection, and global null hypotheses for the case of three endpoints. Unlike the case of three-group comparison, the intersection null hypotheses do not lead to the global null hypothesis and they had to be tested as well. The table shows the raw P-values for the elementary hypotheses and Bonferroni adjusted P-values for the global and intersection null hypotheses. Based on the universal null hypothesis that all $H_{(1)}$, $H_{(2)}$, and $H_{(3)}$ were true, $H_{(1)}$ was rejected at 0.05/3 level although $H_{(2)}$ and $H_{(3)}$ were not. Alternatively, one can say that the Bonferroni adjusted P-value for this global hypothesis was the minimum of the three raw P-values multiplied by 3, or $3 \times 0.013 = 0.039$. Similarly, the two intersection hypotheses concerning $H_{(1)}$ and either $H_{(2)}$ or $H_{(3)}$ had adjusted P-values $2 \times 0.013 = 0.026$, and the intersection hypothesis concerning $H_{(2)}$ and $H_{(3)}$ had adjusted P-values $= 2 \times 0.019 = 0.038$. Under the close testing procedure, one could conclude that the elementary null hypothesis concerning pre-term delivery could be rejected at the 0.05 level, as all hypotheses containing $H_{(1)}$, including $H_{(1)}$, $H_{(1)} \cap H_{(2)}$, $H_{(1)} \cap H_{(3)}$, and $H_{(1)} \cap H_{(2)} \cap H_{(3)}$ had been rejected at the 0.05 level. The P-value was the largest of all the hypotheses involved, that is, the maximum of 0.039, 0.026, 0.026, and 0.013. Hence, the P-value from the close testing is 0.039. Similarly, the other two elementary hypotheses could also be rejected by the close testing, with both P-values being 0.039.

Using Equation (14.2), the Holm adjusted P-values were also 0.039 and gave the same conclusions as the Bonferroni-based close testing procedure. This is no coincidence. These two methods are identical.

TABLE 14.5

Comparisons of Three Major Endpoints in Standard SP and AZI-SP Groups in LAIS

Hypothesis		Raw P-Value	Bonferroni-Based Close Testing
$H_{(1)} \cap H_{(2)} \cap H_{(3)}$			0.039[*]
$H_{(1)} \cap H_{(2)}$			0.026[†]
$H_{(1)} \cap H_{(3)}$			0.026[§]
$H_{(2)} \cap H_{(3)}$			0.038[**]
$H_{(1)}$	Preterm delivery	$P_{(1)} = 0.013$	0.013
$H_{(2)}$	Very preterm delivery	$P_{(2)} = 0.019$	0.019
$H_{(3)}$	Low birth weight	$P_{(3)} = 0.021$	0.021

Source: Luntamo, M., Kulmala, T., Mbewe, B., et al., 2010, Effect of repeated treatment of preg-nant women with sulfadoxine-pyrimethamine and azithromycin on preterm delivery in Malawi: A randomized controlled trial, *American Journal of Tropical Medicine and Hygiene* 83:1212–1220, Table 2.

[*] $3 \times \min(P_{(1)}, P_{(2)}, P_{(3)})$
[†] $2 \times \min(P_{(1)}, P_{(2)})$
[§] $2 \times \min(P_{(1)}, P_{(3)})$
[**] $2 \times \min(P_{(2)}, P_{(3)})$

As the number of endpoints increases, it would become tedious to apply the Bonferroni-based close testing procedure. The same result can be obtained by using the Holm adjustment. Nevertheless, there are other valid ways to test the global and intersection hypotheses. For instance, one may use a valid multivariate test procedure to test the global null hypothesis (Johnson and Wichern 2002). Therefore, the close testing procedure is more general than the Holm adjustment.

15

Regression Analysis Strategy

15.1 Introduction

In large-scale randomized experiments, the primary analysis usually can be simple. The random allocation of exposure and the large sample size minimize the chance of confounding and provide strong evidence for causal relationship. The primary analysis may require only a simple two-group comparison of one major endpoint, such as a two-sample t-test or a chi-square test. However, in many other situations, the relationships among variables are complex. Multivariable regression analysis is an important tool to answer the research questions.

The process of determining what variables are to be included in the multivariable regression analysis is called *variable selection* or *model building*. The analysts may need to build multiple models and compare the findings in order to fully understand the relationships. In statistics, *parsimony* refers to the use of as few variables as possible to explain the phenomena concerned. Montgomery et al. (2001) said, "One should always maintain a sense of parsimony, that is, use the simplest possible model that is consistent with the data and knowledge of the problem environment." In statistical practice, however, there is often overemphasis about "the data" and too little consideration about "knowledge of the problem environment" (p. 223). The data is a mixture of truth, and systematic and random errors. Model development that focuses exclusively on consistency with sample data may result in models that disagree with the knowledge and failure to reflect the truth in the population.

The choice of analysis strategy should be based on the specific research aims, not the other way around. A change in model reflects a change in research aims. As an example, analysis of child weight adjusted for age is about the phenomenon of underweight. In contrast, analysis of child weight adjusted for height concerns the phenomenon of wasting. It is paramount to have a clear view about the specific research aims before attempting to build a regression model. Otherwise it is easy to give the right answer to the wrong question.

15.2 Rationale of Using Multivariable Regression

Multivariable regression analysis is often used to explore the potential associations between variables and to test hypotheses while adjusting for potential confounders. Inclusion of covariates may also improve precision and statistical power. Some conceptual issues should be noted before considering the variable selection techniques.

15.2.1 Positive and Negative Confounding

As discussed in Chapter 2, positive and negative confounding can arise in observational studies and also in randomized experiments if the sample size is small. Researchers and readers are usually aware of positive confounding. However, it appears that negative confounding is often ignored, as reflected in popular covariate selection strategies. This will be discussed again in a later section.

There are many occasions that the covariates are not confounders. Figure 15.1a,b illustrates situations that Z is not a confounder. Instead, it is part of the causal mechanism, or an *intermediate variable* through which the exposure impacts on the outcomes. In panel (a) X impacts on the outcome Y only via Z. In Figure 15.1b, X impacts on the outcome Y partly through Z. In the assessment of causality, intermediate variables are usually not to be adjusted for. This is because the adjustment would remove the impact of

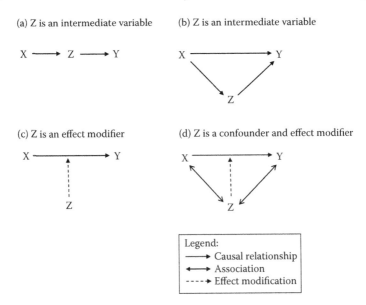

FIGURE 15.1
Examples of relationship and confounding.

the exposure variable. Sometimes the research question is about whether there is another (unknown) mechanism through which the exposure causes an impact. In other words, is Figure 15.1a or Figure 15.1b a better description of the truth? In this case, one may consider estimating and comparing two models, one with and one without adjustment for the intermediate variable Z. If the association between X and Y disappears after the adjustment, the pattern in Figure 15.1a is more plausible. If the association weakens but remains visible, the pattern in Figure 15.1b is more plausible. For this interpretation to work, the measurement of the variables needs to be precise. Otherwise it would be unclear whether the remaining association reflects alternative causal mechanisms or insufficient adjustment.

15.2.2 Baseline Characteristics in Randomized Trials

A contentious issue in randomized trials is whether the analysis should include statistical adjustment for baseline characteristics. If the sample size is large, randomization makes it likely that the groups are comparable at baseline. Imbalance can occur even if the randomization is properly conducted, especially when the sample size is modest. When it occurs, there may be a concern that the comparison is confounded by unbalanced baseline prognostic factors and therefore may mandate the adjustment for them. A fairly common response to the issue is that in a properly conducted randomized trial, imbalance in baseline prognostic factors is a matter of chance. This has been reflected by the P-value in the simple comparison; no adjustment is needed (e.g., Nauta 2010).

Senn (2005) disagreed with this view and made an analogy of rolling two fair dice, one red and one black. The red dice is rolled first, and then the black dice. A researcher is to place a bet on the sum of the scores. One scenario is that the researcher has to place the bet before the two dice are rolled. Another scenario is that the researcher is allowed to place or adjust the bet after the score on the red dice is revealed. In the first scenario, the researcher has no chance to change his or her mind. In the second scenario, it would be odd if the researcher ignores the information from the red dice. Consider the red dice data on baseline prognostic factors, and the black dice the subsequent data. Following the analogy, it would be odd that the researcher ignores the baseline data in the analysis. "One should not use the fact that one has randomized as an excuse for ignoring baseline prognostic information in analyzing a clinical trial" (Senn 2005, p. 26).

Although guidelines from the International Conference on Harmonisation (ICH) of Technical Requirements for Registration of Pharmaceuticals for Human Use do not provide operational details, it is clear that ignoring the baseline imbalances is not preferred. ICH Guideline E9 (1998) states:

> The play of chance may lead to unforeseen imbalances between the treatment groups in terms of baseline measurements not predefined as

covariates in the planned analysis but having some prognostic importance nevertheless. This is best dealt with by showing that an additional
analysis which accounts for these imbalances reaches essentially the
same conclusions as the planned analysis. If this is not the case, the effect
of the imbalances on the conclusions should be discussed (p. 31).

The practical problems include how to define imbalance (there always is
some!), how do we decide on which baseline prognostic factors are important,
and do we adjust for multiple comparisons if there are any. If adjustment for
the baseline prognostic factor is allowed too freely, it offers the opportunity to
commit "data torture." There is no simple solution if adjusted and unadjusted
analyses disagree. Ideally the researchers should focus on only important
prognostic factors and consider what a practically significant imbalance is.
The decisions concerning controlling for baseline imbalances should be made
before seeing the results of the adjusted and unadjusted analysis.

Example 15.1

The Lungwena Antenatal Intervention Study (LAIS) assessed the impact
of sulfadoxine-pyrimethamine (SP) and azithromycin (AZI) on gestational duration (GA) and birth weight. The hypothesis was that prevention
of infectious diseases would prevent preterm birth and therefore improve
birth weight. Although preterm birth was itself the primary outcome variable, it was also the intermediate variable through which the interventions
impacted on birth weight. To estimate the association between the interventions and birth weight with adjustment for GA would underestimate
the true impact. It might be of interest, however, to conduct a secondary
analysis that adjusted for GA to examine whether there was any other
mechanisms through which the intervention affected birth weight.

In the LAIS, parity, malaria parasitemia, and maternal HIV positivity
were important prognostic factors. Despite a fairly large sample size of
about 400 women per group, it was found that the standard care group
had a poorer prognosis in terms of more nulliparous women (25.2%) and
more women with malaria parasitemia (11.3%) at baseline, as compared
to 20.1% and 6.1% in these two variables in the AZI-SP group. The prevalence of the two risk factors differed by about 5%. To a lesser extent,
the standard care group had lower HIV positivity than the monthly SP
group (11.0% versus 14.5%). The study chose to include sensitivity analyses with parity and malaria parasitemia as categorical variables using
generalized linear models. It was reported that the adjusted and unadjusted analyses gave similar results. The difference in HIV positivity
was considered small and not adjusted for.

15.2.3 Reflection of Study Design

Some clinical trials employ stratified randomization (Poccok 1983). Potential
study participants are classified according to a stratification factor measured

at baseline, such as gender or age groups. For each level of the factor, there is one randomization list. This method assures that the intervention groups will be approximately comparable in terms of the stratification factor. The use of stratification implies the researchers' belief that the stratification factor is predictive of the outcome, otherwise there is no reason to use a stratified randomization scheme. Then it is logically consistent to include such variables in the analysis (Senn 1997). Inclusion of independent variables that explains the variation in the outcomes usually though not necessarily improves precision (Robinson and Jewell 1991). Believing that there are predefined predictors of the outcomes and yet not including them in the analysis is to willingly give up statistical power.

15.2.4 Prediction and Prognostic Index

Although a lot of regression analyses are attempts to identify causal relation, there are times that the purpose is to predict. The prediction may identify people who are at high risk of an adverse outcome or in need of a type of service. Then services may be provided for these people accordingly. While a single predictor is sometimes sufficient to identify the at-risk population, the use of multiple predictors can be considered for improving the predictive accuracy.

Example 15.2

Although some forms of early sexual maturation in girls are benign, there was a concern that rapidly progressive central precocious puberty (RPCPP) might have morbid origins. Early recognition of RPCCP followed by more intensive investigations was suggested (Calcaterra et al. 2009). A prognostic index was developed to differentiate RPCPP from other forms of precocious puberty. The prognostic index assigned a score to each of five binary variables, namely, uterine volume ≥5, estradiol ≥50 pmol/L, bone age >2 SD, presence of endometrial thickness, and breast volume ≥0.85 cm^3. The sum of the five component scores gave the RPCPP risk score that ranges from 0 to 7. A girl with a risk score >4 was considered at high risk.

15.2.5 Answering Different Questions

The use of multivariable regression can help us answer multiple questions. In Section 15.2.1 we alluded to this point. By comparison of models with and without adjustment for an intermediate variable we may examine whether there are other unknown causal mechanisms. In the next section we will discuss a typical problem in the studies of growth and development. It sheds light on how to use regression analysis to answer different questions.

15.3 Point Measures, Change Scores, and Unexplained Residuals

The use of point measures, change scores, and unexplained residuals were introduced in Chapter 11. The popularity of unexplained residuals is a relatively new phenomenon. To a certain extent it was a response to the difficulties in the interpretation of multiple regression coefficients in the studies of the fetal origin hypothesis. This section focuses on the situation that the exposure is measured at two ages, one early and one late measure denoted by x_1 and x_2, respectively.

Lucas et al. (1999) reported that many studies of the fetal origin hypothesis showed an association between size at birth (x_1) and adult cardiovascular and metabolic diseases only after regression adjustment for the adult body size (x_2). They discussed several regression models. The *early (late) model* only includes the early (late) size as an independent variable. The *combined model* includes both the early and late sizes as independent variables. They maintained that, having adjusted for late size, the regression coefficient for early size does not represent the effect of early size itself. Equations (15.1) and (15.2) in Figure 15.2 elaborate the rearrangement of the combined model. Equation (15.1) uses the superscript (C) to emphasize that the regression coefficients

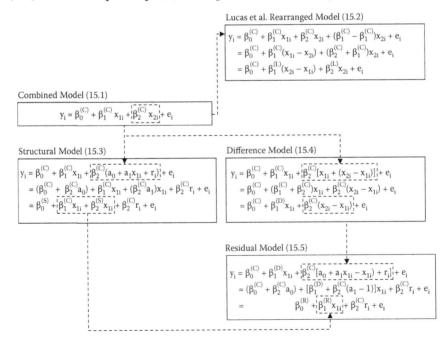

FIGURE 15.2
Multiple ways to formulate a regression model.

are obtained by fitting this combined model equation. It can be seen that the combined model Equation (15.1) is equivalent to Equation (15.2) that estimates the regression coefficient for the change $(x_2 - x_1)$ conditional on late exposure (x_2). The *negative* of the regression coefficients for early size in Equation (15.1) can be interpreted as the regression coefficients for the change $(x_2 - x_1)$, after adjustment for late size, that is,

$$-\beta_1^{(C)} = \beta_1^{(L)}$$

The last line of Equation (15.2) uses a different superscript (L) to emphasize that the regression coefficient is the value from fitting a regression equation with the change score as one of the independent variables. It should be emphasized that the coefficient $\beta_1^{(L)}$ does not indicate the effect of the change, but the effect of the change *after adjustment for the late size*.

Debates on what variables to use and how to interpret the regression findings appear to continue after the discussion by Lucas et al. (1999). In my view, a problem is that it is unclear why the analysis of a change should be adjusted for the later exposure. It is counterintuitive to ask, "Given what will happen later, what is the impact of the present change?" It is much more understandable to ask, "Given what happened earlier, what is the impact of the present change?"

It is more intuitive to rearrange Equation (15.1) to the form of Equation (15.3) and interpret the coefficients accordingly. I call this *structural model* because the concept is the same as *structural equation modeling* (SEM). First, x_2 can be expressed as a linear function of x_1 plus a residual, as described in Chapter 11, Section 11.1. Then we can see that x_1 has a *direct effect* on the outcome, $\beta_1^{(C)}$, and an *indirect effect* via being a determinant of x_2, $(\beta_2^{(C)} a_1)$. The direct effect is direct in the sense that it does not affect the outcome via the late exposure. The indirect effect estimated in an SEM (Sobel 1987) is identical to the product of the regression coefficients $(\beta_2^{(C)} a_1)$. As such, instead of interpreting $\beta_1^{(C)}$ as the regression coefficient for change adjusted for late exposure, it can be interpreted as the direct effect of the early exposure. An important issue in understanding the model is that the late size should not be considered a confounder. It is an intermediate variable.

Li and Gandhi (personal communication, April 2012) suggested to show the rearrangement of the combined model to a *difference model*, as in Equation (15.4), and a *residual model*, as in Equation (15.5). As shown in Figure 15.2, the late exposure x_2 can be seen as a sum of the early exposure x_1 and the change score $\Delta x = (x_2 - x_1)$. Rearranging the elements in Equation (15.4) shows that the regression coefficient for the change score adjusted for early exposure is identical to that for the late exposure variable in Equation (15.1), that is, $\beta_2^{(C)}$. This answers the question, "Given what happened earlier, what is the impact of the present change?" Early growth and development tends to be negatively related to the subsequent change, say, a newborn who suffered intrauterine growth restriction may be light at birth, but then the baby caught up in weight after the release from the growth restricting intrauterine environment. Having controlled for the subsequent change, the regression coefficient for the early exposure, $\beta_1^{(D)}$, answers a hypothetical question of what if the early exposure was not followed by catch-up (or catch-down) growth.

Example 15.3

Cheung et al. (2002) examined the association between weight at birth (z-score), childhood weight gain from birth to 7 years (Δ z-score), and elevated level of psychological distress in adulthood. Having adjusted for sociodemographic and maternal health covariates but not childhood weight gain, one z-score increase in birth weight was associated with a 5% decrease in the odds of elevated psychological distress (OR = 0.95). Having added childhood weight gain into the regression equation, one z-score increase in birth weight was associated with a 10% decrease in the odds (OR = 0.90). The authors concluded that the impact of a lighter birth weight might be compensated by catch-up weight gain during childhood. That is, lighter babies tended to gain more weight during childhood, which might compensate the impact of a lower birth weight. Therefore, analysis of birth weight without adjustment for childhood weight gain showed a weaker association with the outcome (OR = 0.95). Having adjusted for childhood weight gain, the odds ratio for birth weight concerned the question of what if a lighter birth weight was not accompanied by more childhood weight gain. The association became more pronounced (OR = 0.90).

The difference model can be rearranged to become the residual model in Equation (15.5). The change $(x_2 - x_1)$ can be partitioned into two elements: expected change and unexplained residual. This can be shown by replacing x_2 in the last line of Equation (15.4) by a component predicted by x_1 and a residual, r. After this rearrangement, the regression coefficient for the early exposure is now a sum of $\beta_1^{(D)}$ and the effect of the consequential expected change. This may be seen as the *total effect* of the early exposure. The estimates of the direct and indirect effects in a SEM sum to this total effect estimate.

In short, by choosing to use x_1, x_2, r, or Δx in the multivariable regression equation, the analysis can answer different questions.

Example 15.4

In the Simulated Clinical Trial (SCT) dataset, infants with lower birth weight had higher weight gain from birth to 4 weeks. Developmental ability at age 36 months was positively related to both birth weight and change in weight from birth to 4 weeks. Table 15.1 shows the results of ordinary least-squares regression analysis of the early, combined, difference, and residual models. The regression could have included other covariates if that was desired. Without adjustment for any variables, a 1 kilogram increase in birth weight was associated with a 0.971 point increase in the

TABLE 15.1

Regression Analysis of Developmental Score at 36 Months: SCT Data

Variables	Early Model, $y = b_0 + b_1x_1$	Combined m = Model, $y = b_0 + b_1x_1 + b_2x_2$	Difference Model, $y = b_0 + b_1x_1 + b_2\Delta x$	Residual Model, $y = b_0 + b_1x_1 + b_2r$
Birth weight (kg), x_1	0.971	0.614	0.978	0.971
Weight 4 weeks (kg), x_2		0.364		
Change from birth to 4 weeks (kg), Δx			0.364	
Unexplained residuals at 4 weeks (kg), r				0.364

ability outcome. Having adjusted for birth weight, weight at 4 weeks, change in weight, and conditional gain in weight (unexplained residuals) all have the same regression coefficient, $\beta_2^{(C)} = 0.364$, as illustrated in Figure 15.2. Having adjusted for weight at 4 weeks, the association between birth weight and the outcome weakened to $\beta_1^{(C)} = 0.614$. As illustrated in Equation (15.3), this was because this coefficient only reflected the direct effect. Having adjusted for change in weight, which removed the compensatory effect of catch-up growth, the association strengthened to $\beta_1^{(D)} = 0.978$. It answered a somewhat different question about what if there was no catch-up growth. Having adjusted for the conditional gain (unexplained residuals), the regression coefficient remained $\beta_1^{(R)} = 0.971$. The residual model told more than the early model and potentially improves the precision of the regression coefficient of the early exposure variable. So, this model should usually be preferred over the early model.

Display 15.1 shows the Stata codes and outputs for estimating an SEM model for the two-equation system:

$$\text{Developmental score} = b_0 + (b_1 \times \text{Birth weight}) + (b_2 \times \text{Weight 4 weeks})$$

$$\text{Weight 4 weeks} = (a_0 + a_1 \times \text{Birth weight})$$

The intention here was not to discuss SEM, but to reiterate the interpretation of the various regression models. One can see that the early model and the SEM gave the same estimate of the total effect of birth weight (0.971). The combined model and the SEM gave the same direct effect estimate (0.614). The indirect effect was 0.357. One could equivalently obtain the indirect effect from taking the difference between the least-squares regression coefficients 0.971 and 0.614 from the regression models in Table 15.1.

15.4 Issues in Variable Selection

15.4.1 Forced-Entry

Machin et al. (2006) proposed classifying variables as *design variables* (D), *known variables* (K), and *query variables* (Q) according to prior knowledge and assumptions about their associations with the key exposure and outcome variables. A design variable is part of the key feature of the study design. In a randomized trial, the intervention allocated to the participants is a design variable. In an observational study with repeated measurement that aims to estimate how an outcome changes as people aged, age is a design variable. If a randomized trial uses stratified randomization, the stratification factor is also a design variable. A known variable is known or assumed to be a predictor of the outcome or a confounder of the associations concerned.

```
. sem (ability <- bw weight4wk) (weight4wk <- bw)

(output omitted)
```

```
-----------------------------------------------------------------------------
               |                 OIM
               |    Coef.   Std. Err.     z    P>|z|   [95% Conf. Interval]
---------------+-------------------------------------------------------------
Structural     |
  ability <-   |
    weight4wk  | .3644522   .1860463    1.96   0.050   -.0001919   .7290963
           bw  | .6137203   .2325267    2.64   0.008    .1579764   1.069464
        _cons  | 32.67902   .4778981   68.38   0.000    31.74235   33.61568
---------------+-------------------------------------------------------------
  weight4wk <- |
           bw  | .9805748   .0200092   49.01   0.000    .9413575   1.019792
        _cons  |  1.11706   .0597239   18.70   0.000    1.000003   1.234117
---------------+-------------------------------------------------------------
```

```
(output omitted)

. estat teffects

Direct effects
```

```
---------------+-----------------------------------------------------------
               |                 OIM
               |    Coef.   Std. Err.     z    P>|z|   [95% Conf. Interval]
---------------+-----------------------------------------------------------
Structural     |
  ability <-   |
    weight4wk  | .3644522   .1860463    1.96   0.050   -.0001919   .7290963
           bw  | .6137203   .2325267    2.64   0.008    .1579764   1.069464
---------------+-----------------------------------------------------------
```

```
(output omitted)

Indirect effects
```

```
---------------------------------------------------------------------------
               |                 OIM
               |    Coef.   Std. Err.     z    P>|z|   [95% Conf. Interval]
---------------+-----------------------------------------------------------
Structural     |
  ability <-   |
    weight4wk  |        0  (no path)
           bw  | .3573726    .182578    1.96   0.050   -.0004737    .715219
---------------+-----------------------------------------------------------
```

```
(output omitted)

Total effects
```

```
---------------------------------------------------------------------------
               |                 OIM
               |    Coef.   Std. Err.     z    P>|z|   [95% Conf. Interval]
---------------+-----------------------------------------------------------
Structural     |
  ability <-   |
    weight4wk  | .3644522   .1860463    1.96   0.050   -.0001919   .7290963
           bw  |  .971093   .1443616    6.73   0.000    .6881494   1.254037
---------------------------------------------------------------------------
```

DISPLAY 15.1
Stata codes and outputs: Structural equation modeling of the SCT data.

The knowledge or assumption about whether a variable is a known variable usually comes from the literature, but it may also come from the researchers' experience in the subject matter. A query variable is a variable whose influence is unknown and yet we have to assess it. It is obvious that D variables are to be included in the regression model.

Machin et al. (2006) suggested that K variables should usually be included by *forced-entry*, which refers to including the variables in a regression model regardless of the empirical findings from the present dataset. This approach has been underutilized. It is important to know that there is no statistical test for confounding (Pearl 1998). Test for imbalances in baseline characteristics does not reveal confounding (Senn 1997). For variables known or suspected to confound the relationship of interest, forced-entry is needed to remove doubt about the interferences they may cause.

Inclusion of each redundant variable leads to a loss of one degree of freedom, which is like reducing the sample size by one subject. An important consideration in the use of forced-entry is the number of variables. If there is a large number of K variables (relative to the sample size), forced-entry may cause a serious loss in degrees of freedom. In analyses that aim to explore associations, the number of Q variables can be small or large. If it is a small number, I would still prefer to force-enter the variables. Even if there is no practically or statistically significantly association with the outcome, the point estimate and confidence interval may still be research findings worth reporting.

Example 15.3 (Continued)

In the study of adult psychological health in relation to birth weight and childhood weight gain, Cheung et al. (2002) forced-entered gender, age at follow-up, gestational duration, father's social class, mother's marital status, parity, maternal smoking, and housing tenure as K variables in the regression models. The choice of these variables was based on knowledge from the literature. They were specified before conducting the statistical analysis. Since most of these variables were categorical and needed to be expressed as indicator variables, the models included 19 parameters for the K variables. The total sample size was 9731 people. For such a large sample size, including 19 parameters for K variables as forced-entries raised no concern about loss in precision.

15.4.2 Improper Forward Selection

As discussed in Sun et al. (1996), there is a common but improper implementation of forward selection. They called it *bivariate screen*. This is based on a two-step procedure. In step 1, a series of simple regression models are fitted, each using only one of the candidate variables as the independent variable and the statistical significance is calculated. In step 2, all of the candidate variables that showed a statistically significant association with the dependent variable in step 1 are included.

This approach is problematic because it ignores negative confounding. In step 1, independent variables that have a positively confounded association with the dependent variable may show a statistical significance. They are included in step 2, which controls for the other variables. Therefore, step 2 adjusts for the positive confounding effect and shows the truer association. However, independent variables that have a negatively confounded association with the dependent variable may not show a statistical significance in step 1 to begin with. They are excluded from further investigation in step 2. The true associations are not revealed.

Example 15.5

A cross-sectional study examined the social gradient in birth weight in Hong Kong (Cheung and Yip 2001). In a bivariate screen, the mother's education level showed almost no association with birth weight. Based on the bivariate screen method, the mother's education would have been excluded from further investigation.

However, parity was related to both mother's education and birth weight. Mothers with less education were more likely to have more births, that is, not first parity. The first birth tended to be lighter in weight. Figure 15.3 shows the birth weight by mothers' education and parity, having adjusted for other covariates. At the population level, birth weight varied little across education level and showed no clear

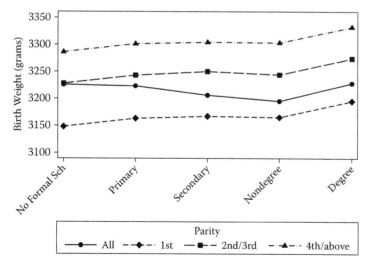

FIGURE 15.3
Birth weight in relation to mother's education and parity. (Data from Cheung, Y. B., and Yip, P. S. F., 2001, Social patterns of birth weight in Hong Kong, 1984–1997, *Social Science & Medicine* 52:1135–1141, Table 3.)

gradient (solid line). Mothers with no formal schooling and mothers with university degrees gave birth to babies with similar weight on average. However, in each stratum defined according to parity, there was a consistent gradient that mothers with no formal schooling gave birth to lighter babies, whereas mothers with university degrees gave birth to heavier babies. The three levels in between did not vary much in mean birth weight. Having adjusted for parity, mother's education was statistically significantly associated with birth weight ($P < 0.01$).

Note that the differences in mean birth weight across education level were small, less than 50 grams. Despite statistical significance, the practical significance of the mother's education appeared to be small in this population. Also note that the interpretation of the findings depends on how we conceptualize the relationship between variables. If we theorize that parity was associated with but not causally determined by education, the analysis showed negative confounding. That is, the association between dependent and independent variables were suppressed by parity. However, if we theorize that education causally influenced parity, we should interpret the lack of association when not adjusting for parity as the total effect. In this conceptual framework, the positive association conditional on parity was one element of the education effect. But education also had an indirect effect on birth weight via lowering parity, which negatively affected birth weight. The sum of the opposite elements roughly canceled out each other in this causal framework.

15.4.3 Further Remarks

I have not intended to discuss in detail statistical means for variable selection. This is because in my opinion the use of subject-matter knowledge and forced-entry is the major approach in a wide range of situations. Readers who are interested in statistical selection of variables are referred to Machin et al. (2006) for a review of conventional methods based on significance tests and change in estimate, and to Hesterberg et al. (2008) and Li and Sillanpää (2012) for reviews of modern methods based on penalized regression.

15.5 Interaction

15.5.1 Concepts

Interaction is also called *effect modification*. It refers to a situation that the strength or direction of an association between an exposure (x_1) and an outcome (y) is affected (or modified) by a third variable (x_2), as depicted in Figure 15.1c.

Example 15.1 (Continued)

In the Lungwena Antenatal Intervention Study (LAIS), it was hypothe-sized *a priori* that the impact of monthly SP as a preemptive treatment for malaria versus standard SP schedule depended on the use of a mosquito bed net at home. For illustration purposes I will ignore the AZI-SP group here. It was hypothesized that the intervention effect would be small among bed net users, because they were less likely to get malaria any-way. The effect of monthly SP was hypothesized to be stronger among nonusers of bed nets. The trial showed that, among nonusers, monthly SP reduced preterm birth by 38% (risk ratio = 0.62; 95% CI = 0.38 to 1.02; P = 0.08). Among bed net users, monthly SP showed no benefit (risk ratio = 1.05; 95% CI = 0.72 to 1.52; P = 0.82). Use of a bed net appeared to be an effect modifier.

It is possible that the third variable (x_2) is also an exposure status of inter-est. If x_1 and x_2 individually both have a positive effect on the outcome, but when both are present neither of them have any effect, they are said to be *antagonistic*. In contrast, if x_1 and x_2 individually has no effect on the outcome but their joint presence has a positive effect, they are said to be *synergistic*. For example, there was a concern about improvement of water supplies alone and sanitary facilities alone may have no effect on child growth (Cheung 1999). If the improvement of both water supplies and sanitary facilities are required to make an impact, they are synergistic. It is possible that an effect modifier is also a confounder, as depicted in Figure 15.1d. In the LAIS, parity (Z) was hypothesized to be an effect modifier (dashed arrow). The interven-tions were suspected to give more benefit to women who had no previous pregnancy. By chance, there was some baseline unbalance in parity between the intervention groups (two-sided arrows linking X and Z). Furthermore, parity was suspected to be related to preterm delivery (two-sided arrows linking Z and Y). So it had a complex role in the web of relationship.

15.5.2 Analysis

Although there is no statistical test for confounding, interaction can be tested. In the simple case of one exposure variable (x_1) and one potential effect modifier (x_2), a regression model in the following form can be fitted:

$$g(y) = \alpha + \beta_1 x_1 + \beta_2 x_2 + \beta_3 x_3 \tag{15.6}$$

where $x_3 = (x_1 \times x_2)$ and $g(.)$ is a link function in a generalized linear model. A regression coefficient β_3 that is statistically significant supports the presence of interaction. The statistical significance can be calculated by using either the Wald test or the likelihood ratio test. If x_1 and x_2 are both binary variables coded as 0 for absence and 1 for presence, β_1 and ($\beta_1 + \beta_3$)

represent the associations between g(y) and x_1 in the absence and presence of x_2, respectively. The strength of β_3 shows the degree of effect modification.

Example 15.4 (Continued)

The SCT dataset resembled the LAIS and included two intervention groups (A and B) and a control group (C). The comparison of the three groups can be represented by two indicator variables A (1 for group A, 0 otherwise) and B (1 for group B, 0 otherwise). Previous pregnancy was a potential effect modifier (PP = 0 for no previous pregnancy, 1 otherwise). Display 15.2 fitted the following logistic regression model about the log odds of preterm birth:

$$\log(\text{odds}) = \alpha + \beta_1 A + \beta_2 B + \beta_3 PP + \beta_4 (A \times PP) + \beta_5 (B \times PP) \qquad (15.7)$$

where the log odds ratio β_4 and β_5 concerns whether the impact of A and B were dependent on a previous pregnancy of the participants. The "xi:", "i.", and "*" operators generated the categorical and interaction variables.

The results show no statistically significant evidence of a previous pregnancy modifying the effect of treatments A (P = 0.266) or B (P = 0.473). Substituting PP = 0 into Equation (15.7), it can be seen that the effect of,

```
. * interaction
.
. xi: logit preterm i.group*i.prevpreg

(output omitted)
```

```
------------------------------------------------------------------------------
     preterm |    Coef.  Std. Err.     z    P>|z|   [95% Conf. Interval]
-------------+----------------------------------------------------------------
   _Igroup_1 | -.0917085  .2056838  -0.45   0.656   -.4948413   .3114243
   _Igroup_2 | -.5970034  .2231349  -2.68   0.007    -1.03434   -.159667
 _Iprevpreg_1 | -.3197704  .2172655  -1.47   0.141   -.745603   .1060622
_IgroXpre_1_1 | -.3593771  .3229304  -1.11   0.266   -.992309   .2735548
_IgroXpre_2_1 |  .2380153  .3319545   0.72   0.473  -.4126036   .8886343
        _cons | -1.098612  .1421338  -7.73   0.000  -1.377189  -.8200351
------------------------------------------------------------------------------
```

```
. lincom _Igroup_2+ _IgroXpre_2_1

 (1) [preterm]_Igroup_2 + [preterm]_IgroXpre_2_1 = 0
```

```
------------------------------------------------------------------------------
     preterm |    Coef.  Std. Err.     z    P>|z|   [95% Conf. Interval]
-------------+----------------------------------------------------------------
         (1) | -.358988  .2457735  -1.46   0.144   -.8406953   .1227193
------------------------------------------------------------------------------
```

DISPLAY 15.2
Stata codes and outputs: Analysis of interaction in the SCT data.

for example, B (versus C) on the log odds of preterm birth among women with no previous pregnancy was given by the coefficient β_2, which was −0.597 (P = 0.007). By substituting PP = 1 into Equation (15.7), the effect of B among women with a previous pregnancy was given by ($\beta_2 + \beta_5$). In Display 15.2 it was −0.597 + 0.238 = −0.359.

The "lincom" command of Stata can provide not only the stratum specific regression coefficient but also its confidence interval and P-value. Here, B was not statistically significant among women with a previous pregnancy (P = 0.144) despite the statistical significance in the no previous pregnancy stratum (P = 0.007). The 95% confidence interval (CI) of the log ORs in the two strata overlapped substantially: (−1.03 to −0.16) and (−0.84 to 0.12), demonstrating again a lack of statistically significant evidence to show effect modification.

The variables involved in the interaction may be quantitative instead of categorical. It is possible to categorize the quantitative variables and then analyze the interaction using the categorical variables, but it loses power and therefore is usually not recommended. If there is a significant interaction between two quantitative variables, one may want to tell what the effects (and 95% CI and P-value) of x_1 at different levels of x_2 are. Note that it is the same interpretation of Equation (15.6) that β_1 represents the effect of x_1 when $x_2 = 0$ no matter if the variables are quantitative or not. One way to chart the effect of x_1 at various levels of x_2 is to calculate the first, second, and third quartiles of x_2, create new x_2 and interaction variables that are centered at the quartiles, and then perform the regression analyses. Since in each step of the analyses the new variables for x_2 take on the value 0 when the original x_2 is at the quartiles, the regression coefficients (and CI and P-values) for x_1 now represent the effect at the three levels of x_2.

Most studies are planned to have sufficient power to examine the effect of an exposure status in the whole sample. Very few studies are sufficiently powered to test hypotheses about interaction. As such, the use of a significance test in determining whether an interaction term should be included in the model is usually not helpful. On the other hand, if the interactions between many variables are explored, the play of chance can produce statistically significant findings about interaction. It is desirable to identify interaction that is known from the literature or develop hypotheses about interaction prior to data analysis. Significance tests are then performed according to the preselected interaction terms. For important effect modifiers, no matter if the interaction is statistically significant or not, provides stratified analyses for each stratum of the effect modifier.

Example 15.1 (Continued)

In the LAIS, it was hypothesized *a priori* that parity, maternal HIV status, and maternal bed net use were effect modifiers. The likelihood ratio test was used to examine the interaction between each of these three variables and the intervention variable. Despite a lack of statistical significance, comparisons of the three intervention groups in each stratum of parity, maternal HIV status, and bed net use were conducted and reported. Despite the lack of statistical significance, the findings were broadly in line with the prior knowledge and were discussed in the context of what was known in the literature.

15.5.3 Model Dependency

Interaction is a statistical phenomenon that needs to be interpreted with care. The presence or absence of an interaction may depend on the metrics. Table 15.2 shows the real percentage of preterm births in the standard and monthly SP groups of LAIS among women who did not use mosquito bed nets. The percentages among women who did use mosquito bed nets were hypothetical for illustration purposes. The risk difference in each stratum was 8%. Fitting the following generalized linear model for binary outcome with an identity link

$$\text{risk} = \alpha + \beta_1 \text{MonthlySP} + \beta_2 \text{BedNetUse} + \beta_3 (\text{MonthlySP} \times \text{BedNetUse})$$

would give $\beta_1 = -0.08$, $\beta_2 = -0.1$, and $\beta_3 = 0$. It shows no interaction. In contrast, the risk ratios in the two strata were 60% and 20%. Fitting the following generalized linear model for binary outcome with a log link:

$$\log(\text{risk}) = \alpha + \beta_1 \text{MonthlySP} + \beta_2 \text{BedNetUse} + \beta_3 (\text{MonthlySP} \times \text{BedNetUse})$$

would give $\beta_1 = -0.511$, $\beta_2 = -0.693$, and $\beta_3 = -1.099$. The model would give RR for monthly SP = exp(−0.511) = 0.6 in the bed net nonusers and exp(−0.511 − 1.099) = 0.2 in nonusers. That $\beta_3 = -1.099$ suggests interaction. Whether this interaction is statistically significant depends on the sample size.

In short, exactly the same data can be concluded as showing interaction or not depending on how the regression model is formulated. There is no one-size-fits-all definition of which model or metric is most suitable for an analysis.

TABLE 15.2

Illustration of Interaction Being Dependent on Metrics

Bed Net Use	Antenatal Care		Risk Difference	Risk Ratio
	Standard SP Schedule	Monthly SP		
Nonusers	20%	12%	8%	60%
Users	10%	2%	8%	20%

15.6 Role of Prior Knowledge

Statistical analysis requires knowledge about the subject matter. First, it guides the analysts on the selection of confounders to control for and the inclusion of interaction and stratified analysis for effect modifiers. It is uncommon that the selection is based purely on statistical means. Prior knowledge may mandate forced entry of variables or stratified analysis. Second, it facilitates the interpretation of analysis results. In Section 15.3, we have discussed that the interpretation of regression models of early size adjusted or not adjusted for later size requires us to recognize that later size is potentially an intermediate variable on the causal pathway. This consideration applies not only to anthropometric variables but also generally to any variables that may form a web of causal relationships (Victora et al. 1997). Other knowledge such as the timing of events may also help to decide the analysis strategy.

Example 15.6

Hernán et al. (2002) examined the impact of folic acid supplementation in early pregnancy (exposure) on the risk of neural tube defects (outcome). The data included a binary variable of whether the pregnancy ended in stillbirth or therapeutic abortion (pregnancy result). A logistic regression model with and without adjustment for this binary variable resulted in a more than 10% difference in the estimate. Nevertheless, they reasoned that it was inappropriate to adjust for this variable in the estimate of the impact of the exposure on the outcome, because a pregnancy result cannot have influenced the folic acid supplementation in early pregnancy. Therefore, it could not be a confounder.

It is not easy for one person to master both statistical analysis and the subject matter. However, some basic understanding, such as the issue of timing of events in the preceding example, is useful for data analysis and yet possible to be obtained by statistical analysts with limited background in the research topic. Furthermore, a truly collaborative effort among statistical analysis experts and subject-matter experts can provide a synergism that improves the quality of statistical analysis.

References

Aalen, O. O. 1978. Nonparametric inference for a family of counting processes. *Annals of Statistics* 6:701–726.

Aalen, O. O. 1988. Heterogeneity in survival data analysis. *Statistics in Medicine* 7:1121–1137.

Abubakar, A., Holding, P., Van de Vijver, F., et al. 2010. Developmental monitoring using caregiver reports in a resource-limited setting: The case of Kilifi, Kenya. *Acat Paediatrica* 99:291–297.

Aickin, M., and Gensler, H. 1996. Adjusting for multiple testing when reporting research results. The Bonferroni vs Holm methods. *American Journal of Public Health* 86:726–728.

Akl, E. A., Oxman, A. D., Herrin, J., et al. 2011. Using alternative statistical formats for presenting risks and risk reductions. *Cochrane Database Systematic Reviews* 3:CD006776.

Altman, D. G. 1993. Construction of age-related reference centiles using absolute residuals. *Statistics in Medicine* 12:917–924.

Andersen, E. B. 1973. A goodness of fit for the Rasch model. *Psychometrika* 1:123–140.

Andersen, L. G., Angquist, L., Eriksson, J. G., et al. 2010. Birth weight, childhood body mass index and risk of coronary heart disease in adults: Combined historical cohort studies. *PLoS One* 5:e14126.

Armstrong, R. D., Frome, E. L., and Kung, D. S. 1979. Algorithm 79-01: A revised simplex algorithm for the absolute deviation curve fitting problem. *Communications in Statistics: Simulation and Computation* 8:175–190.

Ashworth, A. 1998. Effects of intrauterine growth retardation on mortality and morbidity in infants and young children. *European Journal of Clinical Nutrition* 52:S34–42.

Babbie, E. 2010. *The Practice of Social Research*, 12th ed. Belmont, CA: Wadsworth.

Barker, D. J. P. 1998. *Mothers, Babies and Health in Later Life*, 2nd ed. Edinburgh: Churchill Livingstone.

Bartlett, M. S. 1937. The statistical concept of mental factors. *British Journal of Psychology* 28:97–104.

Bauer, P. 1991. Multiple testing in clinical trials. *Statistics in Medicine* 10:871-890.

Beattie, R. B., and Johnson, P. 1994. Practical assessment of neonatal nutrition status beyond birthweight: An imperative for the 1990s. *British Journal of Obstetrics and Gynaecology* 101:842-846.

Beddo, V. C., and Kreuter, F. 2004a. A Handbook of Statistical Analyses using SPSS [book review]. *Journal of Statistical Software* 11(2).

Beddo, V. C., and Kreuter, F. 2004b. A Handbook of Statistical Analyses using Stata [book review]. *Journal of Statistical Software* 11(3).

Benjamini, Y., and Hochberg, Y. 1995. Controlling the false discovery rate: A practical and powerful approach to multiple testing. *Journal of the Royal Statistical Society, Series B* 57:290–300.

Benn, R. T. 1971. Some mathematical properties of weight-for-height indices used as measures of adiposity. *British Journal of Preventive and Social Medicine* 25:42–50.

Benoit, T. C., Jocelyn, L. J., Moddemann, D. M., et al. 1996. Romanian adoption. The Manitoba experience. *Archives of Pediatrics and Adolescent Medicine* 150:1278–1282.

Bensch, G. W., Greos, L. S., Gawchik, K. S., et al. 2011. Linear growth and bone maturation are unaffected by 1 year of therapy with inhaled flunisolide hydrofluoroalkane in prepubescent children with mild persistent asthma: A randomized, double-blind, placebo-controlled trial. *Annals of Allergy, Asthma and Immunology* 107:323–329.

Berger, R. L., and Hsu, J. C. 1996. Bioequivalence trials, intersection-union tests and equivalence confidence sets. *Statistical Sciences* 11:283–319.

Bergsjø, P., Denman, D. W., Hoffman, H. J., et al. 1990. Duration of human single-ton pregnancy: A population study. *Acta Obstetricia et Gynecologica Scandinavica* 69:197–207.

Berk, L. E. 2003. *Child Development*, 6th ed. Boston: Pearson.

Bhandari, N., Bahi, R, Taneja, S., et al. 2002. Growth performance of affluent Indian children is similar to that in developed countries. *Bulletin of the World Health Organization* 80:189–195.

Binder, D. A. 1983. On the variances of asymptotically normal estimators from com-plex surveys. *International Statistical Review* 51:279–292.

Black, M., and Matula, K. 1999. *Essentials of Bayley Scales of Infant Development II Assessment*. New York: Wiley.

Bland, J. M., and Altman, D. G. 1986. Statistical methods for assessing agreement between two methods of clinical measurement. *Lancet* 1(8476):307–310.

Bland, J. M., and Altman, D. G. 1997. Cronbach's alpha. *British Medical Journal* 314:572.

Bollen, K. A., and Curran, P. J. 2006. *Latent Curve Models: A Structural Equation Approach*. Hoboken, NJ: Wiley.

Box, G. E. P., and Norman, R. D. 1987. *Empirical Model-Building and Response Surfaces*. New York: Wiley.

Bradburn, N. M. 1992. Presidential address: A response to the non-response problem. *Public Opinion Quarterly* 56:391–397.

Calcaterra, V., Sampaolo, P., Klersy, C., et al. 2009. Utility of breast ultrasonography in the diagnostic work-up of precocious puberty and proposal of a prognostic index for identifying girls with rapidly progressive central precocious puberty. *Ultrasound in Obstetrics and Gynecology* 33:86–91.

Cameron, N., Preece, M. A., and Cole, T. J. 2005. Catch-up growth or regression to the mean? Recovery from stunting revisited. *American Journal of Human Biology* 17:412–417.

Chee, M. W. L., Chen, K. H. M., Zheng, H., et al. 2009. Cognitive function and brain structure correlations in healthy elderly East Asians. *NeuroImage* 46:257–269.

Cheung, Y. B. 1999. The impact of water supplies and sanitation on growth in Chinese children. *Journal of the Royal Society for the Promotion of Health* 119:89–91.

Cheung, Y. B. 2000. Marital status and mortality in British women: A longitudinal study. *International Journal of Epidemiology* 29:93–99.

Cheung, Y. B. 2007. A modified least-squares approach to the estimation of risk differ-ence. *American Journal of Epidemiology* 166:1337–1344.

Cheung, Y. B., Albertsson-Wikland, K., Luo, Z. C., et al. 2002. Benn Index is related to postnatal linear growth. *Journal of Pediatric Endocrinology and Metabolism* 15:1161–1166.

Cheung, Y. B., and Ashorn, P. 2009. Linear growth in early life is associated with sui-cidal ideation in 18-year-old Filipinos. *Acta Paediatrica* 99:1719–1723.

Cheung, Y. B., Gladstone, M., Maleta, K., et al. 2008. Comparison of four statistical approaches to score child development: A study of Malawian children. *Tropical Medicine & International Health* 13:987–993.

Cheung, Y. B., Jalil, F., Yip, P. S., et al. 2001a. Association between size at birth, paediatric diarrhoeal incidence and postnatal growth. *Acta Paediatrica* 90:1309–1315.

Cheung, Y. B., Khoo, K. S., Karlberg, J., et al. 2002. Association between psychological symptoms in adults and growth in early life: Longitudinal follow up study. *British Medical Journal* 325:749–751.

Cheung, Y. B., and Yip, P. S. F. 2001. Social patterns of birth weight in Hong Kong, 1984–1997. *Social Science & Medicine* 52:1135–1141.

Cheung, Y.B., Yip, P. S. F., and Karlberg, J. P. E. 2001b. Fetal growth, early postnatal growth and motor development in Pakistani infants. *International Journal of Epidemiology* 30:66–72.

Cheung, Y. B., Yip, P. S. F., and Karlberg, J. P. E. 2001c. Parametric modeling of neonatal mortality in relation to size at birth. *Statistics in Medicine* 20:2455–2466.

Chia, A., Chua, W. H., Cheung, Y. B., et al. 2012. Atropine for the treatment of childhood myopia: Safety and efficacy of 0.5%, 0.1% and 0.01% doses (ATOM2). *Ophthalmology* 119:347–354.

Chitty, L. S., and Altman, D. G. 2003. Charts of fetal size: Kidney and renal pelvis measurements. *Prenatal Diagnosis* 23:891–897.

Clayton, D., and Hills, M. 1993. *Statistical Models in Epidemiology*. Oxford: Oxford University Press.

Cleveland, W. S. 1979. Robust locally weighted regression and smoothing scatterplots. *Journal of the American Statistical Association* 74:829–836.

Cohen, J. 1988. *Statistical Power Analysis for the Behavioral Sciences*, 2nd ed. Hillsdale, NJ: Lawrence Erlbaum Associates.

Colcombe, S. J., and Kramer, A. F. 2003. Fitness effects on the cognitive function of older adults: A meta-analytic study. *Psychological Science* 14:125–130.

Cole, T. J. 1988. Fitting smoothed centile curves to reference data (with discussion). *Journal of the Royal Statistical Society, Series A* 151:358–418.

Cole, T. J. 2004. Children grow and horses race: Is the adiposity rebound a critical period for later obesity? *BMC Pediatrics* 4:6.

Cole, T. J., and Green, P. J. 1992. Smoothing reference centile curves: The LMS method and penalized likelihood. *Statistics in Medicine* 11:1305–1319.

Cole, T. J., Freeman, J. V., and Preece, M. A. 1995. Body mass index reference curves for the UK, 1990. *Archives of Disease in Childhood* 73:25–29.

Cole, T. J., Freeman, J. V., and Preece, M. A. 1998. British 1990 growth reference centiles for weight, height, body mass index and head circumference fitted by maximum penalized likelihood. *Statistics in Medicine* 17:407–429.

Cole, T. J., Henson, G. L., Tremble, J. M., et al. 1997. Birthweight for length: Ponderal index, body mass index or Benn index? *Annals of Human Biology* 24:289–298.

Cox, D. R. 1972. Regression models and life tables (with discussion). *Journal of the Royal Statistical Society, Series B* 34:187–220.

Craik, F. I., Bialystok, E., and Freedman, M. 2010. Delaying the onset of Alzheimer disease: Bilingualism as a form of cognitive reserve. *Neurology* 75:1726–1729.

Cronbach, L. J. 1951. Coefficient alpha and the internal structure of test. *Psychometrika* 16:297–334.

Davis, E. P., Glynn, L. M., Schetter, C. D., et al. 2007. Prenatal exposure to maternal depression and cortisol influences infant temperament. *Journal of American Academy of Child & Adolescent Psychiatry* 46:737–746.

De Jager, P. L., Shulman, J. M., Chibnik, L. B., et al. 2012. A genome-wide scan for common variants affecting the rate of age-related cognitive decline. *Neurobiology of Aging* 33:1017.

de Onis, M., and Habicht, J. P. 1996. Anthropometric reference data for international use: Recommendations from a World Health Organization expert committee. *American Journal of Clinical Nutrition* 64:650–658.

Der, G., and Everitt, B. S. 2009. *A Handbook of Statistical Analyses Using SAS*, 3rd ed. Boca Raton, FL: Chapman & Hall/CRC.

Donner, A. 1986. A review of inference procedures for the intraclass correlation coefficient in the one-way random effects model. *International Statistical Review* 54:67–82.

Drachler, M. L., Marshall, T., and de Carvalho Leite, J. C. 2007. A continuous-scale measure of child development for population-based epidemiological surveys: A preliminary study using item response theory for the Denver Test. *Paediatric and Perinatal Epidemiology* 21:138–153.

Drewett, R. 2007. *The Nutritional Psychology of Childhood*. Cambridge: Cambridge University Press.

Dubios, J., Benders, M., Borradori-Tolsa, C, et al. 2008. Primary cortical folding in the human newborn: An early marker of later functional development. *Brain* 131:2028.

Dunlap, W. P., Dietz, J., and Cortina, J. M. 1997. The spurious correlation of ratios that have common variables: A Monte Carlo examination of Pearson's formula. *The Journal of General Psychology* 124:182–193.

Durbán, M., Harezlak, J., Wand, M. P., et al. 2005. Simple fitting of subject-specific curves for longitudinal data. *Statistics in Medicine* 24:1153-1167.

Efron, B., and Tibshirani, R. J. 1993. *An Introduction to the Bootstrap*. New York: Chapman & Hall/CRC.

Embretson, S. E., and Reise, S. P. 2000. *Item Response Theory for Psychologists*. Mahwah, NJ: Lawrence Erlbaum.

Engle, W. A., and American Academy of Pediatrics Committee on Fetus and Newborn. 2004. Age terminology during the perinatal period. *Pediatrics* 114:1362–1364.

Everitt, B. S. 2001. *A Handbook of Statistical Analyses Using S-PLUS*. Boca Raton, FL: Chapman & Hall/CRC.

Everitt, B. S., and Hothorn, T. 2009. *A Handbook of Statistical Analyses Using R*, 2nd ed. Boca Raton: Chapman & Hall/CRC.

Fairclough, D. 2010. *Design and Analysis of Quality of Life Studies in Clinical Trials*, 2nd ed. Boca Raton, FL: Chapman & Hall/CRC.

Fayers, P. M., and Machin, D. 2007. *Quality of Life: The Assessment, Analysis and Interpretation of Patient-Reported Outcomes*, 2nd ed. Chichester, UK: Wiley.

Feng, L., Yap, P. L. K., Lee, T. S., et al. 2009. Neuropsychiatric symptoms in mild cognitive impairment: A population-based study. *Asia-Pacific Psychiatry* 1:23–27.

Feng, Z., McLerran, D., and Grizzle, J. 1996. A comparison of statistical methods for clustered data analysis with Gaussian error. *Statistics in Medicine* 15:1793–1806.

Fergusson, D. M., Horwood, L. J., and Ridder, E. M. 2005. Show me the child at seven II: Childhood intelligence and later outcomes in adolescence and young adulthood. *Journal of Child Psychology and Psychiatry* 46:850–858.

Fleiss, J. L., and Cohen, J. 1973. The equivalence of weighted kappa and the intraclass correlation coefficient as measures of reliability. *Educational and Psychological Measurement* 33:613–619.

Fok, T. F., So, H. K., Wong, E., et al. 2003. Updated gestational age specific birth weight, crown-heel length, and head circumference of Chinese newborns. *Archives of Disease in Childhood-Fetal and Neonatal Edition* 88:F229–236.

Fomon, S. J., and Nelson, S. E. 2002. Body composition of the male and female reference infants. *Annual Review of Nutrition* 22:1–17.

Frankenburg, W. K., Bodds, J. B., and Fandal, A. W. 1991. *Denver Developmental Screening Test, Singapore. Test Manual, 1991*. Singapore: Ministry of Health.

Frison, L., and Pocock, S. J. 1992. Repeated measures in clinical trials: Analysis using mean summary statistics and its implications for design. *Statistics in Medicine* 11:1685–1704.

Gandhi, M., Ashorn, P., Maleta, K., et al. 2011. Height gain during early childhood is an important predictor of schooling and mathematics ability outcomes. *Acata Paediatrica* 100:1113–1118.

Gandhi, M., Teivaanmäki, T., Maleta, K., et al. 2013. Child development at 5 years of age predicted mathematics ability and schooling outcomes in Malawian adolescents. *Acta Paediatrica* 102:58–65.

Giedd, J. N. 2004. Structural magnetic resonance imaging of the adolescent brain. *Annals of the New York Academy of Sciences* 1021:77–85.

Gladstone, M., Lancaster, G. A., Umar, E., et al. 2010. The Malawi Developmental Assessment Tool (MDAT): The creation, validation, and reliability of a tool to assess the child development in rural African settings. *PLoS Medicine* 7:e1000273.

Gladstone, M. J., Lancaster, G. A., Jones, A. P., et al. 2008. Can Western developmental screening tools be modified for use in a rural Malawian setting? *Archives of Disease in Childhood* 93:23–29.

Gleason, J. R. 1997. Computing intraclass correlations and large ANOVAs. *Stata Technical Bulletin Reprints* 6:167–176.

Goldstein, L. J., Lay, D. C., and Schneider, D. I. 1993. *Calculus and Its Applications*, 6th ed. Englewood Cliffs, NJ: Prentice Hall.

Gómez, G., Espinal, A. W., and Lagakos, S. 2003. Inference for a linear regression model with an interval-censored covariate. *Statistics in Medicine* 22:409–425.

Gopnik, A., Meltzoff, A. N., and Kuhl, P. K. 1999. *The Scientist in the Crib: Minds, Brains, and How Children Learn*. New York: Harper Collins.

Gould, W. W. 1993. sg11.1: Quantile regression with bootstrapped standard errors. *Stata Technical Bulletin Reprints* 2:137–139.

Gould, W. W. 1997. Better numerical derivatives and integrals. *Stata Technical Bulletin Reprints* 6:8–12.

Gould, W. W. 1998. sg70: Interquantile and simultaneous-quantile regression. *Stata Technical Bulletin Reprints* 7:167–176.

Gould, W., and Sribney, W. 1999. *Maximum Likelihood Estimation with Stata*®. College Station, TX: Stata Press.

Grantham-McGregor, S., Cheung, Y. B., Cueto, S., et al. 2007. Developmental potential in the first 5 years for children in developing countries. *Lancet* 369:60–70.

Greenwood, A. M., Armstrong, J. R., Byass, P., et al. 1992. Malaria chemoprophylaxis, birth weight and child survival. *Transactions of the Royal Society of Tropical Medicine and Hygiene* 86:483–485.

Griffins, J., Fraser, C., Gras, L., et al. 2006. The effect on treatment comparisons of different measurement frequencies in human immunodeficiency virus observational databases. *American Journal of Epidemiology* 163:676–683.

Griffiths, R., and Huntley, M. 1996. *Griffiths Mental Development Scales–Revised: Birth to Two Years*. Oxford: Hogrefe.

Gujarati, D. N. 1995. *Basic Econometrics*, 3rd ed. New York: McGraw Hill.

Hadlock, F. P., Harrist, R. B., Deter, R. L., et al. 1982. Fetal femur length as a predictor of menstrual age: Sonographically measured. *American Journal of Roentgenology* 135:875–878.

Hand, D. 1994. Deconstructing statistical questions. *Journal of the Royal Statistical Society, Series A* 157:317–356.

Hao, L., and Naiman, D. Q. 2007. *Quantile Regression*. Thousand Oaks, CA: Sage.

Hardin, J., and Hilbe, J. 2001. *Generalized Linear Models and Extensions*. College Station, TX: Stata Press.

Hardouin, J. B. 2007. Rasch analysis: Estimation and tests with raschtest. *The Stata Journal* 7:22–44.

Hardouin, J. B., and Mesbah, M. 2007. The macro program%AnaQol to estimate the parameters of IRT models. *Communications in Statistics–Simulation and Computation* 36:437–453.

Hastie, T., and Tibshirani, R. 1987. Generalized additive models: Some applications. *Journal of the American Statistical Association* 82:371–386.

Hayes, R. J., and Moulton, L. H. 2009. *Cluster Randomised Trials*. Boca Raton, FL: CRC Press.

Heeren, T., and R. D'Agostino. 1987. Robustness of the two independent samples t-test when applied to ordinal scaled data. *Statistics in Medicine* 6:79–90.

Hennekens, C. H., and Buring, J. E. 1987. *Epidemiology in Medicine*. Boston: Little, Brown and Company.

Hernán, M. A., Hernández-Fiaz, S., Werler, M. M., et al. 2002. Causal knowledge as a prerequisite for confounding evaluation: An application to birth defects epidemiology. *American Journal of Epidemiology* 155:176–184.

Hertz-Picciotto, I., and Rockhill, B. 1997. Validity and efficiency of approximation methods for tied survival times in Cox regression. *Biometrics* 53:1151–1156.

Hesterberg, T., Cho, N. H., Meier, L., et al. 2008. Least angle and L1 penalized regression: A review. *Statistics Surveys* 2:61–93.

Hill, A. B. 1965. The environment and disease: Association or causation? *Proceedings of the Royal Society of Medicine* 58:295-300.

Hochberg, Y., and Tamhane, A. C. 1987. *Multiple Comparison Procedures*. New York: Wiley.

Hoijtink, H., and Boomsma, A. 1995. On person parameter estimation in the dichotomous Rasch model. In *Rasch Models, Foundations, Recent Developments and Applications*, ed. G. H. Fisher and I. W. Molenaar, 53–68. New York: Springer.

Hosmer, D. W., and Lemeshow, S. 2000. *Applied Logistic Regression*, 2nd ed. New York: Wiley.

Hsieh, F. Y. 1995. A cautionary note on the analysis of extreme data with Cox regression. *The American Statistician* 49:226–228.

Hu, F. B., Goldberg, J., Hedeker, D., et al. 1998. Comparison of population-averaged and subject-specific approaches for analyzing repeated binary outcomes. *American Journal of Epidemiology* 147:694–703.

International Conference on Harmnonisation of Technical Requirements for Registration of Pharmaceuticals for Human Use (ICH). 1998. *ICH harmonised tripartite guideline. Statistical principles for clinical trials. E9*. http://www.ich.org, accessed April 28, 2012.

Johnsen, S. L., Rasmussen, S., Wilsgaard, T., et al. 2006. Longitudinal reference ranges for estimated fetal weight. *Acta Obstetricia et Gynecologica Scandinavica* 85:286–297.

Johnson, R. A., and Wichern, D. W. 2002. *Applied Multivariate Statistical Analysis*, 5th ed. Upper Saddle River, NJ: Prentice Hall.

Jukes, M. C. H., Pinder, M., Grigorenko, E. L., et al. 2006. Long-term impact of malaria chemoprophylaxis in early childhood on cognitive abilities and educational attainment: Follow-up of a controlled trial in The Gambia, West Africa. *PLoS Clinical Trials* 1:e19.

Kado, D. M., Prenovost, K., and Crandall, C. 2007. Narrative review: Hyperkyphosis in older persons. *Annals of Internal Medicine* 147:330–338.

Karlberg, J. 1987. On the modelling of human growth. *Statistics in Medicine* 6:185–192.

Karlberg, J., Albertsson-Wikland, K., Kwan, E. Y., et al. 1997. The timing of early postnatal catch-up growth in normal, full-term infants born short for gestational age. *Hormone Research* 48(Suppl. 1):17-24.

Karlberg, J., Ashraf, R. N., Saleemi, M., et al. 1993. Early child health in Lahore, Pakistan: XI. Growth. *Acta Paediatrica* 390(Suppl):119–149.

Karlberg, J., Jalil, F., and Lindblad, B. S. 1998. Longitudinal analysis of infantile growth in an urban area of Lahore, Pakistan. *Acta Paediatrica Scandinavica* 77:392–401.

Karlberg, J., Jalil, F., Lam, B., et al. 1994. Linear growth retardation in relation to the three phases of growth. *European Journal of Clinical Nutrition* 48(Suppl. 1):S25–44.

Katzman, R., Terry, R., De Teresa, R., et al. 1988. Clinical, pathological, and neurochemical changes in dementia: A subgroup with preserved mental status and numerous neocortical plaques. *Annals of Neurology* 23:138–144.

Kavšek, M. 2004. Predicting later IQ from infant visual habituation and dishabituation. *Applied Developmental Psychology* 25:369–393.

Keijzer-Veen, M. G., Euser, A. M., van Montfoort, N., et al. 2005. A regression model with unexplained residuals was preferred in the analysis of the fetal origins of adult diseases hypothesis. *Journal of Clinical Epidemiology* 58:1320–1324.

Koenker, R., and Bassett, G. Jr. 1982. Robust tests for heteroscedasticity based on regression quantiles. *Econometrica* 50:43–61.

Krakauer, N. Y., and Krakauer, J. C. 2012. A new body shape index predicts premature mortality hazards independently of body mass index. *PLOS One* 7:e39504.

Kramer, M. S., Platt, R. W., Wen, S. W., et al. 2001. A new and improved population-based Canadian reference for birth weight for gestational age. *Pediatrics* 108:E35.

Kronmal, R. 1993. Spurious correlation and the fallacy of the ratio standard revisited. *Journal of the Royal Statistical Society, Series A* 156:379–392.

Kuczmarski, R. J., Ogden, C. L., Guo, S. S., et al. 2002. 2000 CDC growth charts for the United States: Methods and development. *National Center for Health Statistics. Vital and Health Statistics* 11(246).

Kuhl, P. K., Tsao, F. M., and Liu, H. M. 2003. Foreign-language experience in infancy: Effects of short-term exposure and social interaction on phonetic learning. *Proceedings of the National Academy of Science* 100:9096–9101.

Kulich, M., Rosenfeld, M., Campbell, J., et al. 2005. Disease-specific reference equations for lung function in patients with cystic fibrosis. *American Journal of Respiratory and Critical Care Medicine* 172:885–891.

Kwok, M. K., Leung, G. M., Lam, T. H., and Schooling, C. M. 2011. Early life infections and onset of puberty: Evidence from Hong Kong's children of 1997 birth cohort. *American Journal of Epidemiology* 173:1440–1452.

Kwon, J. W., Song, Y. M., Park, H., et al. 2008. Effects of age, time period, and birth cohort on the prevalence of diabetes and obesity in Korean men. *Diabetes Care* 31:255–260.

Laird, N. M., and Wang, F. 1990. Estimating rates of change in randomized clinical trials. *Controlled Clinical Trials* 11:405–419.

Landau, S. and Everitt, B. S. 2004. *A Handbook of Statistical Analyses Using SPSS*. Boca Raton, FL: Chapman & Hall/CRC.

Landis, J. R., and Koch, G. G. 1977. The measurement of observer agreement for categorical data. *Biometrics* 33:159–174.

Landis, S. H., Anath, C. V., Lokomba, V., et al. 2009. Ultrasound-derived fetal size nomogram for a sub-Saharan African population: A longitudinal study. *Ultrasound in Obstetrics and Gynecology* 34:379–386.

Lawless, J. F., and Babineau, D. 2006. Models for interval censoring and simulation-based inference for lifetime distributions. *Biometrika* 93:631–686.

Le Mare, L., and Audet, K. 2006. A longitudinal study of the physical growth and health of post-institutionalized Romanian adoptees. *Paediatrics and Child Health* 11:85–91.

Lebel, C., and Beaulieu, C. 2011. Longitudinal development of human brain wiring continues from childhood into adulthood. *Journal of Neuroscience* 2011:10937–10947.

Leon, D. A., Smith, G. D., Shipley, M., and Strachan, D. 1995. Adult height and mortality in London: Early life, socioeconomic confounding, or shrinkage? *Journal of Epidemiology and Community Health* 49:5–9.

Li, Z., and Sillanpää, M. J. 2012. Overview of LASSO-related penalized regression methods for quantitative trait mapping and genomic selection. *Theoretical and Applied Genetics* 125:419–435.

Linacre, J. M. 1998. Detecting multidimensionality: Which residual data-type works best? *Journal of Outcome Measurement* 2:266–283.

Linacre, J. M. 2002. What do infit and outfit, mean-square and standardized mean? *Rasch Measurement Transactions* 16:878.

Linacre, J. M., and Wright, B. D. 1994. Chi-square fit statistics. *Rasch Measurement Transactions* 8:350.

Lindsey, J. K. 1998. A study of interval censoring in parametric regression models. *Lifetime Data Analysis* 4:329–354.

Litière, S., Alonso, A, and Molenberghs, G. 2007. Type I and type II error under random-effects misspecification in generalized linear mixed models. *Biometrics* 63:1038–1044.

Liu, Y. X., Jalil, F., and Karlberg, J. 1998. Growth stunting in early life in relation to the onset of the childhood component of growth. *Journal of Pediatric Endocrinology and Metabolism* 11:247–260.

Liu, Y. X., Wikland, K. A., and Karlberg, J. 2000. New reference for the age at childhood onset of growth and secular trend in the timing of puberty in Swedish. *Acta Paediatrica* 89:637–643.

Lohr, S. L. 1999. *Sampling: Design and Analysis*. Pacific Grove, CA: Duxbury Press.

Long, J. S., and Freese, J. 2001. Scalar measures of fit for regression models. *Stata Technical Bulletin Reprints* 10:197–205.

Lucas, A., Fewtrell, M. S., and Cole, T. J. 1999. Fetal origins of adult disease—The hypothesis revisited. *British Medical Journal* 319:245–249.

Lundgren, E. M., Cnattingius, S., Jonsson, B., et al. 2001. Intellectual and psychological performance in males born small for gestational age with and without catch-up growth. *Pediatric Research* 50:91–96.

Luntamo, M., Kulmala, T., Cheung, Y. B., et al. 2013. The effect of antenatal monthly sulfadoxine-pyrimethamine, alone or with azithromycin, on foetal and neonatal growth faltering in Malawi. *Tropical Medicine and International Health*, E-pub ahead of print.

Luntamo, M., Kulmala, T., Mbewe, B., et al. 2010. Effect of repeated treatment of pregnant women with sulfadoxine-pyrimethamine and azithromycin on preterm delivery in Malawi: A randomized controlled trial. *American Journal of Tropical Medicine and Hygiene* 83:1212–1220.

Machin, D., Campbell, M., Tan, S. B., et al. 2009. *Sample Size Tables for Clinical Studies*, 3rd ed. Oxford: Wiley-Blackwell.

Machin, D., Cheung, Y. B., and Parmar, M. K. B. 2006. *Survival Analysis: A Practical Approach*, 2nd ed. Chichester, UK: Wiley.

Madden, D. J., Gottlob, L. R., Denny, L. L., et al. 1999. Aging and recognition memory: Changes in regional cerebral blood flow associated with components of reaction time distributions. *Journal of Cognitive Neuroscience* 11:511–520.

Marcus, R., Peritz, E., and Gabriel, K. R. 1976. On closed testing procedures with special reference to ordered analysis of variance. *Biometrika* 63:655–660.

Marquardt, D. W. 1970. Generalized inverses, ridge regression, biased linear estimation, and nonlinear estimation. *Technometrics* 12:591–612.

Martorell, R., Horta, B. L., Adair, L. S., et al. 2010. Weight gain in the first two years of life is an important predictor of schooling outcomes in pooled analyses from five birth cohorts from low- and middle-income countries. *Journal of Nutrition* 140:348–354.

Michaelsen, K. F., Skov, L., Badsberg, J. H., et al. 1991. Short-term measurement of linear growth in preterm infants: Validation of a hand-held knemometer. *Pediatric Research* 30:464–468.

Midthjell, K., Kruger, O., Holmen, J., et al. 1999. Rapid changes in the prevalence of obesity and known diabetes in an adult Norwegian population. *Diabetes Care* 22:1813–1820.

Montgomery, D. C., Peck, E. A., and Vining, C. G. 2001. *Introduction to Linear Regression Analysis*, 3rd ed. New York: Wiley.

Motulsky, H. 2010. *Intuitive Biostatistics*. Oxford: Oxford University Press.

Moyé, L. A. 1998. P-value interpretation and alpha allocation in clinical trials. *Annals of Epidemiology* 8:351–357.

Mueller, W. H., and Martorell, R. 1988. Reliability and accuracy of measurement. In *Anthropometric Standardisation Reference Manual*, T. G. Lohman, A. F. Roche, and R. Martorell, ed., 83–86. Champaign, IL: Human Kinetics Books.

Nagy, Z., Westerberg, H., and Klingberg, T. 2004. Maturation of white matter is associated with the development of cognitive functions during childhood. *Journal of Cognitive Neuroscience* 285:2094–2100.

Nauta, J. 2010. *Statistics in Clinical Vaccine Trials*. Berlin: Springer-Verlag.

Nelder, J. A., and Wedderburn, R. W. M. 1972. Generalized linear models. *Journal of the Royal Statistical Society, Series A* 135:370–384.

Nelson, W. 1972. Theory and applications of hazard plotting for censored failure data. *Technometrics* 14: 945–966.

Neuhaus, J. M., Hauck, W. W., and Kalbfleisch, J. D. 1992. The effects of mixture distribution misspecification when fitting mixed-effects logistic models. *Biometrika* 79:755–762.

Neuhaus, J. M., Kalbfleisch, J. D., and Hauck, W. W. 1991. A comparison of cluster-specific and population-averaged approaches for analyzing correlated binary data. *International Statistical Review* 59:25–35.

Niklasson, A., and Albertsson-Wikland, K. 2008. Continuous growth reference from 24th week of gestation to 24 months by gender. *BMC Pediatrics* 8:8.

Niklasson, A., Ericson, A., Fryer, J. G., et al. 1991. An update of the Swedish reference standards for weight, length and head circumference at birth for given gestational age (1977–1981). *Acta Paediatrica Scandinavica* 80:756–762.

Nowakowski, R. S. 1987. Basic concepts of CNS development. *Child Development* 58:568-595.

O'Brien, R. M. 2007. A caution regarding rules of thumb for variance inflation factors. *Quality & Quantity* 41:673–690.

Oken, E., Kleinman, K. P., Rich-Edwards, J., et al. 2003. A nearly continuous measure of birth weight for gestational age using a United States national reference. *BMC Pediatrics* 3:6.

Orenstein, W.A., Bernier, R. H., and Hinman, A. R. 1988. Assessing vaccine efficacy in the field. Further observations. *Epidemiological Review* 10:212-241.

Pan, H., and Cole, T. 2005. *User's Guide to LMSChartMaker*. UK: Medical Research Council.

Pascalis, O., de Schonen, S., Morton, J., et al. 1995. Mother's face recognition by neonates: A replication and an extension. *Infant Behavior and Development* 18:79–85.

Pearl, J. 1998. Why there is no statistical test for confounding, why many think there is, and why they are almost right. UCLA Technical Report R-256. Los Angeles: UCLA.

Perini, T. A., de Oliveira, G. L., Santos Ornellas, J., et al. 2005. Technical error of measurement in anthropometry. *Revista Brasileira de Medicina do Esporte* 11:86–90.

Perneger, T. V. 1998. What's wrong with Bonferroni adjustments. *British Medical Journal* 316:1236–1238.

Phuka, J. C., Gladstone, M., Maleta, K., et al. 2012. Developmental outcomes among 18-month-old Malawians after a year of complementary feeding with lipid-based nutrient supplements or corn soy flour. *Maternal and Child Nutrition* 8:239–248.

Poccok, S. J. 1983. *Clinical Trials: A Practical Approach*. Chichester, UK: Wiley.

Pregibon, D. 1980. Goodness of link tests for generalized linear models. *Applied Statistics* 29:15–24.

PROMIS. 2005. PROMIS SCC Data Analysis Plan. Available at http://www.nihpromis.org/documents/PROMIS_Statistical_Analysis.pdf (accessed July 17, 2012).

Pujol, J., Lopez-Sala, A., Sebastian-Gelles, N., et al. 2004. Delayed myelination in children with developmental delay detected by volumetric MRI. *Neuoimage* 22:897–903.

Rabe-Hesketh, S., and Everitt, B. S. 2007. *A Handbook of Statistical Analyses Using Stata*, 4th ed. Boca Raton: Chapman & Hall/CRC.

Rabe-Hesketh, S., and Skrondal, A. 2008. *Multilevel and Longitudinal Modeling Using Stata*, 2nd ed. College Station, TX: Stata Press.

Raiche, G. 2005. Critical eigenvalue sizes in standardized residual principal component analysis. *Rasch Measurement Transactions* 19:1012.

Randolph, C., Tierney, M. C., Mohr, E., et al. 1998. The Repeatable Battery for the Assessment of Neuropsychological Status (RBANS): Preliminary clinical validity. *Journal of Clinical and Experimental Neuropsychology* 20:310–319.

Rasch, G. 1960. *Probabilistic Models for Some Intelligence and Attainment Tests*. Chicago: University of Chicago Press.

Raven, J. 2000. The Raven's progressive matrices: Change and stability over culture and time. *Cognitive Psychology* 41:1–48.

Raven, J., Raven, J. C., and Court, J. H. 2003. *Manual for Raven's Progressive Matrices and Vocabulary Scales.* San Antonio, TX: Pearson Assessment.

RCO&G. 2002. *The Investigation and Management of the Small-for-Gestational-Age Fetus. Guideline No. 31.* London: Royal College of Obstetricians and Gynaecologists.

Rice, J. A. 1995. *Mathematical Statistics and Data Analysis,* 2nd ed. Belmont, CA: Duxbury Press.

Rigby, R. A., and Stasinopoulos, D. M. 2004. Smooth centile curves for skew and kurtotic data modeled using the Box-Cox power exponential distribution. *Statistics in Medicine* 23:3053–3076.

Rijken, M. J., Rijken, J. A., Papageorghiou, A. T., et al. 2011. Malaria in pregnancy: The difficulties in measuring birthweight. *British Journal of Obstetrics and Gynaecology* 118:671–678.

Robinson, G. K. 1991. That BLUP is a good thing: The estimation of random effects. *Statistical Science* 6:15–51.

Robinson, L. D., and Jewell, N. P. 1991. Some surprising results about covariate adjustment in logistic regression models. *International Statistical Review* 59:227–240.

Rogers, W. H. 1994. sg11.2: Calculation of quantile regression standard errors. *Stata Technical Bulletin Reprints* 3:77–78.

Rolland-Cachera, M. F., Deheeger, M., Maillot, M., et al. 2006. Early adiposity rebound: Causes and consequences for obesity in children and adults. *International Journal of Obesity (London)* 30(Suppl. 4): S11–17.

Roseboom, T. J., van der Meulen, J. H. P., Osmond, C., et al. 2000. Coronary heart disease after prenatal exposure to the Dutch famine, 1944–45. *Heart* 84:595–598.

Rothman, K. J. 1990. No adjustments are needed for multiple comparisons. *Epidemiology* 1:43–46.

Rothman, K. J., Greenland, S., and Lash, T. L. 2008. *Modern Epidemiology,* 3rd ed. Philadelphia: Lippincott, Williams & Wilkins.

Royston, P. 1995. Calculation of unconditional and conditional reference intervals for foetal size and growth from longitudinal measurements. *Statistics in Medicine* 14:1417–1436.

Royston, P. 2004. Multiple imputation of missing values. *The Stata Journal* 4:227–241.

Royston, P. 2007. Multiple imputation of missing values: Further update of ice, with an emphasis on interval censoring. *The Stata Journal* 7:445–464.

Royston, P., and Altman, D. G. 1994. Regression using fractional polynomials of continuous covariates: Parsimonious parametric modeling (with discussion). *Applied Statistics* 43:429–467.

Royston, P., Altman, D. G., and Sauerbrei, W. 2006. Dichotomizing continuous predictors in multiple regression: A bad idea. *Statistics in Medicine* 25:127–141.

Royston, P., and Wright, E. M. 1998. A method for estimating age-specific reference intervals ("normal ranges") based on fractional polynomials and exponential transformation. *Journal of the Royal Statistical Association, Series A* 161:79–101.

Royston, P., and Wright, E. M. 2000. Goodness-of-fit statistics for age-specific reference intervals. *Statistics in Medicine* 19:2943–2962.

Royston, P., and Sauerbrei, W. 2005. Building multivariable regression models with continuous covariates in clinical epidemiology with an emphasis on fractional polynomials. *Methods of Information in Medicine* 44:561–571.

Royston, P., and Sauerbrei, W. 2007. Multivariable modeling with cubic regression splines: A principled approach. *Stata Journal* 7:45–70.

Royston, P., Parmar, M. K., and Altman, D. G. 2008. Visualizing length of survival in time-to-event studies: A complement to Kaplan-Meier plots. *Journal of National Cancer Institute* 100:92–97.

Rubin, D. B. 1987. *Multiple Imputation for Nonresponse in Surveys*. New York: Wiley.

Rubin, D. B. 1996. Multiple imputation after 18+ years (with discussion). *Journal of the American Statistical Association* 91:473–489.

Rutter, M. 1998. Developmental catch-up and deficit following adoption after severe global early deprivation. *Journal of Child Psychology and Psychiatry* 39:465–476.

Rutter, M., O'Connor, T. G., and English and Romanian Adoptees (ERA) Study Team. 2004. Are there biological programming effects for psychological development? Findings from a study of Romanian adoptees. *Developmental Psychology* 40:81–94.

Santrock, J. W. 2008. *A Topical Approach to Life-Span Development*, 4th ed. Boston: McGraw-Hill.

Save the Children. 2012. *Nutrition in the First 1,000 Days. State of the World's Mothers 2012*. Westport, CT: Save the Children.

Schaie, K. W. 2005. *Developmental Influences on Adult Intelligence: The Seattle Longitudinal Study*, 2nd ed. New York: Oxford University Press.

Schernhammer, E. S., Tworoger, S. S., Eliassen, A. H., et al. 2007. Body shape throughout life and correlations with IGFs and GH. *Endocrine-Related Cancer* 14:721–732.

Schlesselman, J. J. 1971. Power families: A note on the Box and Cox transformation. *Journal of the Royal Statistical Society, Series B* 33: 307–311.

Schoenfeld, D. 1982. Partial residuals for the proportional hazards regression model. *Biometrika* 69: 239–241.

Schulz, K. F., and Grimes, D. A. 2005. Multiplicity in randomised trials I: Endpoints and treatments. *Lancet* 365:1591–1595.

Searle, S. R., Casella, G., and McCulloch, C. E. 1992. *Variance Components*. New York: Wiley.

Seed, P. 2001. Comparing several methods of measuring the same quantity. *Stata Technical Bulletin Reprints* 10:73–82.

Senn, S. 1997. *Statistical Issues in Drug Development*. Chichester, UK: Wiley.

Senn, S. 2005. Baseline balance and valid statistical analyses: Common misunderstanding. *Applied Clinical Trials* 14:24–26.

Senn, S., and Julious, S. 2009. Measurement in clinical trials: A neglected issue for statisticians? *Statistics in Medicine* 28:3189–3209.

Shah, P. S., Ye, X. Y., Synnes, A., et al. 2012. Prediction of survival without morbidity for infants born at under 33 weeks gestational age: A user-friendly graphical tool. *Archives of Disease in Childhood—Fetal and Neonatal Edition* 97:F110–115.

Shuttleworth-Edwards, A. B., Kemp, R. D., Rust, A. L., et al. 2004. Cross-cultural effects on IQ test performance: A review and preliminary normative indications on WAIS-III test performance. *Journal of Clinical and Experimental Neuropsychology* 26:903–920.

Silverwood, R. J., De Stavola, B. L., Cole, T. J., et al. 2009. BMI peak in infancy as a predictor for later BMI in the Uppsala Family Study. *International Journal of Obesity* 33: 929–937.

Sinclair, J.C., and Bracken, M. B. 1994. Clinically useful measures of effect in binary analyses of randomised trials. *Journal of Clinical Epidemiology* 47:881–889.

Singh, D. 2002. Female mate value at a glance: Relationship of waist-to-hip ratio to health, fecundity and attractiveness. *Neuroendocrinology Letters* 23(Suppl. 4):81–91.

Smith, G. C., and Pell, J. P. 2003. Parachute use to prevent death and major trauma related to gravitational challenge: Systematic review of randomised controlled trials. *British Medical Journal* 327:1459–1461.

Sobel, M. E. 1987. Direct and indirect effects in linear structural equation models. *Sociological Methods Research* 16:155–176.

Spiegelman, D., and Hertzmark, E. 2005. Easy SAS calculations for risk or prevalence ratios and differences. *American Journal of Epidemiology* 162:199–200.

Stacy, E. W. 1962. A generalization of the gamma distribution. *Annals of Mathematical Statistics* 33:1187–1192.

Stasinopoulos, D. M., and Rigby, R. A. 2007. Generalized additive models for location, scale and shape in R. *Journal of Statistical Software* 23:7.

StataCorp. 2011. *Stata: Release 12.* College Station, TX: StataCorp LP.

Stewart, P. W., Reihman, J., Lonky, E., and Pagano, J. 2012. Issues in the interpretation of associations of PCBs and IQ. *Neurotoxicology and Teratology* 34:96–107.

Stovring, H., Gyrd-Hansen, D., Kristiansen, I. S., et al. 2008. Communicating effectiveness of intervention for chronic diseases: What single format can replace comprehensive information? *BMC Medical Informatics and Decision Making* 8:25.

Strauss, J., Witoelar, F., Sikoki, B., and Wattie, A. M. 2009. *The Fourth Wave of the Indonesia Family Life Survey (IFLS4): Overview and Field Report. WR-675/1-NIA/ NICHD.* Santa Monica, CA: RAND.

Sullivan, L. M., and D'Agostino, R. B. 2003. Robustness and power of analysis of covariance applied to ordinal scaled data as arising in randomized controlled trials. *Statistics in Medicine* 22:1317–1334.

Sun, G. W., Shook, T. L., and Kay, G. L. 1996. Inappropriate use of bivariable analysis to screen risk factors for use in multivariable analysis. *Journal of Clinical Epidemiology* 49:907–916.

Tanaka, C., Matsui, M., Uematsu, A., et al. 2012. Developmental trajectories of the fronto-temporal lobes from infancy to early adulthood in healthy individuals. *Developmental Neuroscience*, Epub ahead of print.

Tanner, J. M. 1989. *Foetus into Man,* 2nd ed. Ware, UK: Castlemead Publications.

Terry, M. B., Wei, Y., and Esserman, D. 2007. Maternal, birth and early-life influences on adult body size in women. *American Journal of Epidemiology* 166:5–13.

Therneau, T. M., and Grambsch, P. M. 2000. *Modeling Survival Data: Extending the Cox Model.* New York: Springer.

Thompson, J. R. 1998. Invited commentary: Re: "Multiple comparisons and related issues in the interpretation of epidemiologic data." *American Journal of Epidemiology* 147:801–806.

Thompson, R. A., and Nelson, C. A. 2001. Developmental science and the media: Early brain development. *American Psychologist* 56:5–15.

Toth, M. J., Tchernof, A., Sites, C. K., et al. 2000. Menopause-related changes in body fat distribution. *Annals of the New York Academy of Sciences* 904:502–506.

Ulijaszek, S. J., and Kerr, D. A. 1999. Anthropometric measurement error and the assessment of nutritional status. *British Journal of Nutrition* 82:165–177.

United Nations Development Programme. 2003. *Human Development Report 2003.* New York: Oxford University Press.

van Buuren, S. 2007. Multiple imputation of discrete and continuous data by fully conditional specifications. *Statistical Methods in Medical Research* 16:219–242.

van Buuren, S. 2013. Growth charts of human development. *Statistical Methods in Medical Research,* Epub ahead of print.

van Buuren, S., Brand, J. P. L., Groothuis-Oudshoorn, C. G. M., et al. 2006. Fully conditional specification in multivariate imputation. *Journal of Statistical Computation and Simulation* 76:1049–1064.

van Buuren, S., Boshuizen, H. C., and Knook, D. L. 1999. Multiple imputation of missing blood pressure covariates in survival analysis. *Statistics in Medicine* 18:681–694.

van Buuren, S., and Fredriks, M. 2001. Worm plot: A simple diagnostic device for modeling growth reference curves. *Statistics in Medicine* 20:1259–1277.

van der Linden, W. 1998. Optimal assembly of psychological and educational tests. *Applied Psychological Measurement* 22:195–211.

Verney, S. P., Granholm, E., Marshall, S. P., et al. 2005. Culture-fair cognitive ability assessment. Information processing and psychophysiological approaches. *Assessment* 12:303–319.

Victora, C. G., de Onis, M., Hallal, P. C., et al. 2010. Worldwide timing of growth faltering: Revisiting implications for interventions. *Pediatrics* 125:e473–480.

Victora, C. G., Huttly, S. R., Fuchs, S. C., et al. 1997. The role of conceptual frameworks in epidemiological analysis: A hierarchical approach. *International Journal of Epidemiology* 26:224–227.

Vidmar, S., Carlin, J., Hesketh, K., and Cole, T. 2004. Standardising anthropometric measures in children and adolescents with new functions for egen. *Stata Journal* 4:50–55.

Villar, J., and Belizan, J. M. 1982. The timing factor in the pathophysiology of the intrauterine growth retardation syndrome. *Obstetrics & Gynecological Survey* 37:499–506.

Wang, M., and Long, Q. 2011. Modified robust variance estimator for generalized estimating equations with improved small-sample performance. *Statistics in Medicine* 30:1278–1291.

Wei, L. J., and Glidden, D. V. 1997. An overview of statistical methods for multiple failure time data in clinical trials. *Statistics in Medicine* 16:833–839.

Wei, Y., Pere, A., Koenker, R., et al. 2006. Quantile regression methods for reference growth charts. *Statistics in Medicine* 25:1369–1382.

White, I. R., Royston, P., and Wood, A. M. 2011. Multiple imputation using chained equations: Issues and guidance for practice. *Statistics in Medicine* 30:377–399.

WHO Expert Consultation. 2004. Appropriate body-mass index for Asian populations and its implications for policy and intervention strategies. *Lancet* 363:157–163.

WHO Multicentre Growth Reference Study Group. 2006a. Reliability of anthropometric measurements in the WHO Multicentre Growth Reference Study. *Acta Paediatrica* 450:38–46.

WHO Multicentre Growth Reference Study Group. 2006b. WHO Motor Development Study: Windows of achievement for six gross motor development milestones. *Acta Paediatrica* Suppl 450:86–95.

Wilcox, A. J. 2001. On the importance—and the unimportance—of birth weight. *International Journal of Epidemiology* 30:1233–1241.

Williams, R. L. 2000. A note on robust variance estimation for cluster-correlated data. *Biometrics* 56:645–646.

Wilson, R. S., Mendes De Leon, C. F., Barnes, L. L., et al. 2002. Participation in cognitively stimulating activities and risk of incident Alzheimer disease. *Journal of the American Medical Association* 287:742–748.

World Health Organization. 2006. *WHO Child Growth Standards. Length/Height-for-Age, Weight-for-Age, Weight-for-Length, Weight-for-Height, and Body Mass Index-for-Age. Methods and Development*. Geneva: WHO Press.

World Health Organization. 2007. *WHO Child Growth Standards. Head Circumference-for-Age, Arm Circumference-for-Age, Triceps Skinfold-for-Age and Subscapular Skinfold-for-Age. Methods and Development*. Geneva: WHO Press.

World Health Organization Working Group. 1986. Use and interpretation of anthropometric indicators of nutritional status. *Bulletin of the World Health Organization* 64:929–941.

Wright, B. D. 1996. Local dependency, correlations and principal components. *Rasch Measurement Transactions* 10:509–511.

Wright, E. M., and Royston, P. 1997. Age-specific reference intervals ("normal ranges"). *Stata Technical Bulletin Reprints* 6:91–104.

Wright, E. M., and Royston, P. 1999. Two methods for assessing the goodness-of-fit of age-specific reference interval. *Stata Technical Bulletin Reprints* 8:100–108.

Xu, Y., Cheung, Y. B., Lam, K. F., et al. 2012. Estimation of summary protective efficacy using a frailty mixture model for recurrent event times. *Statistics in Medicine* 31:4023–4039.

Yang, S., Tilling, K., Martin, R., et al. 2011. Pre-natal and post-natal growth trajectories and childhood cognitive ability and mental health. *International Journal of Epidemiology* 40:1215–1226.

Yaqoob, M., Ferngren, H., Jalil, F., et al. 1993. Early child health in Lahore, Pakistan: XII. Milestones. *Acta Paediatrica* 390(Suppl):151–157.

Yehuda, R., Engel, M. S., Brand, S. R., et al. 2005. Transgenerational effects of post-traumatic stress disorder in babies of mothers exposed to the World Trade Center attacks during pregnancy. *Journal of Clinical Endocrinology & Metabolism* 90:4115–4118.

Yen, W. M. 1984. Effects of local item dependence on the fit and equating performance of the three-parameter logistic model. *Applied Psychological Measurement* 8:125–145.

Yusuf, S., Hawken, S., Ounpuu, S., et al. 2005. Obesity and the risk of myocardial infarction in 27000 participants from 52 countries: A case-control study. *Lancet* 366:1640–1649.

Zeger, S. L., Liang, K. Y., and Albert, P. S. 1988. Models for longitudinal data: A generalized estimating equation approach. *Biometrics* 44:1049–1060.

Zhu, J., and Chen, H. Y. 2011. Utility of inferential norming with smaller sample sizes. *Journal of Psychoeducational Assessment* 29:570–580.

Zou, G. 2004. A modified Poisson regression approach to prospective studies with binary data. *American Journal of Epidemiology* 159:702–706.

Appendix A:
Stata Codes to Generate Simulated Clinical Trial (SCT) Dataset

```
/******************** Data Description ********************
Variable       Variable_label         Value        Value_label
id             ID                     1 - 1500
group          Group                  0            C (Control)
                                      1            A
                                      2            B
male           Gender                 0            Male
                                      1            Female
prevpreg       Previous               0            No
               pregnancy              1            Yes
gestweek       Gestational age (weeks)
preterm        Preterm birth          0            No
                                      1            Yes
bw             Birthweight (kg)
lbw            Low birthweight        0            No
                                      1            Yes
age4wk         Age at postnatal visit,
               target age 4 weeks
weight4wk      Weight at postnatal
               visit (kg)
age36mo        Age at development
               assessment, target age
               36 months
m1 to m12      Dev. milestones        0            Fail
                                      1            Pass
ability        Underlying trait
*************************************************************/

clear
version 12.1
set seed 123
set more off

** generate group allocation
set obs 1500
gen id = _n
gen group = int((id-1)/500)
label variable group "Intervention groups"
label define grouplab 0 "C (Control)" 1 "A" 2 "B"
label value group grouplab
```

```
** generate demographics
gen prevpreg = rbinomial(1,.5)
gen male = rbinomial(1,.51)
label variable prevpreg "Previous pregnancy"
label define prevpreglab 0 "No" 1 "Yes"
label value prevpreg prepreglab
label variable male "Gender"
label define malelab 0 "Female" 1 "Male"
label value male malelab

** generate gestational duration
gen term = cond(prevpreg,rbinomial(1,invlogit(2.0)),///
  rbinomial(1,invlogit(1.0+.5*group)))
gen gestweek = cond(term,34+8*rbeta(3.5,1.7),///
30+8*rbeta(3.5,1.7))
label variable gestweek "Gestational age (in weeks)"
recode gestweek (37/max = 0 "Not preterm")///
  (min/37 = 1 "Preterm"), gen(preterm)

** generate birthweight
gen bwgram = round(3000+70*male+110*(gestweek-39)-///
  12*(gestweek-39)^2+rnormal(0,(340+5*(gestweek-39))))
gen bw = round(bwgram/1000,0.001)
recode bwgram (2500/max = 0 "No") (min/2500 = 1 "Yes"), gen(lbw)
label variable bw "Birthweight (kg)"
label variable lbw "Low birthweight"

** generate postnatal assessment
gen age4wk = round(28+rnormal(0,1)*4,1)
gen weight4wk = 4.0+0.15*male+0.8*(bw-3)-.12*(bw-3)^2+///
  0.05*(gestweek-39)+0.025*(age4wk-28)+rnormal(0,0.3)
label variable age4wk "Actual age at postnatal assessment"
label variable weight4wk "Weight at postnatal assessment (kg)"

** generate developmental milestones here
gen age36mo = round(round(365.25*3+rnormal(0,10))/30.4375,.001)
gen ability = -3.5+age36mo+1*bw+.5*(weight4wk-bw)+rnormal(0,2.5)
forvalues i = 1(1)10{
gen m`i' = rbinomial(1,invlogit(-5-`i'*.1+.15*ability))
}

* create a different slope item and
* an item with gender difference
gen m11 = rbinomial(1,invlogit(-10+.3*ability))
gen m12 = rbinomial(1,invlogit(-5+(.15+0.1*male)*ability))

save SCTdata,replace
```

Appendix B:
Stata Codes to Generate Simulated Longitudinal Study (SLS) Dataset

```
/******************** Data Description ********************
Variable     Variable_label              Value      Value_label
id           ID                          1 - 200
SES          Socioeconomic status        0          Lower
                                         1          Higher
male         Gender                      0          Female
                                         1          Male
ttd          Time-to-death
death        Death/censoring             0          Censored
                                         1          Death
visit        Visit, bi-monthly           0,2,...,24
visitage     Actual age at visits
heightcm     Height in cm
haz          Height-for-age z-score
stunt        haz < -2 z-score            0          Not stunted
                                         1          Stunted
tts_midage   Time-to-stunting,
             midpoint age estimate
tts_lbage    Time-to-st, upper bound
tts_ubage    Time-to-st, lower bound
**********************************************************/

clear
version 12.1
set more off
set seed 123

** set sample size, 100 per group, 2 groups
set obs 200
gen id=_n

** generate covariates
gen SES=rbinomial(1,.4)
label variable SES "Socio-Econ Status"
label define SESlab 0 "Low" 1 "High"
label value SES SESlab
gen male=rbinomial(1,.51)
label variable male "Gender"
label define malelab 0 "Female" 1 "Male"
label value male malelabe
```

```
** generate time-to-death data according to a Weibull
** distribution, censored at 24 mo or last visit
** actual age varied from target age with SD = 3 days
** generate time-to-death data, censored at 24 mo or last visit
gen lastv=min(round(2+runiform()*200,2),24)
gen lastvage=lastv+rnormal(0,3)/30.4375
gen agedead=((-ln(runiform())/exp(-5-2*SES))^(1/0.5))/30.4375
gen ttd=min(agedead,lastvage)
gen death=cond(agedead>lastvage,0,1)

** generate haz data, in long format
** age window 0.1 SD
gen freq=int(ttd/2+1) if death==1
replace freq=round(ttd/2)+1 if death==0
expand freq
bys id: gen visit = (_n-1)*2
gen haz = rnormal(0,.5) if visit==0
bys id (visit): replace haz=haz[_n-1]-0.3+0.1*SES+rnormal(0,.1)///
  if visit>0 & visit< = 12
bys id (visit): replace haz=haz[_n-1]-0.1+0.05*SES+rnormal(0,.1)///
  if visit > 12
gen stunt = haz< = -2
by id: gen stunt_ever = sum(stunt)
by id: replace stunt_ever = stunt_ever[_N] > = 1

** generate age at visits
gen dvisit = -visit
bys id (dvisit) : gen visitage=lastvage if death==0 & _n==1
bys id (dvisit) : replace visitage=visit+rnormal(0,.1)        ///
  if visit!=0 & _n>1
bys id (dvisit) : replace visitage=visit+abs(rnormal(0,.1))   ///
  if visit==0 & _n>1
bys id (dvisit) : replace visitage=visit+rnormal(0,.1)        ///
  if death==1 & visit! = 0 & _n==1
bys id (dvisit) : replace visitage=visit+abs(rnormal(0,.1))   ///
  if death==1 & visit==0 & _n==1
replace visitage=(visit+ttd)/2 if death==1 & ttd<visitage

** lower and upper bound of age at stunting
sort id visit
stset visitage, fail(stunt) id(id) exit(time.)
stgen event_order=nfailures()
replace event_order=event_order+1
sort id visit
by id: gen lastvisit=1 if _n==_N
gen tts_lbage=_t0 if _d==1 & event_order==1
replace tts_lbage=_t if stunt_ever==0 & lastvisit==1 &       ///
  event_order==1
gen tts_ubage=_t if _d==1 & event_order==1
replace tts_ubage=. if stunt_ever==0 & lastvisit==1 &        ///
  event_order==1
sort id tts_lbage
by id: replace tts_lbage=tts_lbage[1] if _n!=1
by id: replace tts_ubage=tts_ubage[1] if _n!=1
```

```
sort id visit
stset, clear

** midpoint
gen tts_midage=(tts_lbage+tts_ubage)/2
replace tts_midage=tts_lbage if tts_midage==.

** back-transform (approximate) cm data
gen     heightcm=(42.98-2.63*((visitage+1)^-1)      ///
   +8.69*((visitage+1)^.5))*(haz*0.034+1) if male = =0
replace heightcm=(48.73-7.08*((visitage+1)^-.5)     ///
   +8.07*((visitage+1)^.5))*(haz*0.034+1) if male==1

** clean up; long format;
** only use one record for time-to-event analysis
foreach var in ttd visitage tts_ubage tts_lbage tts_midage ///
  haz heightcm {
  replace `var'=round(`var',.001)
  }
drop lastv lastvage lastvisit agedead freq dvisit event_order
sort id visit
save SLSdata, replace
```

Appendix C:
Stata Program for Detrended Q-Q Plot

```
*** Generate detrended q-q plot
*** Syntax: wormplot varname

capture program drop wormplot
program define wormplot
 version 12.1
 syntax varname [if]
 marksample touse
 tempvar z rank q d ub lb
 qui {
  egen `rank' = rank(`varlist') if `touse'
  sum `varlist' if `touse'
  local N = r(N)
  gen `z' = (`varlist'-r(mean))/r(sd) if `touse'
  gen `q' = invnormal(`rank'/(`N'+1))
  gen `d' = `z'-`q'
  gen `ub' = 1.96*(1/normalden(`q'))*sqrt(normal(`q')*    ///
    (1-normal(`q'))/`N')
  gen `lb' = -`ub'
  twoway (scatter `d' `q' if `touse')                     ///
   (rline `ub' `lb' `q' if `touse', sort mcolor(black)    ///
    msymbol(none) lcolor(black) lpattern(dash))           ///
    , ytitle(Deviation) yline(0,lpattern(vshortdash))     ///
    ylabel(, angle(horizontal))                           ///
   xtitle(Normal quantile) legend(off)
 }
end
```

Appendix D:
Stata Program for Weighted Maximum Likelihood Estimation for Binary Items

Tina Ying Xu,

Duke-NUS Graduate Medical School

```
** Save as: WMLE.do
clear
set mem 500m
set more off
clear
version 12.1
************************************************************************
* users specify:                                                      *
* 1. name of folder containing the data files and sub-routines (*.ado) *
* 2. name of data file that contains ICC parameters (one row per item) *
* 3. name of data file that contains the data (one row per person)    *
* 4. confidence interval (e.g. 95% CI)                                *
* 5. items to be included in the scoring                              *
************************************************************************

cd "F:\HumanGrowthDev\"
local file_item_par "item_par"
local file_item_response "item_response"
local level_CI = 95

local item_include "score1-score3"
************************************************************************

run IRT_WMLE_main.ado
run IRT_WMLE_d2.ado

local itemIDvar "item"
local itemPvarA "parA"
local itemPvarB "parB"

tempfile Est_file1
use "`file_item_response'",clear

tempvar pattern latent_Est latent_SD latent_lower_CI latent_upper_CI
tempname pattern_level wle_est wle_sd wle_lower_CI wle_upper_CI
```

```
qui {
egen `pattern' = group(`item_include'), missing
sort `pattern'
order `pattern'

cap drop `latent_Est'
cap drop `latent_SD'
cap drop `latent_lower_CI'
cap drop `latent_upper_CI'
gen double `latent_Est' =.
gen double `latent_SD' =.
gen double `latent_lower_CI' =.
gen double `latent_upper_CI' =.
save `Est_file1', replace
}

qui sum `pattern'
local `pattern_level' = r(max)

forv i = 1/``pattern_level'' {
qui{
use `Est_file1', clear
keep if `pattern' = =`i'
keep if _n = =1
keep `item_include'
stack `item_include', into(score) clear
rename _stack `itemIDvar'
merge `itemIDvar' using `file_item_par', sort uniqusing

**** program IRT_WMLE
count if ~missing(score)
if(r(N)>1) {
IRT_WMLE_main score `itemPvarB' if ~missing(score),///
  level(`level_CI') nocons offset(`itemPvarA')
local `wle_est' = _b[`itemPvarB']
local `wle_sd' = _se[`itemPvarB']
local `wle_lower_CI' =///
  _b[`itemPvarB']-invnormal(0.5+`level_CI'/200)*_se[`itemPvarB']
local `wle_upper_CI' =///
  _b[`itemPvarB']+invnormal(0.5+`level_CI'/200)*_se[`itemPvarB']
****di ``wle_est'' "***" ``wle_sd''///
"***"``wle_lower_CI'' "***"``wle_upper_CI''

use `Est_file1', clear
replace `latent_Est' = ``wle_est'' if `pattern' = =`i'
replace `latent_SD' = ``wle_sd'' if `pattern' = =`i'
replace `latent_lower_CI' = ``wle_lower_CI'' if `pattern' = =`i'
replace `latent_upper_CI' = ``wle_upper_CI'' if `pattern' = =`i'
save `Est_file1', replace
}
}
if(`i' = =1) {
nois _dots 0, title(Loop running) reps(``pattern_level'')
}
nois _dots `i' 0
}
```

```
use `Est_file1', clear
rename `latent_Est' Estimate
rename `latent_SD' SE
rename `latent_lower_CI' LB
rename `latent_upper_CI' UB
format Estimate SE LB UB%9.4f
```

Appendix D1: Subroutine 1 for WMLE

```
** Save as: IRT_WMLE_main.ado

cap program drop IRT_WMLE_main
program define IRT_WMLE_main, eclass sortpreserve byable(recall)
    version 9.1
    if replay() {
        if ("`e(cmd)'" ! = "IRT_WMLE_main") error 301
        syntax [, Level(cilevel) OR]
    }
    else {
        syntax varlist(numeric) [if] [in]    ///
        [, noLOg noCONStant OFFset(varname numeric) Level(cilevel) OR *]
        gettoken lhs rhs: varlist
        local k : word count `rhs'
        if(`k'>1) di in red "More than one predictors specified"
        if(`k'>1) exit
        if(`k' ==0) di in red                 ///
            "Please specify the predictor"
        if(`k' ==0) exit
        if(`rhs' == `offset') di in red       ///
            "The predictor and the offset term can not be the same"
        if(`rhs' == `offset') exit
        marksample touse
        mlopts mlopts, `options'
        if "`log'" != "" local qui quietly
        if "`offset'" != "" local offopt "offset(`offset')"
        markout `touse' `offset'

        if "`constant'" == "" {
        summarize `lhs' if `touse', mean
        if r(mean) == 0 local b0=-20
        else if r(mean) == 1 local b0=20
        else local b0=logit(r(mean))
        local n=r(N)
        local initopt init(_cons=`b0') search(quietly)
                        }

        global wle_b `rhs'
        `qui' di as txt _n "Fitting the full model:"
        ml model d0 IRT_WMLE_d2 (xb: `lhs'=`rhs', `constant' `offopt')///
         if `touse', `mlopts' `modopts' `initopt' missing maximize   ///
         crittype("Log of weighted-likelihood") `log'

         ereturn local title "Weighted likelihood maximum estimates"
         ereturn local depvar `lhs'
         ereturn local cmd "IRT_WMLE_main"
            }
    ml display, level(`level') `or'
end
```

Appendix D2:
Subroutine 2 for WMLE

```
** Save as: IRT_WMLE_d2.ado

capture program drop IRT_WMLE_d2
program define IRT_WMLE_d2
 version 9.1
 args todo b lnf g negH g1
 tempvar xb lj se2 tempL
 tempname tmpname0

 mleval `xb' = `b'
     quietly {
         generate double `lj' = invlogit( `xb') if $ML_y1 == 1
         replace         `lj' = invlogit(-`xb') if $ML_y1 == 0
         mlsum `lnf' = ln(`lj')

         generate double `se2' = `lj'*(1-`lj')
         generate double `tempL' = 1
         foreach var of global wle_b {
             replace `tempL' = `tempL'*`se2'*`var'*`var'
         }
         sum `tempL'
         local `tmpname0'=r(sum)
         scalar define `lnf' = `lnf' + ln(``tmpname0'')/2
     }
end
```

Index

Printed and bound by CPI Group (UK) Ltd, Croydon, CR0 4YY

24/10/2024

01778301-0011